Statistics of Extremes

E. J. Gumbel
Columbia University

DOVER PUBLICATIONS, INC.
Mineola, New York

Bibliographical Note

This Dover edition, first published in 2004, is an unabridged republication of the edition published by Columbia University Press, New York, 1958.

Library of Congress Cataloging-in-Publication Data

Gumbel, Emil Julius, 1891-
    Statistics of extremes / E.J. Gumbel.
        p. cm.
    Originally published: New York : Columbia University Press, 1958.
    Includes bibliographical references and index.
    ISBN 0-486-43604-7 (pbk.)
      1. Mathematical statistics. 2. Maxima and minima. I. Title.

QA276.G82 2004
519.5—dc22

                                                                                                     2004043937

Manufactured in the United States of America
Dover Publications, Inc., 31 East 2nd Street, Mineola, N.Y. 11501

*A la mémoire de ma femme
bien chérie qui quatre fois
de suite a du écrire ce livre*

PREFACE

THE THEORY of extreme values is mentioned in most of the recent textbooks of statistics. However, this seems to be the first book devoted exclusively to these problems. It is meant for statisticians, and statistically minded scientists and engineers. To spread the application of the methods, the author has tried to keep it on an elementary level. Graphical procedures are preferred to tedious calculations. Special cases and easy generalizations are given as exercises. Therefore, the book may even meet the requirements of a textbook.

Since the line of thought pursued here is new, the reader should not expect complete solutions. There are bound to be shortcomings and gaps in a first treatment. Despite the usual prescriptions of academic habits, unsolved problems are here clearly stated. Perhaps a carping critic might maintain that more problems are raised than solved. However, this is the common fate of science.

This book is written in the hope, contrary to expectation, that humanity may profit by even a small contribution to the progress of science.

The opportunity to start this book presented itself when the necessary free time was imposed upon the author. A first draft of the manuscript was written in 1949 under a grant by the Lucius N. Littauer Foundation. This draft was revised in 1950 under a grant from the John Simon Guggenheim Memorial Foundation. The work continued while the author was engaged as a Consultant at Stanford University and later as an Adjunct Professor at Columbia University, partially under a grant from the Higgins Foundation, and under contract with the Office of Ordnance Research. Finally, a further grant from the Guggenheim Foundation made it possible to engage Professor Frank Lee (Columbia University) for the drawing of the graphs. The author wishes to state his sincere appreciation to these agencies for their help, without which he would not have had the chance to finish this book.

He is indebted to Mr. E. N. Munns, formerly Chief, Division of Forest Influence, U. S. Department of Agriculture, for valuable office help. Dr. Wolfgang Wasow of the University of California (Los Angeles) suggested changes in Chapters 1 and 3 and contributed to section 8.1. P. G. Carlson (Arthur Andersen and Co.) gave the proof for 3.2.3(7). The idea underlying 6.1.6 is due to Dr Julius Lieblein of the D. Taylor Model Basin. Paragraphs 6.2.4, 6.2.5, and 6.2.7 have been written by B. F. Kimball (N. Y.

State Public Service Commission). Most of the work in section 8.2 was done in collaboration with the late Roger D. Keeney. The Watson Scientific Computing Laboratory (Columbia University) calculated the figures given in Table 6.2.3 and put at the author's disposal an unpublished table of the probability points and the first two differences for $\Phi = .00001$ (.00001) .00600 (.00010) .08000 (.00100) .84000 (.00010) .98400 (.00001) .99995. These data facilitated much numerical work. Table 7.2.3 was calculated by Gladys R. Garabedian of Stanford University. The author takes this occasion to express his thanks for these important contributions.

Mr. Arnold Court (University of California, Berkeley) suggested applications to climatology, and Professor S. B. Littauer (Dept. of Industrial Engineering, Columbia University) proposed valuable improvements.

The author profited greatly from his collaboration with Professor A. M. Freudenthal (Columbia University) in their common work on fatigue of metals (see 7.3.5), and from the continuous interest which the Statistical Engineering Laboratory of the National Bureau of Standards took in his work.

The whole manuscript was revised and corrected by Mr. L. H. Herbach (New York University), P. G. Carlson, C. Derman (Columbia University), and A. Court. The bibliography was checked by Mr. Taro Yamane (New York University). It gives the author great pleasure to thank these collaborators, and in particular Mr. Carlson, for their friendly assistance. They certainly have contributed whatever may be good in this book. The rest should be charged against the author.

New York,
1 October, 1957

CONTENTS

Chapter One: AIMS AND TOOLS   1

1.0. Aims   1
    1.0.1. Conditions   1
    1.0.2. History   2
    1.0.3. The Flood Problem   4
    1.0.4. Methodology   5
    1.0.5. Arrangement of Contents   5

1.1. General Tools   7
    1.1.1. Linear Transformations   7
    1.1.2. Other Transformations   7
    1.1.3. Symmetry   8
    1.1.4. Measures of Dispersion   9
    1.1.5. Moments   10
    1.1.6. Generating Function   11
    1.1.7. Convolution   13
    1.1.8. The Gamma Function   15
    1.1.9. The Logarithmic Normal Distribution   16

1.2. Specific Tools   20
    1.2.0. Problems   20
    1.2.1. The Intensity Function   20
    1.2.2. The Distribution of Repeated Occurrences   21
    1.2.3. Analysis of Return Periods   26
    1.2.4. "Observed" Distributions   28
    1.2.5. Construction of Probability Papers   28
    1.2.6. The Plotting Problem   29
    1.2.7. Conditions for Plotting Positions   32
    1.2.8. Fitting Straight Lines on Probability Papers   34
    1.2.9. Application to the Normal Distribution   38

Chapter Two: ORDER STATISTICS AND THEIR EXCEEDANCES   42

2.1. Order Statistics   42
    2.1.0. Problems   42
    2.1.1. Distributions   42

CONTENTS

2.1.2. Averages	43
2.1.3. Distribution of Frequencies	46
2.1.4. Asymptotic Distribution of mth Central Values	47
2.1.5. The Order Statistic with Minimum Variance	51
2.1.6. Control Band	52
2.1.7. Joint Distribution of Order Statistics	53
2.1.8. Distribution of Distances	55
2.2. The Distribution of Exceedances	57
2.2.0. Introduction	57
2.2.1. Distribution of the Number of Exceedances	58
2.2.2. Moments	61
2.2.3. The Median	63
2.2.4. The Probability of Exceedances as Tolerance Limit	64
2.2.5. Extrapolation from Small Samples	67
2.2.6. Normal and Rare Exceedances	69
2.2.7. Frequent Exceedances	72
2.2.8. Summary	73
Chapter Three: EXACT DISTRIBUTION OF EXTREMES	**75**
3.1. Averages of Extremes	75
3.1.0. Problems	75
3.1.1. Exact Distributions	75
3.1.2. Return Periods of Largest and Large Values	78
3.1.3. Quantiles of Extremes	79
3.1.4. Characteristic Extremes	82
3.1.5. The Extremal Intensity Function	84
3.1.6. The Mode	85
3.1.7. Moments	87
3.1.8. The Maximum of the Mean Largest Value	89
3.2. Extremal Statistics	94
3.2.0. Problems	94
3.2.1. Absolute Extreme Values	94
3.2.2. Exact Distribution of Range	97
3.2.3. The Mean Range	100
3.2.4. The Range as Tolerance Limit	103
3.2.5. The Maximum of the Mean Range	106
3.2.6. Exact Distribution of the Midrange	108
3.2.7. Asymptotic Independence of Extremes	110
3.2.8. The Extremal Quotient	111

CONTENTS

Chapter Four: ANALYTICAL STUDY OF EXTREMES — 113

4.1. The Exponential Type — 113
    4.1.0. Problems — 113
    4.1.1. Largest Value for the Exponential Distribution — 113
    4.1.2. Order Statistics for the Exponential Distribution — 116
    4.1.3. L'Hôpital's Rule — 118
    4.1.4. Definition of the Exponential Type — 120
    4.1.5. The Three Classes — 122
    4.1.6. The Logarithmic Trend — 123
    4.1.7. The Characteristic Product — 125

4.2. Extremes of the Exponential Type — 126
    4.2.0. Problems — 126
    4.2.1. The Logistic Distribution — 126
    4.2.2. Normal Extremes, Numerical Values — 129
    4.2.3. Analysis of Normal Extremes — 136
    4.2.4. Normal Extreme Deviates — 140
    4.2.5. Gamma Distribution — 143
    4.2.6. Logarithmic Normal Distribution — 146
    4.2.7. The Normal Distribution as a Distribution of Extremes — 147

4.3. The Cauchy Type — 149
    4.3.0. Problems — 149
    4.3.1. The Exponential Type and the Existence of Moments — 149
    4.3.2. Pareto's Distribution — 151
    4.3.3. Definition of the Pareto and the Cauchy Types — 152
    4.3.4. Extremal Properties — 153
    4.3.5. Other Distributions without Moments — 154
    4.3.6. Summary — 155

Chapter Five: THE FIRST ASYMPTOTIC DISTRIBUTION — 156

5.1. The Three Asymptotes — 156
    5.1.0. Introduction — 156
    5.1.1. Preliminary Derivation — 156
    5.1.2. The Stability Postulate — 157
    5.1.3. Outline of Other Derivations — 162
    5.1.4. Interdependence — 164

5.2. The Double Exponential Distribution — 166
    5.2.0. Introduction — 166
    5.2.1. Derivations — 166
    5.2.2. The Methods of Cramér and Von Mises — 170

5.2.3. Mode and Median	172
5.2.4. Generating Functions	173
5.2.5. Standard and Mean Deviations	174
5.2.6. Probability Paper and Return Period	176
5.2.7. Comparison with Other Distributions	179
5.2.8. Barricelli's Generalization	184
5.3. Extreme Order Statistics	187
5.3.0. Problems	187
5.3.1. Distribution of the mth Extreme	187
5.3.2. Probabilities of the mth Extreme	189
5.3.3. Generating Functions	192
5.3.4. Cramér's Distribution of mth Extremes	194
5.3.5. Extreme Distances	197
5.3.6. The Largest Absolute Value and the Two Sample Problem	198

Chapter Six: USES OF THE FIRST ASYMPTOTE — 201

6.1. Order Statistics from the Double Exponential Distribution	201
6.1.0. Problems	201
6.1.1. Maxima of Largest Values	201
6.1.2. Minima of Largest Values	203
6.1.3. Consecutive Modes	206
6.1.4. Consecutive Means and Variances	208
6.1.5. Standard Errors	212
6.1.6. Extension of the Control Band	216
6.1.7. The Control Curve of Dick and Darwin	218
6.2. Estimation of Parameters	219
6.2.0. Problems	219
6.2.1. Exponential and Normal Extremes	219
6.2.2. Use of Order Statistics	223
6.2.3. Estimates for Probability Paper	226
6.2.4. Sufficient Estimation Functions, by B. F. Kimball	229
6.2.5. Maximum Likelihood Estimations, by B. F. Kimball	231
6.2.6. Approximate Solutions	232
6.2.7. Asymptotic Variance of a Forecast, by B. F. Kimball	234
6.3. Numerical Examples	236
6.3.0. Problems	236
6.3.1. Floods	236
6.3.2. The Design Flood	238
6.3.3. Meteorological Examples	241

CONTENTS

6.3.4. Application to Aeronautics	245
6.3.5. Oldest Ages	246
6.3.6. Breaking Strength	248
6.3.7. Breakdown Voltage	249
6.3.8. Applications to Naval Engineering	251
6.3.9. An Application to Geology	254

Chapter Seven: THE SECOND AND THIRD ASYMPTOTES — 255

7.1. The Second Asymptote	255
7.1.0. Problems	255
7.1.1. Fréchet's Derivation	255
7.1.2. The Cauchy Type	259
7.1.3. Averages and Moments	264
7.1.4. Estimation of the Parameters	266
7.1.5. The Increase of the Extremes	269
7.1.6. Generalization	270
7.1.7. Applications	271
7.1.8. Summary	272
7.2. The Third Asymptote	272
7.2.0. Introduction	272
7.2.1. The Von Mises Derivation	273
7.2.2. Other Derivations	276
7.2.3. Averages and Moments of Smallest Values	280
7.2.4. Special Cases	285
7.2.5. The Increase of the Extremes	287
7.2.6. The 15 Relations Among the 3 Asymptotes	288
7.3. Applications of the Third Asymptote	289
7.3.0. Problems	289
7.3.1. Estimation of the Three Parameters	289
7.3.2. Estimation of Two Parameters	293
7.3.3. Analytical Examples	298
7.3.4. Droughts	299
7.3.5. Fatigue Failures	302

Chapter Eight: THE RANGE — 306

8.1. Asymptotic Distributions of Range and Midrange	306
8.1.0. Problems	306
8.1.1. The Range of Minima	307
8.1.2. Generating Function of the Range	308

8.1.3. The Reduced Range	309

       8.1.3. The Reduced Range 309
       8.1.4. Asymptotic Distribution of the Midrange 311
       8.1.5. A Bivariate Transformation 312
       8.1.6. Asymptotic Distribution of the Range 316
       8.1.7. Boundary Conditions 320
       8.1.8. Extreme Ranges 321
       8.1.9. Summary 324
  8.2. Extremal Quotient and Geometric Range 324
       8.2.0. Problems 324
       8.2.1. Definitions 325
       8.2.2. The Extremal Quotient 326
       8.2.3. The Geometric Range 327
  8.3. Applications 331
       8.3.0. Problems 331
       8.3.1. The Midrange 331
       8.3.2. The Parameters in the Distribution of Range 333
       8.3.3. Normal Ranges 336
       8.3.4. Estimation of Initial Standard Deviation 338
       8.3.5. Climatological Examples 340

Summary 345

Bibliography 349

Index 373

TABLES

1.2.9.	Normal Standard Deviation σ_n as Function of Sample Size n	39
2.1.2.	Probabilities of Modal mth Values	45
2.1.6.	Reduced Standard Errors $\sigma(\hat{z}_m)\sqrt{n}$ for Unit Standard Deviation	52
2.2.1.	The Probabilities $w(5,m,5,x)$	60
2.2.7.	The Six Limiting Distributions of Exceedances	73
4.1.6.	The Three Classes of Distributions of the Exponential Type	125
4.2.2(1).	Normal Extremes	130
4.2.2(2).	Characteristic Normal Extremes u_n for Large Samples	133
4.2.2(3).	Normal Extremes and Intensity	133
5.1.1.	The Three Asymptotic Probabilities of Largest Values	157
5.1.3(1).	Asymptotic Probabilities of *Initial* Variates	162
5.1.3(2).	Asymptotic Probabilities of Extreme Values	164–165
5.2.4.	Semivariants of the First Asymptote	174
5.2.6.	The Three Curvatures of Largest Values	178
5.2.7.	Selected Probabilities for Normal and Largest Values	180
5.3.2.	Asymptotic Probabilities of the Three Largest Values	190
5.3.3.	Characteristics of mth Extremes	194
6.1.1.	Extreme Values for the Double Exponential Distribution	202
6.1.2(1).	Characteristic Minima	204
6.1.2(2).	Probabilities of Maxima and Minima of Largest Values	205
6.1.2(3).	The Iterated Natural Logarithm	205
6.1.5.	Standard Errors of Reduced Central Values	214
6.1.6.	The Extreme Control Band	218
6.2.2.	The Weights in Lieblein's Estimator for $N = 4, 5, 6$	225
6.2.3.	Means and Standard Deviations of Reduced Extremes	228
6.3.2.	The Standardized Flood z_ω for Some Rivers	240
6.3.5.	Modal Values and Characteristic Oldest Ages, U.S.A., 1939–41	248
7.2.3.	Reduced Averages and Third Moment of the Third Asymptote	282
7.2.4.	The Four Pseudosymmetrical Cases	285
7.2.6.	Asymptotic Probabilities of Iterated Extremes	288
7.3.4.	Analysis of Droughts	301
7.3.5.	Parameters for Fatigue Failures of Nickel	304
8.1.5.	Asymptotic Distribution and Probability of ξ	316

8.1.6.	Characteristics of Reduced Extremes, Range, and Midrange	319
8.1.8.	Modal Extreme Reduced Ranges	323
8.2.3.	Transformations of Extremes to Extremal Statistics	330
8.3.1(1).	Population Standard Deviation of the Midrange	331
8.3.1(2).	Observed Midranges of Lives of Bulbs (in Ten Hours)	332
8.3.2.	Population Mean and Standard Deviation of the Range	334
8.3.4.	Characteristics for Normal Ranges	340
8.3.5(1).	Frequencies of Dew Points, Washington, D.C., 1905–1945	343
8.3.5(2).	Annual Range of Temperature in $C°$, Bergen, Norway, 1878–1926	343
8.3.5(3).	Annual Range of Temperature in $F°$, New Haven, Connecticut, 1907–1948	343
8.3.5(4).	Estimation of the Parameters for the Four Ranges	344

GRAPHS

1.1.9.	Characteristics for the Logarithmic Normal Distribution	19
1.2.1.	Intensity Functions $\mu(x)$	22
1.2.2(1).	Control Intervals for the Return Period	24
1.2.2(2).	Design Factor versus Design Quotient	25
1.2.5(1).	The First Laplacean and the Normal Probability	30
1.2.5(2).	The First Laplacean and the Normal Probability on Normal Paper	31
1.2.6.	Plotting Positions on Normal Paper	33
1.2.9(1).	Increase of Normal Standard Deviation as Function of the Sample Size	39
1.2.9(2).	Annual Mean Discharges, Colorado River at Bright Angel Creek, 1922–1939	40
2.1.4.	Asymptotic and Exact Standard Errors of mth Normal Values	50
2.1.6.	Control Band for Normal Central Values	54
2.2.4.	Probabilities $W(10,m,10,x)$ of Exceedances	65
2.2.5.	Sample Size n as Function of Percentage q and Probability $F_1(q)$	68
2.2.6.	Distributions and Probabilities of Rare Exceedances	71
3.1.3(1).	Return Periods of Largest Values	80
3.1.3(2).	Calculation of Quantiles from the Median Extreme	81
3.1.8.	Mean Largest Standardized Values	92
3.2.1.	Probability Paper for Normal Deviations	94
3.2.3.	The Mean Range Interpreted as Area	101
3.2.4.	The Observed Range as Tolerance Limit	105
3.2.6.	Standard Error of the Center as Function of Sample Size	109
4.1.1(1).	Averages of Largest Exponential Values	114
4.1.1(2).	Distributions of Largest Exponential Values	115
4.1.6.	The Three Classes of the Exponential Type	125
4.2.2(1).	Probabilities of Normal Extremes	129
4.2.2(2).	Quantiles of Normal Extremes	131
4.2.2(3).	Distributions of Normal Extremes for 2, 3, 5, and 10 Observations	132
4.2.2(4).	Averages of Normal Extremes in Natural Scale	134
4.2.2(5).	Occurrence Interval and Sample Size	134
4.2.2(6).	Standard Deviations for Normal Extremes, Deviates, and Ranges	135

GRAPHS

4.2.2(7).	Moment Quotients for Largest Normal Values and Deviates as Functions of Sample Size	136
4.2.2(8).	Beta Diagrams for Normal Extremes, Deviates, and Ranges	137
4.2.3.	Modal Largest Normal Value for $2n$ Observations	139
4.2.4.	Probabilities of Studentized Normal Extreme Deviate	142
4.2.5.	Characteristic Largest Value for Gamma Distribution	145
4.2.6.	Asymptotic Parameters for the Logarithmic Normal Distribution	147
4.3.4.	The Three Classes for the Cauchy Type	154
5.2.6.	Extremal Probability Paper	177
5.2.7(1).	Extreme and Normal Distributions	180
5.2.7(2).	Probabilities for Extreme and Normal Standardized Values	181
5.2.7(3).	Probabilities of Standardized Extremes, Range, and Midrange	182
5.2.7(4).	Normal and Extremal Distribution as Function of the Probability	183
5.2.7(5).	First Asymptotic Probability Traced on Logarithmic Normal Paper	184
5.3.1.	Asymptotic Distributions of mth Reduced Extremes	188
5.3.2(1).	Asymptotic Probabilities of mth Extremes Traced on Normal Paper	190
5.3.2(2).	The Convergence of the Median to the Mode for mth Extremes	191
5.3.2(3).	Asymptotic Mode and Median of mth Extremes	192
5.3.4.	Cramér's Probabilities of Transformed mth Extremes	196
5.3.6.	Correction for Obtaining the Composite Mode for Two Samples	199
6.1.2.	Probabilities of Average Extreme Values	206
6.1.3.	Modal mth Extremes $\tilde{y}_m(N)$	207
6.1.4.	Mean mth Extremes $\bar{y}_m(N)$	211
6.1.5.	Homogeneity Test for the Most Probable Annual Flood	213
6.2.1(1).	Probabilities of Extreme Exponential Values	220
6.2.1(2).	Probabilities of Extreme Values for the Logistic Variate	220
6.2.1(3).	Probabilities of Normal Extremes and Extreme Deviates	221
6.2.1(4).	Calculated and Asymptotic Distributions of Normal Extremes for 10, 100, 1,000 Observations	222
6.2.3.	Population Mean $\bar{y}_N$ and Standard Deviation σ_N as Function of N	227
6.2.6.	Correction B_N for Maximum Likelihood Method	233
6.3.1.	Floods. Mississippi River, Vicksburg, Mississippi, 1898–1949	237

GRAPHS

6.3.2.	Design Flood for Given Risk	240
6.3.3(1).	Annual Maxima of Temperatures and Their Differences from the Annual Mean. Bergen, Norway, 1857–1926	242
6.3.3(2).	Largest One-day Precipitations. Barakar Catchment Area, India	243
6.3.6.	Survivorship Function for Rubber Specimen	250
7.1.2(1).	Fréchet's Probability (II) for Different Values of k	261
7.1.2(2).	The Second Asymptotic Distribution of Largest Values	262
7.1.2(3).	The Generalized Probability $V(y)$	263
7.1.3.	Averages for the Second Asymptote	265
7.1.4(1).	Estimate of k from Coefficient of Variation	266
7.1.4(2).	Estimate of k from the Reciprocal Moments	267
7.2.1.	Probabilities III of Extreme Values, Traced on Normal Paper	275
7.2.2(1).	Probabilities III, Traced on Logarithmic Extremal Paper	277
7.2.2(2).	Influence of k on the Shape of the Distribution III of Smallest Values	278
7.2.3.	Averages of the Third Asymptote	280
7.3.1(1).	Parameter k and the Standardized Distances $A(k)$ and $B(k)$ as Functions of the Skewness	290
7.3.1(2).	Standardized Distances from Lower Limit	291
7.3.2(1).	Estimation of k from the Coefficient of Variation	294
7.3.2(2).	Criterion for Lower Limit Zero	295
7.3.3(1).	Parameters v_n and k_n for Normal Extremes	299
7.3.3(2).	Approximations of the Probabilities of Normal Extremes by the Third Asymptote	300
7.3.4.	Droughts: Colorado and Connecticut Rivers	301
7.3.5.	Fatigue Failures of Nickel	303
8.1.4.	Asymptotic Probabilities of Extremes, Range and Midrange	312
8.1.6(1).	Asymptotic Distribution of the Range	318
8.1.6(2).	Asymptotic Probabilities of Standardized Extremes, Range and Midrange	319
8.1.8(1).	Averages of Extreme Reduced Ranges	322
8.1.8(2).	The Range of the Range	324
8.2.2.	Asymptotic Probabilities and Distributions of the Extremal Quotient for the Cauchy Type	326
8.2.3(1).	Asymptotic Probabilities $\Psi'^{(2)}(\xi)$ and $\Psi'^{(3)}(\xi)$ for the Geometric Range	328
8.2.3(2).	Estimation of k from the Coefficient of Variation of the Geometric Range	329
8.3.1(1).	Population Standard Deviation of Midrange	332
8.3.1(2).	Midranges of Lives of Bulbs Traced on Logistic Paper	333

8.3.2.	Population Mean Range and Standard Deviation as Function of Sample Size N	335
8.3.3(1).	Probabilities of Normal Ranges	337
8.3.3(2).	Characteristics of the Normal Range	338
8.3.5(1).	Range of Dew Points: Washington, D.C., 1905–1945	341
8.3.5(2).	Annual Range of Temperatures: Bergen, Norway, and New Haven, Connecticut	342

Statistics
of
Extremes

Chapter One: AIMS AND TOOLS

*"Blessed is he who expects nothing
for he shall never be disappointed."*

1.0. AIMS

1.0.1. Conditions. The aim of a statistical theory of extreme values is to analyze observed extremes and to forecast further extremes. The extremes are not fixed, but are new statistical variates depending upon the initial distribution and upon the sample size. However, certain results which are distribution-free may be reached.

Statistical studies of extreme values are meant to give an answer to two types of questions: 1) Does an individual observation in a sample taken from a distribution, alleged to be known, fall outside what may reasonably be expected? 2) Does a series of extreme values exhibit a regular behavior? In both cases, "reasonable expectation" and "regular behavior" have to be defined by some operational procedure.

The essential condition in the analysis is the "clausula rebus sic stantibus." The distribution from which the extremes have been drawn and its parameters must remain constant in time (or space), or the influence that time (or space) exercises upon them must be taken into account or eliminated. Another limitation of the theory is the condition that the observations from which the extremes are taken should be independent. This assumption, made in most statistical work, is hardly ever realized. However, if the conditions at each trial are determined by the outcome of previous trials as, e.g., in Pólya's and Markoff's schemes, some classical distributions are reached. Therefore, the assumption that distributions based on dependent events should share the asymptotic properties of distributions based on independent trials is not too far-fetched.

One of the principal notions to be used is the "unlimited variate." Here "common sense" revolts at once, and practical people will say: "Statistical variates should conform to physical realities, and infinity transcends reality. Therefore this assumption does not make sense." The author has met this objection when he advocated his theory of floods (6.3.1), at which time this issue was raised by people who applied other unlimited distributions without realizing that their methods rely on exactly the same notion.

This objection is not as serious as it looks, since the denial of the existence of an upper or lower limit is linked to the affirmation that the probability for extreme values differs from unity (or from zero) by an amount which becomes as small as we wish. Distributions currently used have this property. The exploration of how unlimited distributions behave at infinity is just part of the common general effort of mathematics and science to transgress the finite, as calculus has done since Newton's time for the infinite, and nuclear physics is doing for the infinitesimal.

References: Borel, Chandler.

1.0.2. History. The founders of the calculus of probabilities were too occupied with the general behavior of statistical masses to be interested in the extremes. However,† as early as 1709 Nicolaus Bernoulli considers an actuarial problem: n men of equal age die within t years. What is the mean duration of life of the last survivor? He reduces this question to the following: n points lie at random on a straight line of length t. Then he calculates the mean largest distance from the origin.

The first researches pertaining to the theory of largest values started from the normal distribution. This was reasonable in view of its practical importance. The astronomers who dealt with repeated observations of the same object, say the diameter of a star, were the first to be interested in establishing a criterion for the acceptance or rejection of an outlying value. In 1852, Peirce stated such a test. Another widespread method is due to Chauvenet (1878). A systematic study of this problem was made by P. Rider (1933), who also gave an extensive bibliography.

Extreme values are linked to small probabilities. In this connection, Poisson's law must be mentioned, since it deals with small probabilities. It gives the number of rare events, while the theory of extreme values considers their size. For sixty years Poisson's derivation was nothing but a mathematical curiosity, until L. von Bortkiewicz (1898) showed its statistical meaning and importance. This author was also the first (1922) to study extreme values. In the following year, R. von Mises introduced the fundamental notion of the characteristic largest value (without using this name) and showed its asymptotic relation to the mean of the largest normal values. In 1925, L. H. C. Tippett calculated the probabilities of the largest normal values for different sample sizes up to 1000, and the mean normal range for samples from 2 to 1000. Tippett's tables are the fundamental tools for all practical uses of largest values from normal distributions.

The fact that most of these studies started from this distribution

---

† I am indebted to Mr. J. Dutka, Rutgers University, for having drawn my attention to this example given by Todhunter (p. 195).

hampered the development, since none of the fundamental theorems of extreme values are related, in a simple way, to the normal distribution. The first study of largest values for other distributions was made by E. L. Dodd in 1923. His work is based on "asymptotic" values which are again similar to the characteristic largest values.

The first paper based on the concept of a type of initial distributions different from the normal one is due to M. Fréchet (1927). He was also the first to obtain an asymptotic distribution of the largest value. More important, he showed that largest values taken from different initial distributions sharing a common property may have a common asymptotic distribution. He introduced the stability postulate according to which the distribution of the largest value should be equal to the initial one, except for a linear transformation. Fréchet's paper, published in a remote journal, never gained the recognition it merited. This was due to the fact that the type of initial distributions considered is not very frequent. In addition, R. A. Fisher and L. H. C. Tippett published, in the next year, the paper which is basic for work on extreme values. They used the same stability postulate, and found in addition to Fréchet's asymptotic distribution two others valid for other initial types. The authors stressed the slow convergence of the distribution of the largest normal value toward its asymptote, and thus showed the reason for the relative sterility of all previous endeavors.

In 1936, R. von Mises classified the initial distributions possessing asymptotic distributions of the largest value, and gave sufficient conditions under which the three asymptotic distributions are valid. In 1943, B. Gnedenko gave necessary and sufficient conditions. In 1948, G. Elfving and the present author derived, almost at the same time, the relations of the asymptotic distribution of the normal range to certain Bessel functions. In 1953 the National Bureau of Standards published *Probability Tables for the Analysis of Extreme Value Data* and in 1954 this author's brochure *Statistical Theory of Extreme Values and Some Practical Applications*, where many numerical examples may be found. More details, especially concerning the contributions of E. S. Pearson and his school, will be given in the text.

In his thesis (1954) R. A. da Silva Leme gave a systematic expository treatment of the asymptotic distributions of extreme values and their applications to engineering problems, expecially the safety of structures. The latter problem was studied previously (1953) in the same connection and in great detail by Arne L. Johnson.

1.0.3. The Flood Problem.

"However big floods get, there will always be a bigger one coming; so says one theory of extremes, and experience suggests it is true."

(PRESIDENT'S WATER COMM., p. 141.)

The oldest problems connected with extreme values arise from floods. Their economic importance was early realized since ancient agrarian economies were based exclusively on water flow, and waterways provided the main system of communication. Their importance has increased in industrial economies. Through the construction of hydro-electric plants, water becomes a permanent source of energy. Its flow is used for reservoirs, irrigation, and the fight against erosion. Finally, life and property should be protected against the damages caused by inundations. The social importance of these aspects of flood control which entail adequate provision against future floods cannot be overemphasized.

An inundation happens when water flows where it ought not to flow. For the statistical treatment, consider the mean daily discharge of a river through a given profile at a specific station, measured in cubic meters (or cubic feet) per second. Among the 365 (or 366) daily discharges during a year, there is one measure which is the largest. This discharge is called the annual flood. Hence within each year there is one such flood, which may have occurred on more than one day, and which need not be an inundation. Until recent years, engineers employed purely empirical procedures for the analysis of hydrological data and, especially, the evaluation of the frequencies of floods and droughts. These studies led practically nowhere until the statistical nature of the problem was recognized. Even then the results were meager and disappointing. Hence the claim that "sound engineering judgment" is superior to statistics, a statement which may have been true as long as the adequate statistical method was unknown. At present the statistical nature of these problems has been realized and the empirical procedures are slowly being replaced by methods derived from the theory of extreme values.

Similar stationary time series may easily be obtained for annual droughts, largest precipitations, snowfalls, maxima and minima of atmospheric pressures and temperatures, and other meteorological phenomena.

The question of what constitutes an extreme value seems simple enough. Yet, in meteorology, difficulties may arise from the fact that the year is a natural unit of periodicity which automatically determines the number $n = 365$ as the sample size from which the extreme values of daily observations are taken. The arbitrary starting point of January 1 spreads the winter season over two years. Suppose the minimum temperature of one year occurs during December, and the minimum temperature in the next year in January. These two values are the first and the second smallest temperatures for one single season. To avoid the artificial separation introduced by the calendar years, some meteorological publications use "water years" starting in October, thus obtaining only one smallest temperature value within one season. The highest temperatures, which in our climate only occur in summer, are not affected by this procedure.

References: Am. Soc. Civ. Eng. Handbook, Barrow, Baur, Beard (1942), Chow (1951), Coutagne (1930, 1934, 1951, 1952a, b, c), Creager, Eggenberger, E. E. Foster (1942), H. A. Foster (1933), Gibrat (1932, 1934), Goodrich, Grant, Hazen, Hurst, Lane and Kai Lei, Linsley, Moran, President's Water Resources Policy Comm., L. K. Sherman, Slade (1936a,b), Todd, Urquhart, Wisler and Brater.

1.0.4. Methodology. Although the problem of extreme values is a quite recent one, it has attracted the attention of many scientists working in very different fields. Problems related to extreme values exist in astronomy, meteorology, naval engineering, oceanography, quality control, in connection with breaking strength of material and the building code, in population statistics (the oldest age), and in economics.

To satisfy such diversified demands, we have studied the methodology common to all these problems in order to construct a systematic theory. No attempt was made to cover the whole range of publications. The rapid development of statistics in recent years would render such an attempt futile. The criterion was to include whatever seemed of general value. The book is based mainly on the fundamental papers of Fréchet, Fisher and Tippett, Von Mises, and Elfving. In addition the author could not avoid repeating some of his own work. Only the future can tell whether he has succeeded in his choice. To satisfy the demand for completeness, a bibliography is given, and it is hoped that no relevant publication is missing therefrom.

1.0.5. Arrangement of Contents. The systematic treatment of extreme values starts with methods which are distribution-free. After some interesting initial distributions have been considered, more general methods are developed, which hold for certain types of distributions; finally the asymptotic theory is constructed.

The book comprises eight chapters. Each chapter consists of several sections divided into paragraphs. Each section starts with a statement of the problems. The first chapter introduces those statistical tools which will be frequently used, especially the concept of intensity function, taken from actuarial statistics, and return period, taken from engineering practices.

The patient reader is asked to believe that these notions are necessary. He will be convinced of it from the third chapter onward.

The second chapter introduces distribution-free methods and culminates in the proof that a forecast of the number of exceedances of the extreme is more reliable than a similar forecast for the median. The third chapter shows the general properties of extremes which do not require the knowledge of the initial distribution; in particular it sets bounds on the increase of mean extremes with sample size.

The fourth chapter is devoted to exact distributions of the extremes. Here the initial distribution, the parameters contained therein, and the number of observations must be known. Two types of initial unlimited distribution are worked out: the usual exponential type, and the Cauchy type, which possesses a longer tail. Since the distributions of the extremes depend upon the behavior of the initial distribution at large absolute values of the variate, the two types possess different properties with respect to the extremes. A third type consists of certain limited distributions.

As long as small samples are considered, no new parameters enter into the distribution of extremes. This agreeable situation changes as soon as large samples are approximated by the three asymptotic theories which are studied in Chapters 5 to 7. Here two new parameters appear, which are connected with the intensity function and return period.

The first asymptotic distribution of extreme values valid for the exponential type of initial distributions seems to be the most important. Chapter 5 is devoted to the study of this theory. Applications especially to floods are given in Chapter 6. The second and third asymptotic distributions valid for the Cauchy type, and for certain limited distributions, are shown in Chapter 7. Certain empirical formulae derived by engineers for the breaking strength of materials are special cases or generalizations of the third asymptotic distribution of the extremes.

The last chapter deals with functions of extremes, their sum, difference, quotient, and geometric mean. The last two statistics are believed to be new, and it is not known whether they will have any practical application. They are included since they are special cases of an important general theorem.

Although the author does not share the usual pragmatic outlook which judges a theory by its practical value, the presentation chosen is such that practical conclusions can be drawn from all the theorems given. In addition some typical applications are shown. No distribution is stated without an explanation of how the parameters may be estimated, even at the risk that the methods used will not stand up to the present rigorous requirements of mathematically minded statisticians, who are welcome to find better estimates.

1.1. GENERAL TOOLS†

1.1.1. Linear Transformations. We deal mainly with continuous statistical variates. The probabilities that the variates are less than a certain value are designated by capital letters, and the derivatives of the probabilities by the corresponding small letters. These densities of probability are called *distributions*.

Instead of the usual standardization of variates we often use a linear transformation

(1) $$z = \alpha(x - \beta)$$

where β is an average of the same dimension as x and $1/\alpha$ is of the dimension x. Consequently z has no dimension. The transformation (1) leads from a distribution $f(x)$ of the variate x to the distribution

(2) $$\varphi(z) = \frac{1}{\alpha} f\left(\beta + \frac{z}{\alpha}\right)$$

of the reduced variate z. In general, all distributions where the parameters may be eliminated by a linear transformation are written in a reduced form. However, for all practical purposes, we have to return to the initial variate.

1.1.2. Other Transformations. From a distribution $f(x)$ we may pass to a distribution $\varphi(z)$ of a new variate z defined by the transformation

(1) $$z = h(x)$$

by means of the equation

(2) $$\varphi(z) = f(x) \left|\frac{dx}{dz}\right|$$

The indicated absolute value is necessary since $\varphi(z)$ cannot be negative. The probabilities, $\Phi(z)$ and $F(x)$ are equal

(3) $$\Phi(z) = F(x)$$

provided that the derivative $h'(x)$ is positive. Consequently the quantiles $x(F)$ defined by the probability $F = q\%$ and especially the median of the variate z are obtained from the quantiles of x by the transformation (1). The mode $\tilde{z}$, however, if it exists, has to be calculated from (2).

† A reader informed in statistics may omit this section and return to it when the necessity arises.

The transformation $z = x^{-1}$ called *reciprocal* may be applied to a non-negative variate. The transformed distribution $\varphi(z)$ is from (2)

(4) $$\varphi(z) = z^{-2} f\left(\frac{1}{z}\right).$$

For $z = 1$ we have $x = 1$ and $\varphi(1) = f(1)$. If the initial distribution has one and only one mode, the transformed distribution need not have a mode and vice versa. The probability $\Phi(z)$ of a value less than z is

(5) $$\Phi(z) = 1 - F(x) ; \quad \Phi(1) = 1 - F(1)$$

Therefore the median $\check{z}$ of the reciprocal variate is the reciprocal of the initial median $\check{x}$. From (4) it follows that

$$\varphi(\check{z}) \gtrless f(\check{x}) \quad \text{if} \quad \check{x} \gtrless 1.$$

The density of probability at the median is conserved if and only if the median is equal to unity. The order of the quartiles is reversed by the transformation. The interquartile range remains the same.

Exercises: 1) Show that the distribution obtained from a logarithmic transformation of a normal variate is invariant under a reciprocal transformation. 2) How is the mode of the transformed distribution related to the transformation of the initial mode?

References: Rietz (1922).

1.1.3. Symmetry. A distribution with median zero is called symmetrical if, for all x

(1) $$F(-x) = 1 - F(x).$$

Evidently

(2) $$f(-x) = f(x); \quad f'(-x) = -f'(x).$$

The mean (provided it exists) coincides with the median and the most (or least) probable value if unique. All odd central moments vanish. The probabilities are

(3) $$F(x) = \frac{1 + \theta(x)}{2} ; \quad F(-x) = \frac{1 - \theta(x)}{2} ; \quad \theta(x) = \int_{-x}^{+x} f(z)\, dz.$$

A reduction of the variate has no influence on the symmetry, but shifts its center. Symmetry is a quality invariant under a linear transformation. Non-linear transformations applied to a symmetrical distribution may destroy symmetry. Non-linear transformations applied to an asymmetrical distribution may introduce symmetry. Asymmetrical distributions are called skewed. The skewness is called positive (negative), if the mode precedes (follows) the median. Then the larger part of the distribution is on the right (left) and the steeper slope of the distribution is on the left (right) side of the mode.

Two distributions $f_1(x_1)$ and $f_2(x_2)$ of two variates x_1 and x_2 so that for all x_1 and x_2

(4) $$F_1(-x_1) = 1 - F_2(x_2) \; ; \quad f_1(-x_1) = f_2(x_2)$$

are called *mutually symmetrical*. The means, modes, and medians coincide in size and differ in sign. All even central moments coincide in size and sign. All odd central moments coincide in size and differ in sign. The quartiles of one distribution are the quartiles of the other in reverse order. Let x_1 and x_2 be two mutually symmetrical variates, and let y_1 and y_2 be two transformed variates where the transformations $x_1(y_1)$ and $x_2(y_2)$ are linear, but different. Then the distributions $\varphi_1(y_1)$ and $\varphi_2(y_2)$ are still called mutually symmetrical.

1.1.4. Measures of Dispersion. The population standard deviation σ and the central moment μ_3 are estimated for a sample of size n by

(1) $$s = \sqrt{[1 + 1/(n-1)](\overline{x_0^2} - \bar{x}_0^2)};$$
$$m_3 = [\overline{x_0^3} - 3\overline{x_0^2}\,\bar{x}_0 + 2\bar{x}_0^3]\, n^2/(n-1)(n-2)$$

where the index 0 indicates sample values and where

(2) $$\overline{x_0^\kappa} = \sum_1^n x_\nu^\kappa / n \; ; \quad \nu = 1, 2, \ldots, n.$$

The coefficient of variation

(3) $$V = \sigma/\bar{x}$$

is estimated from the sample by $s/\bar{x}_0$.

We sometimes use the mean deviation ϑ defined by

(4) $$\vartheta = \int_{-\infty}^{+\infty} |x - A| f(x)\, dx$$

where A is an arbitrary fixed value. If

(5) $$\lim_{x = -\infty} x F(x) = 0$$

integration by parts leads from (4) to

(6) $$\vartheta = \bar{x} - A + 2\int_{-\infty}^{A} F(x)\, dx.$$

The mean deviation is a minimum if taken about the median $\check{x}$.

For a given distribution, the evaluation of the integral in (6) leads to ϑ as a function of the parameters, or of one parameter. If the population values $\bar{x}$ and $\check{x}$ are replaced by the sample values $\bar{x}_0$ and $\check{x}_0$, the mean deviations ϑ and ϑ_1 are estimated for the choices of $A = \bar{x}$ and $A = \check{x}$ respectively by

(7) $\quad t = \sum_{1}^{n} |x_\nu - \bar{x}_0|/\sqrt{n(n-1)} \, ; \qquad t_1 = \sum_{1}^{n} |x_\nu - \check{x}_0|/\sqrt{n(n-1)} \, .$

Two mutually symmetrical distributions have the same mean deviation. If the population mean does not exist, the population mean deviation does not exist either, since the integral (6) then diverges.

References: Cadwell (1954b), Reiersol, Rider (1929).

1.1.5. Moments. Sample moments are used to estimate the unknown parameters of a distribution and to calculate the standard errors of such estimates. For non-negative variates, the calculation of the population moments $\overline{x^l}$ about origin and of integral order l (provided that they exist) may be simplified by the introduction of the probability $F(x)$, since

(1) $\quad \overline{x^l} = -\int_0^\infty x^l \, d[1 - F(x)] \, ; \qquad l = 1, 2, \ldots .$

If a distribution lacks population moments from a certain order onward, the sample moments exist, but they cannot be used for the estimation of parameters, since they have no theoretical counterpart. However, we may define *reciprocal moments* $\overline{x^{-l}}$ (provided that they exist) by

(2) $\quad \overline{x^{-l}} = \int_0^{+\infty} x^{-l} f(x) \, dx \, .$

The reciprocal moments occur also if a non-negative value x undergoes the reciprocal transformation (1.1.2). Then the moments of x are related to the moments of z by

(3) $\quad \overline{x^l} = \overline{z^{-l}} \, .$

The reciprocal moments of the reciprocal of a variate are the moments of the original variate, provided that both moments exist.

The most important tool for the estimation of parameters is the asymptotic normal distribution of the sample mean $\bar{x}_0$, median $\check{x}_0$, standard and mean deviations s and t for those distributions with which we are concerned. The expectations in each case are the corresponding population values, and the respective asymptotic variances are

(4) $\quad \begin{aligned} & n\sigma^2(\bar{x}_0) = \sigma^2 \, ; \qquad n\sigma^2(\check{x}_0) = [2f(\check{x})]^{-2} \, ; \\ & n\sigma^2(s) = \sigma^2(\beta_2 - 1)/4 \qquad n\sigma^2(t) = \sigma^2 - \vartheta^2 \, ; \\ & n\sigma^2(a\bar{x}_0 + bs)/\sigma^2 = a^2 + ab\sqrt{\beta_1} + b^2(\beta_2 - 1)/4 \, . \end{aligned}$

An estimate of a parameter from a statistic which has a smaller variance than another one is called more precise.

The moment quotients called Betas used in (4) are defined by

(5) $\beta_1 = \mu_3^2 \mu_2^{-3}$; $\beta_2 = \mu_4 \mu_2^{-2}$; $\beta_3 = \mu_3 \mu_5 \mu_2^{-4}$; $\beta_4 = \mu_6 \mu_2^{-3}$

where the μ_ν are the central moments of order ν. If the distribution contains two or less parameters and can be reduced, the Betas are fixed numbers. If the distribution contains three parameters, β_2 and the higher Betas are functions of the first one.

Exercises: 1) Calculate the asymptotic variances of the sample skewness. 2) Apply the formulae (4) to the exponential and to the normal distributions. 3) Derive the asymptotic variance of a linear function of the sample median and mean deviation.

References: Bennett, Cramér (p. 352), Gumbel and Carlson (1956), Kaplansky, Zeigler.

1.1.6. Generating Function. A simple tool for the calculation of the population moments is the construction of the moment generating function $G_x(t)$ defined by

(1) $$G_x(t) = \int_{-\infty}^{+\infty} e^{xt} f(x)\, dx\,.$$

If this integral exists and can be expressed as a closed function of t, the moments of order l are its lth derivates at $t = 0$. The generating functions $G_1(t)$ and $G_2(t)$ for two mutually symmetrical distributions, 1.1.3(4), are related by

(2) $$G_2(t) = G_1(-t)\,.$$

If we replace t by $-t$ in the generating function, we obtain the moments of a distribution which is symmetrical to the original one.

The generating function $G_z(t)$ of the reduced variate $z = \alpha(x - \beta)$ is related to $G_x(t)$ by

(3) $$G_x(t) = e^{\beta t} G_z\left(\frac{t}{\alpha}\right)\,.$$

Conversely, the generating function $G_x(t)$ of the variate x is obtained from the generating function $G_z(t)$ of a reduced variate z by writing t/α instead of t, and multiplication by $e^{\beta t}$.

For the calculation of the central moments we use the *seminvariant generating function* $L_x(t)$ defined by

(4) $$L_x(t) = -\bar{x}t + \lg G_x(t)\,.$$

The *seminvariants* (or cumulants) λ_ν of order ν are defined as the coefficients of $t^\nu/\nu!$ in the Taylor series

$$(5) \qquad L_x(t) = \sum_{\nu=2}^{\infty} \frac{\lambda_\nu t^\nu}{\nu!}$$

The even (odd) seminvariants for two mutually symmetrical distributions coincide in size and sign (differ in sign). The seminvariants for one distribution determine those for the other one.

The seminvariants λ_ν are related to the central moments μ_ν by

$$(6) \qquad \mu_2 = \lambda_2 \; ; \qquad \mu_3 = \lambda_3 \; ; \qquad \mu_4 = \lambda_4 + 3\lambda_2^2 \, ,$$

$$(7) \qquad \mu_5 = \lambda_5 + 10\lambda_3\lambda_2 \; ; \qquad \mu_6 = \lambda_6 + 15\lambda_4\lambda_2 + 10\lambda_3^2 + 15\lambda_2^3 \, .$$

The Betas and the excess defined by

$$(8) \qquad \varepsilon \equiv \beta_2 - 3$$

are expressed by the seminvariants as

$$(9) \qquad \begin{aligned} &\beta_1 = \lambda_3^2 \lambda_2^{-3} \; ; \qquad \varepsilon = \lambda_4 \lambda_2^{-2} \, , \\ &\beta_3 = \frac{\lambda_3 \lambda_5}{\lambda_2^4} + \frac{10\lambda_3^2}{\lambda_2^3} = \beta_1 \left(\frac{\lambda_5}{\lambda_2 \lambda_3} + 10 \right) , \\ &\beta_4 = \lambda_6/\lambda_2^3 + 10\beta_1 + 15(1 + \varepsilon) \, . \end{aligned}$$

These equations allow the calculation of the Betas from the seminvariant generating function. The Betas are not influenced by linear transformations of the variate, since they have no dimension and since the λ's are not affected by translations. The first two Betas of the mean $\bar{x}_0$ are linked to the Betas themselves by

$$(9') \qquad \beta_1(\bar{x}_0) = \beta_1/n \; ; \qquad \beta_2(\bar{x}_0) - 3 = (\beta_2 - 3)/n$$

The generating function $G_x(t)$ is helpful for logarithmic transformations. For the variate

$$(10) \qquad x = k \lg (z/a)$$

the generating function $G_x(t)$ becomes $G_x(t) = \overline{z^{kt}} a^{-kt}$. If we put $l = kt$, the lth moment and the lth reciprocal moment of the variate z (provided they exist) are obtained from the generating function of the variate x by

$$(11) \qquad \overline{z^l} = a^l G_x\left(\frac{l}{k}\right) ; \qquad \overline{z^{-l}} = a^{-l} G_x\left(-\frac{l}{k}\right) .$$

Let $k = a = 1$. Then the generating function of the logarithm of a variate is the tth moment of the logarithm of the variate:

(12) $$G_{\lg z}(t) = \overline{z^t}.$$

The generating function for $\lg z$ is equal to the generating function for x.

The *geometric moments* g_l of order l for the transformed variate z are defined by

(13) $$\lg \bar{g}_l = \int_0^\infty (\lg z)^l \varphi(z) \, dz,$$

whence

(14) $$\lg \bar{g}_l = \overline{x^l}.$$

The geometric moments for the logarithmically transformed variate z are equal to the usual moments about the origin of the variate x. In particular, the geometric variance σ_g^2 is defined by

(15) $$\sigma_g^2 = \bar{g}_2 - \bar{g}_1^2.$$

The standard deviation of the logarithms is sometimes called the *variability index*.

For a discontinuous non-negative variate with a distribution $w(x)$ we use the factorial moments $\bar{x}_{[k]}$ defined by

(16) $$\bar{x}_{[k]} = \sum_{x=k}^\infty x(x-1)\ldots(x-k+1)w(x)$$

and the factorial moment generating function

(17) $$G_{[x]}(t) = \sum_{x=0}^\infty (1+t)^x w(x)$$

which leads to these moments if $t = 0$ is introduced after differentiation with respect to t.

For the binomial distribution this function is

(18) $$G_{[x]}(t) = (1 + pt)^N$$

and the factorial moments become

(19) $$\overline{x_{[k]}} = N(N-1)\ldots(N-k+1)p^k.$$

Exercise: Calculate the Betas for the (first) Laplacean distribution:

$$f(x) = \frac{e^{-x}}{2}; \quad x \geqq 0; \quad f(x) = \frac{e^x}{2}; \quad x < 0.$$

1.1.7. Convolution. The main use of the generating function is in the convolution of variates. Let $f(x)$ and $g(y)$ be the distributions of two independent variates x and y. Let

(1) $$v \equiv x + y; \quad w \equiv x - y$$

be the sum and the difference of these variates. Then the expectations $\bar{v}$ and $\bar{w}$, and the variances σ_v^2 and σ_w^2, of the sum and the difference are

(2) $\quad \bar{v} = \bar{x} + \bar{y} \;; \qquad \bar{w} = \bar{x} - \bar{y} \;; \qquad \sigma_v^2 = \sigma_w^2 = \sigma_x^2 + \sigma_y^2 \;.$

The distributions $h(v)$ and $l(w)$ are obtained from

(3) $\quad h(v) = \int_{-\infty}^{+\infty} f(x)g(v-x)\,dx \;; \quad l(w) = \int_{-\infty}^{+\infty} f(x)g(x-w)\,dx$

where the actual limits of the integrals are determined from the domains within which the distributions $f(x)$ and $g(y)$ are non-negative. For example, let x and y be non-negative and unlimited toward the right. Then the conditions

$$x \geq 0 \;; \quad v - x \geq 0 \;; \quad x \geq 0 \;; \quad x - w \geq 0$$

lead to

$$0 \leq x \leq v \;; \quad x \geq 0 \,, \quad x \geq w \;.$$

Consequently,

(4) $\quad h(v) = \int_0^v f(x)g(v-x)\,dx; \qquad l(w) = \begin{cases} \int_0^\infty f(x)g(x-w)\,dx; & w < 0 \\ \int_w^\infty f(x)g(x-w)\,dx; & w > 0. \end{cases}$

This process, which is called convolution, enables one to determine the moments of v and w, using only one distribution. However, they may also be obtained from the generating functions $G_v(t)$ and $G_w(t)$ without the explicit knowledge of the distributions $h(v)$ and $l(w)$. The moment generating functions for the sum and the difference are

(5) $\qquad G_v(t) = G_x(t)\,G_y(t) \;; \qquad G_w(t) = G_x(t)\,G_y(-t) \,,$

$\qquad\qquad L_v(t) = L_x(t) + L_y(t) \;; \qquad L_w(t) = L_x(t) + L_y(-t) \,.$

The system (5) allows the calculation of the moments of the sum and of the difference of two variates even if the distribution of the sum and the difference cannot be written explicitly. The even seminvariants of the sum v are equal to the even seminvariants of the difference w, whereas the odd seminvariants $\lambda_{v,2\nu+1}$ and $\lambda_{w,2\nu+1}$ of the sum and difference are

(6) $\quad \lambda_{v,2\nu+1} = \lambda_{x,2\nu+1} + \lambda_{y,2\nu+1} \;; \quad \lambda_{w,2\nu+1} = \lambda_{x,2\nu+1} - \lambda_{y,2\nu+1} \;.$

If the two distributions $f(x)$ and $g(y)$ are identical, the seminvariants of the sum are twice the seminvariants of the original distribution and the odd seminvariants of the difference vanish. If the two distributions $f(x)$ and

$g(y)$ are mutually symmetrical, the generating function of the sum v and the difference w are from 1.1.6(2) and (5).

(7) $$G_v(t) = G_x(t) \cdot G_x(-t) \; ; \qquad G_w(t) = G_x^2(t) \, ,$$
$$L_v(t) = L_x(t) + L_x(-t) \; ; \qquad L_w(t) = 2L_x(t) \, .$$

The even seminvariants of the sum v and the even seminvariants of the difference w are twice the even seminvariants of the variate x or y. The odd seminvariants of the difference w are twice the odd seminvariants of x, and the odd seminvariants of the sum v vanish. Consequently the convolution of two mutually symmetrical distributions creates a symmetrical distribution.

Exercise: Show that the Laplacean variate 1.1.6 may be obtained as the difference of two exponential variates.

1.1.8. The Gamma Function. In connection with generating functions, the Gamma function often occurs. The development of its logarithm is

(1) $$\lg \Gamma(1-t) = \gamma t + \sum_{\nu=2}^{\infty} S_\nu t^\nu / \nu \; ; \qquad |t| < 1$$

where

(2) $$\gamma = 0.5772156649 \ldots .$$

is Euler's constant. The sum S_ν defined by

(3) $$S_\nu = \sum_{\lambda=1}^{\infty} \lambda^{-\nu}$$

exist for $\nu \geq 2$, and for S_1, which diverges, the following convergence holds:

(3') $$\lim_{n=\infty} (S_1 - \lg n) = \gamma$$

The first numerical values of the other S_ν are

(4) $$S_2 = \pi^2/6 \; = 1.64493407 \; ; \qquad S_3 = 1.20205690 \, ,$$
$$S_4 = \pi^4/90 = 1.08232323 \; ; \qquad S_5 = 1.03692776 \, ,$$
$$S_6 = \pi^6/945 = 1.01734306 \; ; \qquad S_7 = 1.00834928 \, .$$

The first few derivatives of the logarithm of $\Gamma(1-t)$ for $t = 0$ are

$$\lg' \Gamma(1) = \gamma \; ; \qquad \lg'' \Gamma(1) = S_2 \, ,$$
$$\lg^{(3)} \Gamma(1) = 2S_3 = 2.4041138 \; ; \qquad \lg^{(4)} \Gamma(1) = 6S_4 = 6.49393939 \, .$$

For $t = 0$, the first derivatives of $\Gamma(1 - t)$ thus are

(5) $\quad \Gamma(1) = 1 \; ; \quad \Gamma'(1) = \gamma \; ; \quad \Gamma^{(2)}(1) = \gamma^2 + \pi^2/6 = \quad 1.978111994$,

$\quad \Gamma^{(3)}(1) = \gamma^3 + \gamma\pi^2/2 + 2S_3 \qquad\qquad\qquad = 5.444874456$,

$\quad \Gamma^{(4)}(1) = \gamma^4 + \gamma^2\pi^2 + 8\gamma S_3 + 3\pi^4/20 \qquad = 23.561457369$.

The quotient $\Gamma(n - t)/\Gamma(n)$ may be written for $n \geq 2$;

(6) $\quad \lg \dfrac{\Gamma(n - t)}{\Gamma(n)} = \lg \Gamma(1 - t) + \sum\limits_{\lambda=1}^{n-1} \lg (1 - t/\lambda)$,

whence

$$\frac{d \lg \Gamma(n - t)/\Gamma(n)}{dt} = \lg' \Gamma(1 - t) - \sum_{\lambda=1}^{n-1} \frac{1}{\lambda - t} .$$

The analytic properties of the Gamma function, its numerical values, and developments different from (1) may be found in the usual mathematical textbooks. For integral values of n the function $\Gamma(1 + n) = n!$ can be approximated by Stirling's formula

(7) $\qquad\qquad\qquad\qquad n! \sim n^n e^{-n} \sqrt{2\pi n}$

If n and m are integers and n is large compared to m, this formula leads to the approximations

(7') $\quad \dfrac{n!}{(n-m)!} \sim \dfrac{(n+m)!}{n!} \sim n^m \; ; \quad \dfrac{(mn)!}{(n!)^m} \sim \dfrac{m^{mn+1/2}}{(2\pi n)^{(m-1)/2}}$.

1.1.9. The Logarithmic Normal Distribution. The distribution which is obtained from the normal one by a logarithmic transformation has often been used in problems connected with extremes. Therefore a study of its properties is necessary. This transformation was first used by Galton. Kolmogoroff has shown that the corresponding distribution holds for the sizes if a certain mass is broken up into small parts by the application of force. It has also an exact basis for the measure of the luminosity of the stars. Grassberger applied it to the daily discharges of a river and, at the same time, to their annual maxima. Gibrat used it for daily discharges, Hazen for floods, and this use is still widespread. Finally, many authors use it for fatigue failures. Some practical statisticians believe that every skewed distribution should be logarithmically normal, and therefore use the logarithmic transformation indiscriminately. It is pointed out that the distribution has no upper limit. Let $f(y)$ and $F(y)$ be the normal functions. Then the distribution $g(z)$ and the probability $G(z)$ obtained from the transformation

(1) $$y = \lg z$$

are

(2) $$g(z) = \frac{1}{z\sqrt{2\pi}} \exp\left(-\frac{\lg^2 z}{2}\right) ; \quad G(z) = F(\lg z).$$

Some authors consider instead

(3) $$g_1(z) = \frac{1}{\sqrt{2\pi}} \exp\left(-\frac{\lg^2 z}{2}\right).$$

This does not result from the transformation (1). Furthermore, $g_1(z)$ is not even a distribution, since the integral taken from zero to infinity is $\sqrt{e}$ rather than unity. Therefore, to make it a distribution, it must be written

(4) $$g_2(z) = \frac{1}{\sqrt{2\pi e}} \exp\left(-\frac{\lg^2 z}{2}\right) ; \quad G_2(z) = F[(\lg z) - 1].$$

This form will not be used in the following.

If a variate x is introduced into (2) by the transformation

(5) $$\lg x = \lg \beta + \delta \lg z$$

the distribution $f_g(x)$ and the probability function $F_g(x)$ are

(6) $$f_g(x) = \frac{1}{\delta\sqrt{2\pi}x} \exp\left[-\frac{\lg^2 (x/\beta)}{2\delta^2}\right] ; \quad F_g(x) = F\left[\frac{\lg (x/\beta)}{\delta}\right].$$

The parameter β may be interpreted either as the median $\check{x}$ or as the geometric mean and should be used to characterize the distribution.

The probability function possesses the symmetry

(7) $$F_g(\beta/x) = 1 - F_g(x/\beta).$$

The mode $\tilde{x}$ is

(8) $$\tilde{x} = \beta \exp(-\delta^2),$$

whence the symmetry relation

(9) $$f_g(k\tilde{x}) = f_g(\tilde{x}/k) ; \quad k > 0.$$

A multiple of the mode has the same probability density as the corresponding fraction. The moments of order l are

(10) $$\overline{x^l} = \beta^l \exp(l^2 \delta^2/2).$$

The mode, median and mean,

(11) $\qquad \tilde{x} = \beta \exp(-\delta^2) < \check{x} = \beta < \bar{x} = \beta \exp(\delta^2/2)$

are related by

(12) $\qquad\qquad\qquad \bar{x}^2 \tilde{x} = \check{x}^3 .$

From (10) the variance and the coefficient of variation are

(13) $\qquad\qquad \sigma^2 = \beta^2 e^{\delta^2}(e^{\delta^2} - 1) ; \qquad V^2 = e^{\delta^2} - 1 .$

The relation between the mean, median, and coefficient of variation obtained from (11) and (13)

(13') $\qquad\qquad\qquad \bar{x}/\check{x} = \sqrt{1 + V^2}$

is traced in Graph 1.1.9. The relation

(14) $\qquad\qquad \sigma^2(\lg x) = \lg(1 + V^2)$

between the logarithmic standard deviation and the coefficient of variation obtained from (5) and (13) traced in the same graph is nearly linear. Consequently, the standard deviation of x is nearly proportional to the product of the mean $\bar{x}$ and the standard deviation of the logarithm of x. The two preceding equations may be combined to

(15) $\qquad\qquad \sigma^2(\lg x) = 2 \lg(\bar{x}/\check{x}) .$

This relation is also traced in Graph 1.1.9. Since from (10) the third central moment is

$$\mu_3 = \beta^3 e^{3\delta^2/2}(e^{3\delta^2} - 3e^{\delta^2} + 2)$$

the skewness expressed by the coefficient of variation also traced in Graph 1.1.9. is

(16) $\qquad\qquad\qquad \sqrt{\beta_1} = V^3 + 3V .$

The relations (12), (13'), and (14) to (16) can serve as criteria for the validity of the theory

Exercises: 1) Trace the β_1, β_2 diagrams. 2) Trace the probability function

(17) $\qquad 1 - F(x) = \exp[-x/(1-x)] ; \qquad 0 \leqq x \leqq 1$

and the corresponding distribution. Calculate the averages, moments about unity, and the first moment quotients.

1.1.9 AIMS AND TOOLS

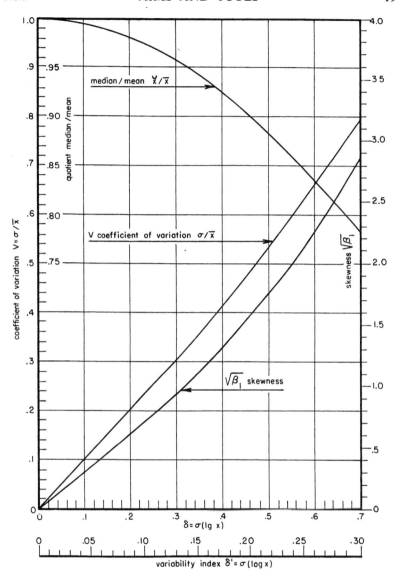

Graph 1.1.9. Characteristics for the Logarithmic Normal Distribution

References: Chow (1954), Cohen, Galton (1875, 1879), Gibrat (1931, 1932, 1934), Gordon, Grassberger (1932), Gumbel (1926), Hazen, Iwai, Jordan, Kalinske, Kapteyn, Kimball (1938), McAllister, Moshman (1953), Quensel, Severo and Olds, Slade (1936b), Weiss (1957).

1.2. SPECIFIC TOOLS

1.2.0. Problems. In the following we introduce two new statistical functions, the *intensity* $\mu(x)$ and the *return period* $T(x)$, indispensable for the analysis of extreme values. One of the parameters in the asymptotic distributions of the extremes (Chapter 5) is a certain value of the intensity function. The return period is of great importance for statistical phenomena where the time factor enters. Engineers who work on flood control are interested in the time intervals between two discharges of a river each of which is equal to or greater than a given one. The mean of these intervals is the return period. The connection between the two notions will be shown in 1.2.3. The return period will be used on probability papers, which are an essential tool in the analysis of extreme values. This leads to the problem of how to plot n individual observations on such a paper and how to fit straight lines.

1.2.1. The Intensity Function. The probability of a value equal to or larger than x is $1 - F(x)$, while the probability of a value between x and $x + dx$ is $f(x)\, dx$. The probability that a value known to be equal to or larger than x be situated between x and $x + dx$ (failure or hazard rate) is

$$(1) \qquad \mu(x)\, dx = \frac{f(x)\, dx}{1 - F(x)} \geq f(x)\, dx \,.$$

The reciprocal $1/\mu(x)$ is called Mill's Ratio. K. Pearson tabulated it for the normal distribution. Graph 1.2.1 shows the function $\mu(x)$ for different variates. In mortality statistics, $f(x)\, dx$ is the probability of a newborn dying between the ages x and $x + dx$, and $1 - F(x)$ is the probability of a newborn reaching age x. Thus $\mu(x)\, dx$ is the probability that a person aged x will die in the next interval dx and it is called the force of mortality. Henceforth it is called the *intensity* function. For a given intensity the probability $F(x)$ is obtained from

$$(2) \qquad 1 - F(x) = (1 - F(x_0)) \exp\left[-\int_{x_0}^{x} \mu(z)\, dz\right],$$

where x_0 is an arbitrary value of x. The intensity $\mu(x)$ may be independent of x. It may increase with x without limit; it may converge toward a constant or it may decrease with x. However, if the variate is unlimited toward the right such that $F(x) < 1$ for any finite x it follows from (2) that *the integral*

$$\int_{x}^{\infty} \mu(z)\, dz$$

must diverge. This condition holds for all intensity functions which increase with x. It still holds for

$$\mu(x) = \kappa/x; \quad \kappa > 0,$$

but it will no longer hold for

$$\mu(x) = \kappa/x^{1+\varepsilon}; \quad \varepsilon > 0.$$

This sets a limit to the possible decrease of the intensity function.

The intensity function is related to the logarithmic derivative of the distribution. Differentiation of the equation $[1 - F(x)]\mu(x) = f(x)$, obtained from (1), leads to

(3) $$\mu'(x)/\mu^2(x) = 1 + f'(x)/[f(x)\mu(x)]; \quad f(x) \neq 0.$$

This equation facilitates the differentiation of the intensity which is often complicated. The intensity function need not be monotonic. See Graph 1.2.1 for the logarithmic normal distribution.

If the distribution has a mode, $\tilde{x}$, we obtain from (3)

(4) $$\mu'(\tilde{x}) = \mu^2(\tilde{x}),$$

provided this equation has a root in the domain of the variate x.

Exercises: 1) Calculate the intensity function for the distributions shown in Graph 1.2.1. 2) Study the distribution corresponding to $\mu(x) = 1 + e^{-x}$.

References: Davis, Gordon, Hooker and Longley-Cook, Komatu, Weiss (1957).

1.2.2. The Distribution of Repeated Occurrences. To derive the notion of return period we construct a dichotomy for a continuous variate. First, we consider the observations equal to or larger than a certain large value x. (This exceedance is the event in which we are interested.) Second, we consider the observations smaller than this value. Let

(1) $$q = 1 - p = F(x)$$

be the probability of a value smaller than x. Observations are made at regular intervals of time, and the experiment stops when the value x has been exceeded once. We ask for the probability $w(v)$ that the exceedance happens for the first time at trial v (geometric distribution).

The variable v is an integer, limited to the left, but unlimited to the right, since the event need not happen. If the event happens at trial v, it must have failed in the first $v-1$ trials. Therefore the probability $w(v)$ is

(2) $$w(v) = pq^{v-1}; \quad v \geq 1.$$

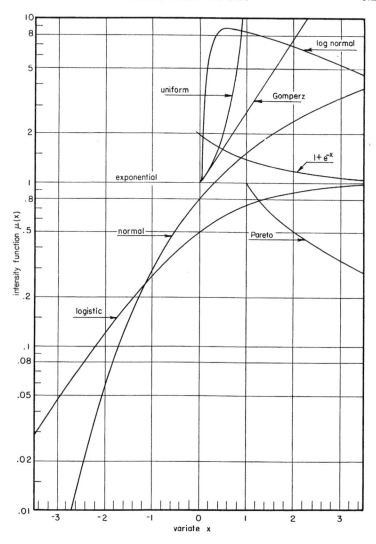

Graph 1.2.1. Intensity Functions $\mu(x)$

It decreases with increasing v. The moment generating function is

(3) $$G_v(t) = p/(e^{-t}-q).$$

The mean $\bar{v}$ for which we write $T(x)$ is

(4) $$\bar{v} = 1/p = T(x) = \frac{1}{1-F(x)} > 1.$$

This result is self-evident: If an event has probability p, we have to make, on the average, $1/p$ trials in order that the event happen once. The mean $T(x)$ will be called the *return period*.

From the generating function (3) the standard deviation becomes

(5) $$\sigma = \sqrt{T^2 - T} \to T - 1/2 \,.$$

The smaller p is, the larger is the spread of the distribution (2).

The cumulative probability $W(v)$ that the event happens before or at the vth trial is

(6) $$W(v) = 1 - q^v \,.$$

Therefore, the cumulative probability $W(T)$ for the event to happen before or at the return period T is from (4)

(7) $$W(T) = 1 - (1 - 1/T)^T \sim 1 - 1/e = 0.63212 \,.$$

The median number of trials $\breve{v} = T_0$ is

(7') $$T_0 = \frac{0.69315}{-\lg(1 - 1/T)} \sim 0.69315T - 0.34657 \,.$$

Of course, T_0 increases with T. The mean is 44% larger than the median. There is as much chance for the event to happen before $.69T$ as after.

If we have N series of this type, the observed distribution $w(v)$ may be compared to the theoretical one given in (2), provided that p is known. If not, p may be estimated from (4). This is also the maximum likelihood estimate. A rough estimate is the observed relative frequency $w(1)$ itself. We may also compare the observed variance with its expectation σ^2 which may be estimated from (5) by replacing p by its estimate. Formula (2) has been applied by P. S. Olmstead to the number of apparatus that functioned v times in the transmission of a signal. Hagstroem showed the role which equation (2) played in the early development of the calculus of probabilities, especially in the controversy between Fermat and Méré.

If x is large then p is small (say $p < .1$), and T is large (say $T > 10$), and the probability (6) may be written

(8) $$W(v) \sim 1 - \exp(-v/T) \,.$$

This exponential probability function valid for large T is traced in Graph 1.2.2(1) on a double logarithmic scale against v expressed in units of T. The probability

(9) $$P = e^{-1/\lambda} - e^{-\lambda}$$

for v to be contained in the interval

$$T/\lambda < v < T\lambda \,; \quad \lambda > 1$$

is also shown in this graph. The interval

$$.4751\ T < v < 2.1050\ T$$

corresponds to the probability $P = 1/2$. If we fix the probability $P = 0.68269$ (or $= 0.95450$) which corresponds to the interval $\bar{x} \pm \sigma$ (or $\pm 2\sigma$) in the normal distribution, the corresponding intervals for v become

(10) $.3196\ T < v < 3.129\ T$; $.04654\ T < v < 21.485\ T$.

For a given distribution and a given value of x corresponding to a large value of T, there is a probability 2/3 that the event happens within the

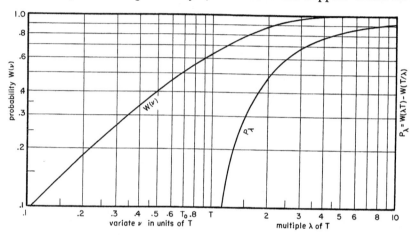

Graph 1.2.2(1). **Control Intervals for the Return Period**

first (one sigma) interval (10) which is called the *control interval*. See Graph 1.2.2(1).

Let the unit interval between observations be years. If T is large, the probability $W(\omega)$ that the event with the return period $T(\omega)$ happens before $N(\omega)$ years is from (8),

(11) $$1 - W(\omega) = \exp[-N(\omega)/T(\omega)].$$

For practical purposes we choose $N(\omega)$, want $W(\omega)$ to be small, and ask for the *design value* x_ω of the variate corresponding to the return period $T(\omega)$. This procedure, proposed by A. Court (1952, p. 45) may be called the *calculated risk*. In first approximation the design duration $T(\omega)$ is

(11') $$T(\omega) \sim N(\omega)/W(\omega).$$

If we want a probability $1 - W(\omega) = .95$ that the unfavorable event will not happen before $N = 10$ years, we have to choose the design value x_ω so large that its return period is 200 years.

1.2.2 AIMS AND TOOLS

The relation (11) is distribution-free. In order to obtain the design value x_ω we have to know the distribution. For example, if the distribution is exponential the value x_ω is, from (11'),

(12) $$x_\omega \sim \lg N(\omega) - \lg W(\omega).$$

The relation between the design factor $z_\omega = (x_\omega - \bar{x})/\sigma$ and the quotient $N(\omega)/W(\omega)$, i.e., the return period $T(\omega)$, is traced in Graph

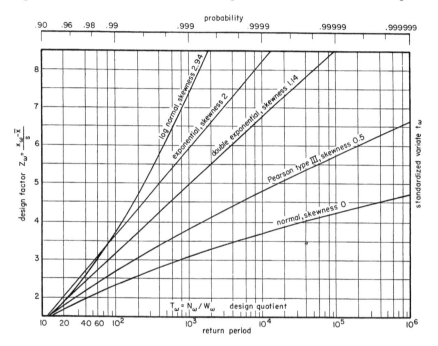

Graph 1.2.2(2). Design Factor versus Design Quotient

1.2.2(2) for different distributions on semilogarithmic paper. (The term double exponential will be explained in 5.2.1.) The choice of $N(\omega)$ and $W(\omega)$ is governed by political, economic, or technical considerations.

Exercises: 1) Calculate T and σ directly from (2) without the help of the generating function. 2) Calculate the higher moments from the generating function. 3) Calculate a table of the design value for the normal distribution.

Problem: What is the distribution of the quotient s^2/σ^2?

References: Hammer, Langbein (1949), Levert, Linsley, Kohler, and Paulhus, H. A. Thomas.

1.2.3. Analysis of Return Periods. The return periods are most interesting if the observations are made at equidistant intervals of time, as is the case in meteorological statistics. Then the return periods are times and, for annual observations, numbers of years. The return period for the median is 2, for the upper quartile it is 4 and so on. If x is smaller than the median, $T(x)$ is smaller than 2 and decreases with decreasing values of x. For a distribution with the initial condition $F(a) = 0$, the return period $T(a)$ is equal to unity.

The value x_D for which the return period of x is doubled is found from 1.2.2(4) as the solution of

(1) $$F(x_D) = [1 + F(x)]/2 .$$

Conversely, the return period of $2x$ is the solution of

(2) $$T(2x) = 1/[1 - F(2x)] .$$

For a given distribution we obtain x_D as a function of x, and $T(2x)$ as a function of $T(x)$, since x is a function of T.

To any continuous distribution there corresponds uniquely a continuous return period, and to any return period there corresponds a unique distribution:

(3) $$f(x) = \frac{1}{T(x)} \frac{d \lg T(x)}{dx} .$$

Finally, the variate x may also be considered as a function of the return period T. The first derivative of this function is

(4) $$\frac{dx}{d \lg T} = \frac{1}{f(x)T(x)} > 0 .$$

Therefore, one should plot the return period $T(x)$ on a logarithmic scale as the abscissa against x traced on an arithmetic scale as the ordinate. From 1.2.1(1) equation (4) becomes

$$\frac{dx}{d \lg T} = \frac{1}{\mu(x)} .$$

Consequently the sign of the second derivative of x as function of $\lg T$ is

(5) $$\text{sign} \frac{d^2 x}{(d \lg T)^2} = -\text{sign} \frac{d\mu(x)}{dx} .$$

If the derivative of the intensity function is always positive (negative), the curve $x = x(\lg T)$ is always concave (convex) downward. If $\mu'(x)$ changes its sign, the curve possesses an inflection. For the usual distribution, the intensity function increases with x. (See Graph 1.2.1.)

Equation (5) may also be expressed in a more practical form. From 1.2.1(3)

$$\text{sign}\frac{d^2x}{(d\lg T)^2} = -\text{sign}\left[\mu(x) + \frac{d\lg f(x)}{dx}\right].$$

The variate x traced as function of $\lg T$ is downward

(6) $\quad\quad\quad\quad\begin{array}{c}\text{concave}\\\text{convex}\end{array}\quad\text{if}\quad \mu(x) \gtreqless -\frac{d\lg f(x)}{dx}.$

If, for large x, the intensity function approaches the logarithmic derivative with negative sign, the variate x as a function of $\lg T$ approaches a straight line.

The problem of how to construct observed return periods will be solved in 1.2.7(2). Engineers who use empirical formulae often violate the initial condition for an unlimited distribution

(7) $$\lim_{x=-\infty} T(x) = 1.$$

For values of the variate which are less than the median another interpretation of the return period is possible: the mean number of observations $_1T(x)$ necessary to obtain once a value less than x is

(8) $$_1T(x) = 1/F(x).$$

If $T(x)$ (and $_1T(x)$) are used for values surpassing (and falling short of) the median, the return periods consist of two branches joining at the median for which both return periods are equal to 2. The return periods increase without limit with the absolute value of x. If for all x

(9) $$F(-x) < 1 - F(x),$$

the return periods of two values, $-x$ and $+x$, are

(9') $$_1T(-x) > T(x).$$

The absolute amount of the negative values increases more slowly with the return period than the positive values. The opposite relation holds if the inequality (9) is reversed.

If the distribution is symmetrical about zero, the two return periods are equal. If two distributions are mutually symmetrical, the return periods $_1T$ for the first one are the return periods T for the second one, and vice versa.

Exercises: 1) Calculate $T(2x)$ as function of $T(x)$ for the exponential and the logistic distribution. 2) Trace $T(2x)$ as function of $T(x)$ for the normal distribution. 3) Analyze the properties of $_1T(x)$. 4) Calculate the distributions of $T(x)$, $_1T(x)$, and of $\log T(x)$.

References: Horton, Levert, Saville, Tiago de Oliveira (1952).

1.2.4. "Observed" Distributions. The simplest testing procedure *seems* to consist in comparing the theoretical and observed distributions. In reality, the theoretical distribution, being a density of probability, is not a counterpart to the observed distribution, and there is no such thing as *the* observed distribution of a continuous variate. To obtain *an* observed distribution, we have to choose certain class intervals and to count the number of observations contained in these cells. This implies *two* arbitrary steps: the choice of the class length, and the choice of the starting point of the division. The influence of the first step is well known but the influence of the second step is seldom mentioned. This is due to the unfortunate procedure of publishing only the grouped data.

The ambiguity of the notion "observed" distributions leads to a serious drawback of the conventional χ^2 test for continuous variates, since we will obtain different values of χ^2 for the same observations, the same theory, the same parameters, and even the same class length, by merely shifting the starting point. In the actual computation of χ^2 only one observed distribution is used. Others will lead to other values of χ^2 which may be different at the chosen significance level. The way out of the ambiguity is to replace the classes of equal length by classes of equal probability. Another criterion which is based on the comparison of each individual observation to its theoretical counterpart will be given in the next paragraph.

References: Anderson and Darling (1952, 1954), Cochran (1952), Gumbel (1943), Mann and Wald, Neyman (1949), Williams.

1.2.5. Construction of Probability Papers. In the following we analyze probability papers which are widely used by engineers. This method avoids the arbitrary division into cells which is basic for the χ^2 test. We are mainly interested in unlimited continuous distributions. The probability papers are constructed to obtain approximate straight lines for the observed (cumulative) frequencies, allow a simple forecast, and lead to a graphical test for the goodness of fit, since the choice of a certain paper is identical with the choice of a distribution. In certain cases of good fit even the estimation of the parameters becomes unnecessary, since the theoretical straight line may be drawn by a ruler.

Let $\Phi(x,\alpha,\beta)$ be the probability function for a variate x unlimited in both directions. Suppose that a reduced variate

(1) $$y = \alpha(x - \beta)$$

exists, so that the probability function

(2) $$F(y) = \Phi(\beta + y/\alpha, \alpha, \beta)$$

for the reduced variate is independent of the parameters α and β.

To construct a probability paper, the variate x is plotted as ordinate, the reduced variate y as abscissa, both in linear scales. On a second horizontal scale we plot the probability $F(y)$. The values $F(y) = 0$ and $F(y) = 1$ do not exist on the paper. The return periods may be traced on an upper line parallel to the abscissa, $_1T(x)$ going to the left, and $T(x)$ going to the right. If the theory holds, the observations plotted on probability paper ought to be scattered closely about the straight line (1). Probability papers may also be constructed for non-linear relations between x and y.

To show the efficiency of probability papers, we compare the symmetrical first Laplacean function with mean zero and unit standard deviation

(3) $\quad F(x) = \tfrac{1}{2} e^{x\sqrt{2}}, x \leqq 0 \, ; \quad F(x) = 1 - \tfrac{1}{2} e^{-x\sqrt{2}} \, ; \quad x \geqq 0$

to the normal one. The first distribution has a peak in the middle, at $x = 0$, where the derivative is discontinuous. It surpasses the second one in the neighborhood of the peak and also for large absolute values of the variate. The two probabilities traced in Graph 1.2.5(1) in the conventional linear way show very small differences, a point which has been stressed by Fréchet and Wilson. On normal probability paper, however, in Graph 1.2.5(2), the part outside the interval $.02 < F < .98$ is enlarged in such a way that a clear distinction between the two curves is possible, provided that the sample size is sufficiently large.

Exercises: 1) Compare Cauchy's probability and the normal one with the same median and the same probable error in the linear way and on normal paper. 2) Construct a Laplacean probability paper and trace the normal, the Laplacean, Cauchy's, and the "logistic" probabilities. 3) Construct a probability paper for Cauchy's distribution and trace the four probabilities. 4) Trace the four probabilities on logistic probability paper.

Reference: Streiff.

1.2.6. The Plotting Problem. The usual problem is not the comparison of two theoretical distributions, but the comparison of an observed series to the theoretical straight line 1.2.5(1). For this purpose we have to decide where to plot the observations x_m, $(m = 1, 2, \ldots, n)$ arranged in increasing magnitude. Our aim is to plot *each* single observation. To arrive at the right answer to this problem, we show first two obvious procedures which will turn out to be unsatisfactory, and a compromise between them which is misleading. After these wrong procedures have been eliminated, the solution will be stated in the next paragraph, and derived in 2.1.3.

If n observations x_m of a continuous variate are arranged in increasing order, the mth observation has the cumulative frequency m/n, and the largest has the frequency 1. If the same observations are arranged in

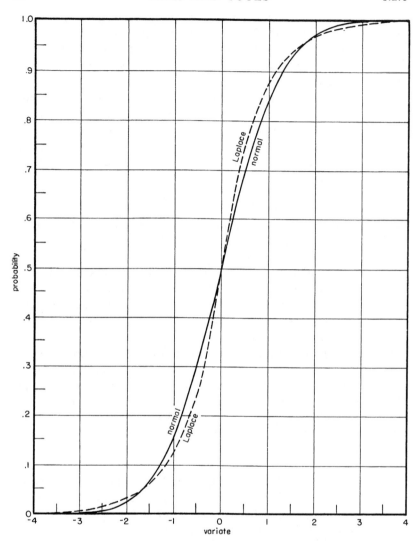

Graph 1.2.5(1). The First Laplacean and the Normal Probability

decreasing magnitude, the mth observation from the top, which is the $n - m + 1$th from the bottom (not the $n - m$th), has the frequency m/n, and the last (which is the smallest) has the frequency zero. Generally we may attribute to the mth observation either the frequency m/n or $(m - 1)/n$, and obtain two series of observed points. Near the median, the two series are very close. But toward the beginning and toward the end the

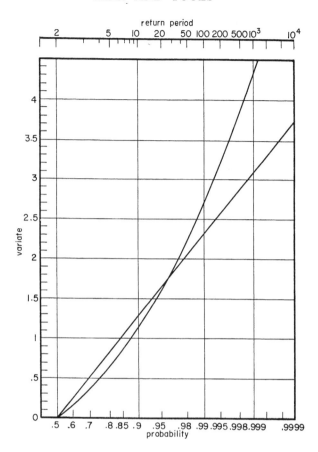

Graph 1.2.5(2). **The First Laplacean and the Normal Probability on Normal Paper**

two series diverge widely. Since the frequencies zero and unity do not exist on probability paper designed for an unlimited variate, the largest (smallest) observation cannot be plotted for the first (second) series.

The observed return periods 1.2.2(4) corresponding to the two frequencies are

(1) $T^{(1)}(x_m) = n/(n-m)$; $T^{(2)}(x_m) = n/(n-m+1)$.

Both increase with the rank m, and are related by

(2) $T^{(2)}(x_{m+1}) = T^{(1)}(x_m) < T^{(2)}(x_m)$.

For the last observation, $T^{(2)}(x_n)$ is equal to the sample size n, while $T^{(1)}(x_n)$ is no longer defined.

One might say: what does it matter if we attribute the frequencies 0.99 or

0.995 to a certain observation? In reality, a difference between return periods of 100 or 200 may be decisive for further course of action. Obviously a method is needed which leads to one and only one series, and which reproduces the smallest as well as the largest observation.

A way out, proposed by Hazen and used by many engineers, is a compromise attributing to x_m neither m/n nor $(m-1)/n$ but their arithmetic mean $(m - 1/2)/n$. It leads to an observed return period,

$$(3) \qquad T^{(3)}(x_m) = n/(n - m + 1/2) ,$$

which is the harmonic mean of $T^{(1)}(x_m)$ and $T^{(2)}(x_m)$. This procedure may be used in the conventional way of plotting $F(x)$ against x, on linear scales, but it cannot be used on probability paper, since the return period of the largest observation $m = n$ becomes $2n$. In other words, the compromise claims that an event which has already happened once in n years will occur, in the mean, in $2n$ years. If the extreme observation has economic consequences, as in the case of floods, the danger factor is heavily underestimated. The compromise is misleading where the plotting problem is of most interest.

To show the differences among the three plotting positions, the 17 yearly mean discharges for the Colorado River at Bright Angel Creek, 1923–1939, measured in 1000 cubic feet per second (Water Supply Paper 879, p. 266), are plotted on normal paper in Graph 1.2.6. The fourth series will be explained in the next paragraph. To each series corresponds another theoretical straight line, and therefore another forecast.

References: Beard (1942), Chernoff and Liebermann (1954), Horton (p. 433), Kimball (1947a).

1.2.7. Conditions for Plotting Positions. To find a way out of this confusion, we must appreciate the statistical nature of the problem. For n observations from a given distribution $f(x)$, the mth value has another distribution depending upon n, m, and $f(x)$. The distribution of the mth value possesses a median and may possess a mode and a mean. This will be studied in the next chapter. At present we are looking for a certain average of the mth value and the corresponding initial probability to be used as plotting positions. This will also fix the observed return periods. As a guidance in the choice we state some postulates:

1) The plotting position should be such that all observations can be plotted.

2) The plotting position should lie between the observed frequencies $(m-1)/n$ and m/n and should be universally applicable, i.e., it should be distribution-free. This excludes the probabilities at the mean, median, and modal mth value which differ for different distributions.

3) The return period of a value equal to or larger than the largest observation, and the return period of a value smaller than the smallest observation, should approach n, the number of observations. This condition need not be fulfilled by the choice of the mean or median mth value.

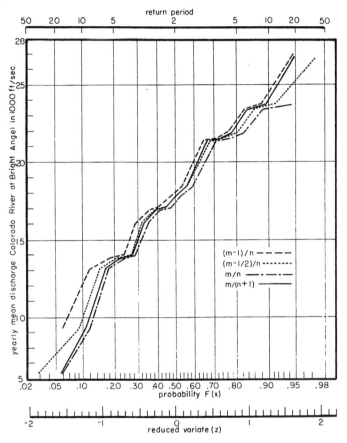

Graph 1.2.6. **Plotting Positions on Normal Paper**

4) The observations should be equally spaced on the frequency scale, i.e., the difference between the plotting positions of the $(m + 1)$th and the mth observation should be a function of n only, and independent of m. This condition is fulfilled for the three procedures outlined but need not be fulfilled for the probabilities at the mean, median, or modal mth values.

5) The plotting position should have an intuitive meaning, and ought to be analytically simple. The probabilities at the mean, modal, or median

mth value have an intuitive meaning. However, the numerical work involved is prohibitive.

In previous writings (1945), the author proposed to use the probabilities at the most probable extremes as plotting positions for the extremes with linear interpolation for the remaining $n-2$ plotting positions. This requires specific tables for different distributions.

For reasons which will be fully explained in the next section, we choose the mean frequency of the mth value

(1) $$\overline{F(x_m)} = m/(n+1)$$

as its plotting position. See Graph 1.2.6. The five conditions are satisfied. The *observed return period* $'T(x_m)$ of the mth values is

(2) $\quad 'T(x_m) = (n+1)/(n-m+1) \,; \quad 'T(x_{n-m+1}) = (n+1)/m \,,$

whence

(3) $\quad\quad 'T(x_1) = 1 + 1/n \,; \quad 'T(x_n) = {}_1'T(x_1) = n+1 \,.$

For large values of n and small values of m, the observed return periods approximate $T^{(1)}(x_m)$ and for large values of m they approximate $T^{(2)}(x_m)$. This solves the questions raised in 1.2.6.

The plotting positions are meant for the individual observations x_m. If the observations are grouped so that the mth up to the $m+k$th observations are represented by a mean value of the variate, we attribute to it a mean rank

(4) $$m' = \sqrt{m(m+k)} \,,$$

because the return period scale on the probability paper is approximately logarithmic. If the smallest or largest observations occur several times, they are plotted individually in order to conserve the return period (3).

The plotting position can also be used for forecasting distributions. Assume n observations have been fitted to a certain distribution $f(x,\alpha,\beta)$. Let a and b be the estimates of the parameters. We want to know the approximate sizes x_m of N future observations. From the plotting positions we derive the N reduced variates $z_m = a(x_m - b)$. The estimates a,b lead to the N values $x_m (m = 1, 2, \ldots, N)$ which constitute the forecast.

References: Beard (1951), Benard and Bos-Levenbach, Gumbel (1945b, 1947a), Haldane, Kimball (1946b).

1.2.8. Fitting Straight Lines on Probability Papers. The advantage of the probability paper compared to the usual ways of plotting lies in the fact that it transforms the theoretical curve $F(x), x$ into a straight line. After the observations have been plotted according to 1.2.7(1), the straight line

1.2.8   AIMS AND TOOLS

may be drawn by a ruler, provided that the scatter of the observations is sufficiently small. The question of the acceptance or rejection of the probability function may be settled by mere inspection. If the fit is good enough, a limited forecast may be made by extending the straight line.

However, in some cases this simple procedure will turn out to be unsatisfactory, and an estimate of the two parameters of the straight line 1.2.5(1) becomes necessary. Any probability paper designed for a variate x which permits a linear reduction

(1) $$x = \beta + y/\alpha \; ; \quad y = \alpha(x - \beta)$$

may be considered as a rectangular grid for y and x. Therefore the classical method of least squares may be applied for the estimation of the parameters β and $1/\alpha$. A small modification of the procedures results in a considerable simplification of the calculation.

The usual procedures consist in minimizing either the vertical distances $\sum (x_m - \beta_1 - y_m/\alpha_1)^2$ or the horizontal distances $\sum (y_m - \alpha_2 x_m + \alpha_2\beta_2)^2$. Since the two postulates differ, we write $b_1, a_1 (b_2, a_2)$ for the solution of the first (second) system. In general, only the first system is used, since it minimizes the errors of the observed values x_m. After introducing the sample means, mean squares, standard deviations, and the cross product

(2) $$\bar{x}_0 = \frac{1}{n} \sum x_m \; ; \quad \overline{x_0^2} = \frac{1}{n} \sum x_m^2 \; ; \quad s_x^2 = (\overline{x_0^2} - \bar{x}_0^2)n/(n-1)$$

$$\bar{y}_n = \frac{1}{n} \sum y_m \; ; \quad \overline{y_n^2} = \frac{1}{n} \sum y_m^2 \; ; \quad \sigma_n^2 = \overline{y_n^2} - \bar{y}_n^2 \; ; \quad (\overline{xy})_n = \frac{1}{n} \sum x_m y_m ,$$

the two systems of normal equations become

$$\bar{x}_0 = b_1 + \bar{y}_n/a_1 \qquad \bar{y}_n = a_2 \bar{x}_0 - a_2 b_2$$
$$(\overline{xy})_n = b_1 \bar{y}_n + \overline{y_n^2}/a_1 \qquad (\overline{xy})_n = a_2 \overline{x_0^2} - a_2 b_2 \bar{x}_0 .$$

The solutions are

(3) $$\frac{1}{a_1} = \frac{(\overline{xy})_n - \bar{x}_0 \bar{y}_n}{\sigma_n^2} \qquad \frac{1}{a_2} = \frac{s_x^2}{(\overline{xy})_n - \bar{x}_0 \bar{y}_n}$$

(4) $$\bar{x}_0 - b_1 = \bar{y}_n/a_1 \qquad \bar{x}_0 - b_2 = \bar{y}_n/a_2 .$$

Each solution requires the knowledge of four averages. The mean $\bar{x}_0$ and the standard deviation s_x of the observations have to be calculated in both cases. The averages $\bar{y}_n$ and σ_n of the reduced variate depend upon the plotting positions, i.e., only upon the number of observations, and can therefore be calculated once for all, if a table of the reduced variate $y(F)$ as function of the probability is available. However, the crossproduct

$(\overline{xy})_n$, which occurs in both solutions, depends upon the observations and the plotting positions and has therefore to be calculated anew for each series of observations.

To eliminate the calculation of the crossproduct, we define new estimates $1/a$ and b by

(5) $$\frac{1}{a} \equiv \sqrt{1/a_1 a_2} \ ; \qquad \bar{x}_0 - b = \sqrt{(\bar{x}_0 - b_1)(\bar{x}_0 - b_2)} \ .$$

Then the equations (3) and (4) lead to

(6) $$1/a = s_x/\sigma_n \ ; \qquad b = \bar{x}_0 - \bar{y}_n/a \ .$$

The estimates (6) require only the knowledge of the sample mean and standard deviation and the population mean $\bar{y}_n$ and standard deviation σ_n.

The new estimates a and $b - \bar{x}_0$ are the geometric means of the estimates a_1, a_2 and $b_1 - \bar{x}_0$, $b_2 - \bar{x}_0$ in the vertical and horizontal systems, respectively.

If the observations are closely scattered about the theoretical line (1), the differences between the estimates of the parameters obtained in the first two systems are small, and the parameters resulting from the new system are approximately the arithmetic means of the parameters resulting from the horizontal and vertical systems. For symmetrical distributions, $\bar{y}_n$ vanishes and b is estimated from the sample mean as usual. For such distributions we need only a table of σ_n. P. G. Carlson has given another interpretation of the equation (6). He constructs a least-squares line

(7) $$y = \alpha(x - \beta)$$

to be fitted to paired observation (x_m, y_m), which does not require the calculation of $\sum x_m y_m$. The derivation consists of minimizing the squared deviations parallel to an arbitrary line and choosing the slope of this line so as to make the coefficient of $\sum x_m y_m$ equal to zero. The line along which the distances are minimized has a slope equal to but opposite in sign to that of the line (7).

The estimates are closely related to the method of moments. The mean and standard deviations for the variate x are

(8) $$\bar{x}_0 = \beta + \bar{y}/\alpha \ ; \qquad s_x = \sigma_y/\alpha \ .$$

If the sample values are used on the left side, the estimates are

(9) $$1/\hat{\alpha} = s_x/\sigma_y \ ; \qquad \hat{\beta} = \bar{x}_0 - \bar{y}/\hat{\alpha} \ .$$

However, there is an inconsistency in this substitution, since $\bar{x}_0$ and s are based on n observations, while the reduced population mean $\bar{y}$ and standard deviation σ_y are obtained by integration over the whole domain of variation. To eliminate this bias, the population values also should be calculated for samples of size n. They are given by equation (2). The equations (6) which are analogous to (9) are therefore logically consistent, since they contain sample and population values for the same sample size n, and converge with increasing sample size to the solution (9).

The estimation of the parameters advocated here aims at a graphical fit which is essential for the forecast. It is a modification of the classical method of moments, but it can also be used if no population moments exist, since the population moments $\bar{x}_n$ and σ_n do exist for any finite sample. The method is also a special case of the least-squares procedure.

Evidently, different methods based on different aims must lead to different results. But the errors of estimation of the parameters are asymptotically normally distributed in the three procedures. It will be shown in 6.2.5 that the maximum likelihood method is very complicated for distributions of extreme values and requires numerical work to an extent which is prohibitive for routine work. This is another reason why the graphical solutions advocated here are preferable.

The problem of a numerical test based on the comparison of the probability and frequency functions for any continuous variate was studied by R. von Mises (1931) and Smirnov (1936, 1939). The probability integral transformation reduces this problem to the question whether observed points are distributed at random on a straight line from zero to unity. Von Mises proposed the test function

(10) $$n\omega^2 = \frac{1}{12n} + \sum_{m=1}^{n} [F(x_m) - (2m-1)/(2n)]^2,$$

while this author proposed in 1942 as a consequence of the plotting position 1.2.7(1)

(10′) $$k^2 = \frac{n+1}{n} \sum_{m=1}^{n} [F(x_m) - m/(n+1)]^2.$$

Smirnov (1936, 1939) obtained the asymptotic probability function for $n\omega^2$. A table was calculated by Anderson and Darling (1952). On the basis of this work Tiago (1955) has shown that k^2 has asymptotically the same distribution as $n\omega^2$. Thus k^2 gives a numerical test for the graphical procedure advocated here.

Sherman proposed the test function

(11) $$\omega_n = \tfrac{1}{2} \sum_{m=1}^{n+1} |F(x_m) - F(x_{m-1}) - 1/(n+1)|$$

with

(11′) $$F(x_0) = 0, F(x_{n+1}) = 1.$$

The values $F(x_m)$ may be read off the probability paper after the theoretical line has been fitted. From the plotting positions the expectation of the difference $F(x_m) - F(x_{m-1})$ is $1/(n+1)$. Sherman proved that the first two moments of ω_n converge with n increasing to

$$\bar{\omega} = 1/e \; ; \qquad \sigma_n^2(\omega) = (2e - 5)/e^2 n$$

and that the standardized value

(12) $$\omega_n^* = 1.51348(e\omega_n - 1)\sqrt{n}$$

is asymptotically normally distributed. In a later article Sherman (1957) proposed a modification of the procedure. The expectation E_n of ω_n is

$$E_n = \left(\frac{n}{n+1}\right)^{n+1}$$

and the dispersion D_n is given by

$$E_n^2 + D_n^2 = \frac{2n^{n+2} + n(n-1)^{n+2}}{(n+2)(n+1)^{n+2}}.$$

Numerical calculations show that the distribution of

$$\omega_n^{(1)} = (\omega_n - E_n)/D_n$$

converges more quickly to normality than the distribution of ω_n^*. Therefore, the test function $\omega_n^{(1)}$ is preferable to the asymptotic expression ω_n^*. This gives a numerical test for goodness of fit reached on the probability paper. Graphical tests will be studied in 2.1.6.

Exercises: 1) Approximate Anderson and Darling's probability function by the logarithmic normal probability function. 2) Calculate tables for E_n and D_n for $n = 20(2) 50(5)100$.

References: Bartholomew (1954), Barton and David, Birnbaum, Daniels, Gumbel (1942a), Hudimoto, Kimball (1947a), Kolmogorov and Hinshin, Kondo, Massey (1951), Schultz, Steffensen.

1.2.9. Application to the Normal Distribution. The estimation of the parameters is now shown for the normal distribution. The reduced values, which we designate in this case by

(1) $$z_m = \alpha(x_m - \beta),$$

are obtained from the plotting positions 1.2.7(1). The parameter β is estimated as the sample mean, as usual, and $1/\alpha$ is estimated from 1.2.8(6).

The standard deviations σ_n are calculated from 1.2.8(2). The symmetry of the normal distribution facilitates this calculation, since for $2n$ and

1.2.9 AIMS AND TOOLS

$2n + 1$ observations we have to calculate only n values z_m^2. In Table 1.2.9 all values z_m^2, already calculated are used again for the subsequent values of n.

Table 1.2.9. Normal Standard Deviation σ_n as Function of Sample Size n

Sample Size n	z_n	σ_n	$\hat{z}_{2,n}$
9	1.2816	.7731	0.7491
19	1.6449	.8638	1.0976
29	1.7507	.8874	1.2128
39	1.9600	.9192	1.3997
49	2.0537	.9319	1.4876
99	2.3263	.9607	1.7589

The values of z_n and $\hat{z}_{2,n}$ in columns 2 and 4 of Table 1.2.9 defined by

$$F(z_n) = n/(n+1) \; ; \qquad F(\hat{z}_{2,n}) = \sigma_n$$

and plotted in Graph 1.2.9(1) lie approximately on a straight line. Therefore,

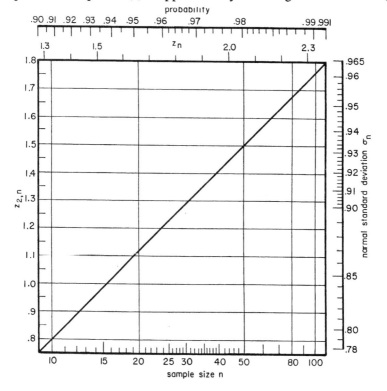

Graph 1.2.9(1). Increase of Normal Standard Deviation as Function of the Sample Size

this relation can be used for linear interpolation of σ_n for values of n not given in the table. To facilitate the interpolation, the sample size n is plotted as abscissa. In addition, z_n and σ_n are also shown on normal probability scales.

B. L. Van der Waerden published a table of the mth normal values $(m = 1, 2 \ldots, n)$ ($2D$) for $n = 6(1)50$, which gives of course also the value of z_n and a table of σ_n^2 ($3D$) for $n = 2(1)150$.

As a numerical example we calculate the straight line for the Graph 1.2.6. The sample mean and standard deviation are

(2) $$\bar{x}_0 = b = 17.75; \quad s = 5.427 .$$

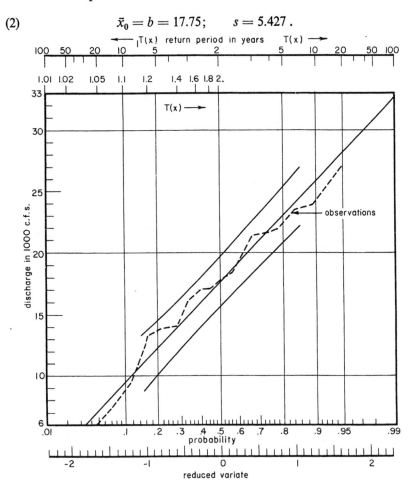

Graph 1.2.9(2). Annual Mean Discharge: Colorado River at Bright Angel Creek, 1923–1939

Interpolation from Table 1.2.9 leads to $\sigma_{17} = .851$. From 1.2.8(6) we obtain the estimate $1/a = 6.38$. The straight line

(3) $$x = 17.75 + 6.38z$$

is drawn in Graph 1.2.9(2). The construction of the control curves will be explained in 2.1.6. Within the next 50 years, up to 1989, i.e., for a total of 67 years the forecast from (3) leads to a maximum of 31,600 c.f.s. for the annual mean discharge and a minimum of 3,900 c.f.s.

The logarithmic normal distribution may also be compared to the observations with the help of probability paper. The ordinate is on a logarithmic scale, the abscissa on a normal probability scale with a parallel to it showing the reduced normal variate y on linear scale. The theory is accepted if the observations are scattered about the line

(4) $$\lg x = \lg \beta + \delta y$$

obtained from 1.1.9(1) and 1.1.9(5). The maximum likelihood method estimates β from the geometric mean and δ from the geometric standard deviation, i.e.,

(5) $$\lg \hat{\beta} = \frac{1}{n} \sum_{v=1}^{n} \lg x_v; \qquad \hat{\delta} = s(\lg x).$$

The graphical method outlined in 1.2.8 replaces the second estimate by

(6) $$\hat{\delta}' = s(\lg x)/\sigma_n,$$

where the values σ_n are given in Table 1.2.9. For the estimation of the parameters there is no need to calculate the higher moments.

Exercise: Calculate σ_n in Table 1.2.9 in such a way that linear interpolation may be used.

Problems: 1) Calculate the standard errors of the estimate (6) and compare it to the standard errors of the usual estimate. 2) Why is the relation between z_n and $\hat{z}_{2,n}$ approximately linear? 3) Trace the diagram 1.2.9(2) for a number of stations of a large river. How do the means and standard deviations increase with increasing size of the watershed?

References: Chernoff and Liebermann (1956).

Chapter Two: ORDER STATISTICS AND THEIR EXCEEDANCES

"*Nec Babylonios temptaris numeros.*"
(*Don't trust random numbers.*)

2.1. ORDER STATISTICS

2.1.0. Problems. The study of order statistics is necessary to justify the plotting positions 1.2.7(1), and also because extreme values are special cases of order statistics. Conversely, certain problems involving order statistics can easily be solved from the theory of extreme values. Under not very restrictive conditions, the distributions of the order statistics in the neighborhood of the median converge toward normality. This property may be used for the estimation of certain parameters and for the construction of control curves for probability papers. Finally, we study the differences among order statistics.

References: Savage, Wilks (1948).

2.1.1. Distributions. The mth ($m = 1, 2, \ldots, n$) among n observations taken in increasing order has a distribution $\varphi_n(x_m)$ which will depend upon the initial distribution $f(x)$, the sample size n, and the order m. If n is odd, the $m = (n+1)/2$th value is the central value which gives an estimate for the population median. If m is of the order $n/2$ we speak of an mth *central value*. The mth value from below is the $n - m + 1$ value from above, and vice versa. Both are called ranked variates or order statistics. A related notion is the (population) quantile, $x(F)$, defined by a certain value of the probability F.

If n increases, m may increase, too, so that m/n remains approximately constant. If m does not increase with n, the quotient m/n decreases. The corresponding values x_m and x_{n-m+1} are called mth extreme values. For $m = 1$ we obtain the extreme values themselves.

The mth value from below x_m (and from above $_m x$) in a sample of size n is such that $m - 1$ values are below (above) the mth, and $n - m$ values are above (below) it. Multiplying the respective probabilities with the probability density for the mth value itself, and taking into account the number of combinations, leads to the distribution $\varphi_n(x_m)$ and $\varphi_n(_m x)$,

(1) $$\varphi_n(x_m) = \frac{n!}{(n-m)!(m-1)} F^{m-1}(1-F)^{n-m} f$$

$$\varphi_n({}_m x) = \frac{n!}{(n-m)!(m-1)} (1-F)^{m-1} F^{n-m} f,$$

where the variable x is omitted.

For symmetrical initial distributions, the two distributions (1) are mutually symmetrical (1.1.3). In the following we concentrate on the first distribution and count m from the bottom. If the initial variate is limited, or unlimited, in one or both directions, the same holds for the mth values.

The distribution $v_n(z_m)$ of the mth value for a transformed variate z is obtained from the distribution $\varphi_n(x_m)$ by the transformation which links x to z.

References: Epstein and Sobel (1952), Hastings *et al.*, Homma (1951), Malmquist, Mosteller (1946), Ogawa, K. Pearson (1931b), Pillai (1951), Renyi, Sarhan and Greenberg, W. R. Thompson (1935), Torabella, Walsh (1946a, b), Wilks (1948), Yang.

2.1.2. Averages. Karl Pearson (1931, b) has studied the distribution of the mth values for several initial distributions. The mean and the generating function $G_m(t)$ of the mth value can be written as integrals. However, the integration can be carried through only if the variable can be expressed in a simple way by the probability function. For most initial distributions, the calculation of the means leads to very complicated formulae. For the normal distribution, the expansion of x as function of F consists of thirteen factors, and the length of such formulae forbids analytical usage.

The *medians* $\check{x}_m$ and ${}_m\check{x}$ of the mth values are the solutions

(1) $$F(\check{x}_m) = \lambda_1(m,n); \qquad F({}_m\check{x}) = \lambda_2(m,n)$$

of

(2) $$\frac{\int_0^{\lambda_1} F^{m-1}(1-F)^{n-m}\, dF}{\int_0^1 F^{m-1}(1-F)^{n-m}\, dF} = \tfrac{1}{2}; \qquad \frac{\int_0^{\lambda_2} F^{n-m}(1-F)^{m-1}\, dF}{\int_0^1 F^{n-m}(1-F)^{m-1}\, dF} = \tfrac{1}{2},$$

where λ_1 and λ_2 can be obtained from Pearson's *Tables of the Incomplete Beta Functions*. Evidently the medians depend upon the initial distribution.

The modes $\tilde{x}_m$ *and* $_m\tilde{x}$ (as far as they exist) are obtained from the logarithms of the distributions 2.1.1(1) as the solutions of

(3) $\quad \dfrac{m-1}{F}f - \dfrac{n-m}{1-F}f + f'/f = 0; \quad -\dfrac{m-1}{1-F}f + \dfrac{n-m}{F}f + f'/f = 0$

or, after the choice of a common denominator,

(3') $\qquad m - 1 - (n-1)F = -f'f^{-2}F(1-F);$

$\qquad\qquad n - m - (n-1)F = -f'f^{-2}F(1-F),$

whence, introducing $\tilde{x}_m$ and $_m\tilde{x}$,

(4) $\qquad F(\tilde{x}_m) = \dfrac{m-1}{n-1} + \dfrac{[f'f^{-2}F(1-F)]}{n-1} \quad (x = \tilde{x}_m);$

$\qquad F(_m\tilde{x}) = 1 - \dfrac{m-1}{n-1} + \dfrac{[f'f^{-2}F(1-F)]}{n-1} \quad (x = {_m\tilde{x}}).$

For symmetrical initial distributions with median zero, the modal mth values from the top and from the bottom differ only in sign. If k is a positive integer, then the modes of the kmth value in a sample of size kn differ from the modes of the mth value in a sample of size n. If the initial distribution is limited to the right or to the left, the distributions of the first few or the last few mth values need not possess maxima.

The initial probabilities $F(\tilde{x}_m)$ at the modal mth values as functions of m and n are given in Table 2.1.2 for some distributions. The return period of the most probable largest value $T(\tilde{x}_n)$ given in the last column is called the *occurrence interval*.

For certain initial distributions, the probabilities at the mode of the smallest value, $m = 1$, vanish. For other distributions, the occurrence interval diverges or is of the order $2n$. Therefore, the probability $F(\tilde{x}_m)$ cannot generally be used as a plotting position. Hazen's plotting position, 1.2.6, which was originally designed for the logarithmic normal distribution which is asymmetrical and unlimited to the right, turns out in the last line of Table 2.1.2 to be the probability of the most probable mth value for a limited symmetrical distribution. This will probably not discourage engineers from using this "practical" formula.

If equation (4) does not lead to a simple solution as for the normal distribution, and if $f'(x) \cdot f^{-2}(x)$ remains within fixed limits, we may use a process of approximation. Equation (4) is written

(5) $\quad F(\tilde{x}_m) = \dfrac{m-1}{n-1} + \dfrac{\Delta(\tilde{x}_m)}{n-1};$

$\qquad \Delta(\tilde{x}_m) = [f'(x)f^{-2}(x)F(x)(1-F(x))]_{(x=\tilde{x}_m)}.$

The use of the first member leads to a first approximation. Its introduction into the correction $\Delta(\tilde{x}_m)$ leads to a second approximation, and so on.

Table 2.1.2. Probabilities of Modal mth Values

Distribution	Domain	Probability $F(x)$	Probability at the Most Probable mth Value $F(\tilde{x}_m)$	Most Probable Rank m [Equation (6)]	Rank for Estimating the Median	Occurrence Interval $T(\tilde{x}_n)$
Uniform	$0 \leq x \leq 1$	x	$\dfrac{m-1}{n-1}$	$(n-1)F+1$	$\dfrac{n+1}{2}$	∞
"Power Function"	$0 \leq x \leq 1$	x^k	$\dfrac{mk-1}{nk-1}$	$nF + (1-F)/k$	$\dfrac{n+1/k}{2}$	∞
Exponential	$x \geq 0$	$1 - e^{-x}$	$(m-1)/n$	$nF+1$	$\dfrac{n+2}{2}$	n
Logistic	$-\infty < x < +\infty$	$(1+e^{-x})^{-1}$	$m/(n+1)$	$(n+1)F$	$\dfrac{n+1}{2}$	$n+1$
Simplified Pareto	$x \geq 1$	$1 - x^{-1}$	$(m-1)/(n+1)$	$(n+1)F+1$	$\dfrac{n+3}{2}$	$\dfrac{n+1}{2}$
Triangular	$0 \leq x \leq 1$	$x(2-x)$	$\dfrac{2m-2}{2n-1}$	$(n-\tfrac{1}{2})F+1$	$\dfrac{2n+1}{4}$	$2n-1$
Sine Square	$0 \leq x \leq \dfrac{\pi}{2}$	$\sin^2 x$	$(m-\tfrac{1}{2})/n$	$nF+\tfrac{1}{2}$	$\dfrac{n+1}{2}$	$2n$

Equation (5) may also be used to derive the rank $\tilde{m}$, henceforth called the *most probable rank*, corresponding to a desired quantile. Then F is given and $\tilde{m}$, obtained from

(6) $$\tilde{m} - 1 = (n-1)F - \Delta(\tilde{x}_m),$$

need not be an integer. Consequently, the estimate of a quantile is obtained by interpolation between two observed values, x_m and x_{m+1}. The most probable ranks for different quantiles are given in columns 5 and 6 of Table 2.1.2. For symmetrical distributions $m = (n+1)/2$ is, of course, the most probable rank of the median.

Exercise: 1) Prove that the sine square probability is the only function which leads to $F(\tilde{x}_m) = (m - \frac{1}{2})/n$. 2) Establish the relation between the moments of the mth values for two mutually symmetrical variables (Lieblein and Zelen, eqn. B24).

Problems: 1) Calculate the modal mth values for the normal and for the Cauchy distributions. 2) Prove the convergence of the procedure advocated in (5).

References: Evans, Goodman (1954), Hastings *et al.*, Mosteller (1946), Teichroew, Thompson (1936), Yang.

2.1.3. Distribution of Frequencies.

If, instead of x_m, we consider F_m as the variate in 2.1.1(1), its distribution $v_n(F_m)$ becomes

(1) $$v_n(F_m) = \binom{n}{m} m F_m^{m-1}(1 - F_m)^{n-m}; \quad 0 \leq F_m \leq 1.$$

This transformation is called the probability integral transformation. The variate F_m is the (cumulative) *frequency of the mth value*. The distributions of the frequencies of the mth values from above and from below are mutually symmetrical. For $m = 1$ ($m = n$), the distribution of the first (last) frequency decreases (increases) with F. For $m \neq 1$ and $m \neq n$ there exist modes $\tilde{F}_m$ which are the solutions of

(2) $$\tilde{F}_m = (m-1)/(n-1).$$

The modes of the frequencies cannot be used as plotting positions, because $F_1 = 0$ and $F_n = 1$ do not exist on a probability paper designed for an unlimited variate. The cumulative frequency of the modes $F(\tilde{x}_m)$ differs from the mode of the frequencies $\tilde{F}_m$. The median $\breve{F}(x_m)$ of the frequencies which differ, of course, from the frequency of the median $F(\breve{x}_m)$, may be calculated from the *Tables of the Incomplete Beta Function*. These values are given by L. G. Johnson, who proposed to use them as plotting positions. Benard obtained the approximation

(2') $$\breve{F}(x_m) = (m - .3)/(n + .4).$$

The reasons which speak against these methods will be given in 3.1.3(3).

The mean frequency $\overline{F(x_m)}$, different from the frequency $F(\bar{x}_m)$ of the mean, is obtained from the moments of order k which are, from (1),

$$(3) \qquad \overline{F_m{}^k} = \frac{n!(m-1+k)!}{(n+k)!(m-1)!}.$$

Hence, for $k = 1$, the mean frequency of the mth value

$$(4) \qquad \overline{F(x_m)} = \overline{F_m} = m/(n+1)$$

differs from its modal frequency (2), from the definition of the quantile and from the frequency $F(\bar{x}_m)$ of the mean mth value. Consequently, the solutions $\hat{x}_m$ of equation (4) differ from the mean mth values. Formula (4) gives the desired solution of the plotting problem since it *fulfills the postulates given in* 1.2.7. Therefore, we advocate its general use, first proposed by W. Weibull (1938). The second moment of the frequencies

$$(5) \qquad \overline{F_m{}^2} = \frac{m(m+1)}{(n+1)(n+2)}$$

leads to the variance of the mth frequencies from below and from above,

$$(6) \qquad \sigma^2(F_m) = \frac{m(n-m+1)}{(n+2)(n+1)^2} = \sigma^2(_mF) = \frac{\overline{F}_m(1-\overline{F}_m)}{n+2}$$

This variance is smallest for $m = 1$, and $m = n$, and increases if we pass from the extremes to the middle of the distribution, contrary to what one might expect.

Exercises: 1) Calculate the skewness and the excess of the distribution of the mth frequency. 2) Prove that $\overline{-\lg F_m} = \sum_{m}^{n} 1/\nu$ (B. F. Kimball). 3) Calculate the standard deviation and the next two moments of $\lg F_m$ from the seminvariant generating function. 4) Calculate the mean and variance of $\sum_{1}^{n}\left(F_m - \frac{m}{n+1}\right)^2$. 5) Calculate the arithmetic and geometric mean return period of the mth value.

References: Dalcher, Gumbel (1942a), Haldane. L. G. Johnson, The median ranks, etc., Industr. Math. 2:1 (1951).

2.1.4. Asymptotic Distribution of mth Central Values.

From the time of Laplace, many authors have proven that the distribution of the central value for the normal distribution is normal. This proof can easily be generalized for other distributions and for other quantiles situated in the neighborhood of the median. We first study the distribution 2.1.3(1) of the frequencies of the mth values, and later the distribution of the mth values themselves. In both cases we assume that, with increasing sample size, the quotient m/n remains in the neighborhood of $\tfrac{1}{2}$. Under this condition, the distributions of the frequencies of the mth values converge toward normal distributions with means and variances given by 2.1.3(4) and (6).

For the proof, let

(1) $$F(x) = \frac{m}{n+1} + z$$

and let p and q be two fixed probabilities of the order $\tfrac{1}{2} \pm O\!\left(\dfrac{1}{n}\right)$, so that approximately

(2) $\quad m = p(n+1); \quad m - 1 = np - q; \quad n - m = nq - p,$

where p and q are small compared to np and nq. Then the distribution $w(z)$ of the variate z becomes, from 2.1.3(1),

$$w(z) = \binom{n}{m} m \left(\frac{m}{n+1} + z\right)^{m-1} \left(1 - \frac{m}{n+1} - z\right)^{n-m}$$
$$= \text{const. exp}\,[(np - q)\lg(1 + z/p) + (nq - p)\lg(1 - z/q)],$$

where the constant stands for all factors independent of z.

In the usual development of the term in brackets

$$(np - q)\left(\frac{z}{p} - \frac{z^2}{2p^2} + \ldots\right) + (nq - p)\left(-\frac{z}{q} - \frac{z^2}{2q^2} - \ldots\right)$$
$$\sim -\frac{nz^2}{2}\left(\frac{1}{q} + \frac{1}{p}\right),$$

we neglect q compared to np, and p compared to nq, a procedure which is reasonable, provided that m and n are large and of the same order of magnitude. Then the asymptotic distribution of z is normal with mean zero and variance

(3) $$\sigma^2(z) = pq/n$$

and the asymptotic distribution of the mth frequency F_m is normal with the mean and variance given in 2.1.3(4) and 2.1.3(6).

The same convergence holds for the mth observations situated in the neighborhood of the median. If n is large, the distribution of the mth value 2.1.1(1) converges toward

$$\varphi(x_m) = \text{const.}\, f(x)\exp\left[-\frac{n}{2pq}(F(x) - F(\hat{x}))^2\right]; \quad x = x_m,$$

2.1.4 ORDER STATISTICS

where $\hat{x}$ is the solution of

(4) $$F(\hat{x}) = \frac{m}{n+1} = p; \quad q = 1 - F(\hat{x}).$$

The expansion of $F(x)$ in the neighborhood of $\hat{x}$ leads to

$$F(x) = F(\hat{x}) + (x - \hat{x})f(\hat{x}) + \frac{(x - \hat{x})^2}{2}f'(\hat{x}) + \dots$$

The third and the following terms may be neglected. Then $f(x) = f(\hat{x})$ is a constant. Consequently, the distribution of the mth values converges toward the normal distribution

(5) $$\varphi(x_m) = \text{const. } \exp\left[-\frac{n(x - \hat{x})^2 f^2(\hat{x})}{2F(\hat{x})(1 - F(\hat{x}))}\right],$$

where $\hat{x}$ is the solution of (4) and the variance σ_m^2 is given by

(6) $$n\sigma_m^2 = \frac{F(\hat{x})(1 - F(\hat{x}))}{f^2(\hat{x})}.$$

If we choose numerical values for $F(\hat{x})$, the right side depends upon $\hat{x}$ only and no longer upon m.

Formula (6) holds asymptotically for symmetrical and slightly asymmetrical distributions where the median is situated near the mode, and for ranks in the neighborhood of the median. It becomes invalid for small probabilities, i.e., for extreme values. Hastings calculated the standard error of the mth normal values for samples of 2 to 10. The asymptotic values obtained from (6) shown in Graph 2.1.4 are not as different from the calculated values as one might expect.

Kawata (1951) has shown that the asymptotic normal distributions of the mth value hold also in the case $n \to \infty$; $m/n \to 1$; $n - m \to \infty$, provided that the probability function is of the form

(7) $\quad 1 - F(x) = x^{-k}h(x)$, where $h(cx)/h(x) \to 1$; $k > 0$; $c > 0$.

Order statistics are often used for the estimation of parameters. The asymptotic errors of estimation are known from the standard errors of order statistics, as long as they may be assumed to be normally distributed. As a general rule we assume this approach to be sufficiently fast within the interval $0.15 \leq F \leq 0.85$, although it depends upon the distribution and the sample size. Order statistics outside of this interval, extremes, are not normally distributed. This fact sets a limit to the use of order statistics. In addition, there is a practical difficulty. The most probable rank $\tilde{m}$ from which the order statistics are obtained is in general not an integer, but is confined in an interval $m < \tilde{m} < m + 1$. If the

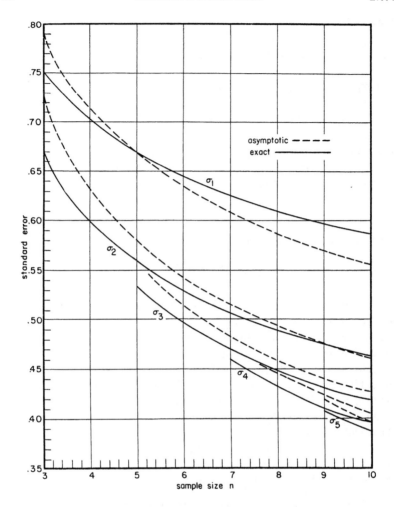

Graph 2.1.4. Asymptotic and Exact Standard Errors of mth Normal Values

corresponding two consecutive order statistics are equal, or if they differ by an amount of the order of the smallest unit which can be observed, there is no practical ambiguity in the estimation. However, if the gap between the two consecutive order statistics is large, we have to interpolate between them according to the values of the probability function, and this interpolation may introduce errors which may outweigh the advantage in the simplicity of procedure or in precision gained by the use of the order statistics relative to other estimates of the parameters.

References: Cramér, Dwass, Eyraud (1935), Fogelson, Godwin, Haag,

Hastings et al., H. L. Jones (1948), Nair (1940), Noether (1948), Smirnov (1935), W. R. Thompson (1936), Tukey (1946), Walsh (1950), Westenberg, Woodruff.

2.1.5. The Order Statistic with Minimum Variance. The variance 2.1.4(6) of the mth value depends upon three factors, $1/n$, $F(\hat{x})$, and $f(\hat{x})$. The first indicates the usual decrease of the variance with increasing number of observations. The second, $F(1 - F)$, is a numerical value depending upon m and n and is maximum for the median. However, this behavior is modified by the third factor, the density of probability $f(\hat{x})$, which will be studied now. To establish the dependency of $\sigma^2(\hat{x}_m)$ upon $\hat{x}_m$, it is sufficient to consider

(1) $$d \lg n\sigma^2(\hat{x}_m)/dF = d[\lg F + \lg (1 - F) - 2 \lg f]/dF.$$

Since the derivative $F'(x) = f(x)$ is always positive we eliminate first the case that $n\sigma^2(\hat{x}_m)$ is a constant. This holds for sine square probability (Table 2.1.2), where

(2) $$n\sigma^2(\hat{x}_m) = \tfrac{1}{4}.$$

If $n\sigma^2(\hat{x}_m)$ depends upon $\hat{x}_m$, differentiation leads to

(1') $$d \lg n\sigma^2(\hat{x}_m)/dF = 1/F - 1/(1 - F) - 2f'/f^2.$$

This expression vanishes if the distribution is symmetrical and has a mode or an antimode (minimum) at the median, i.e., if for the same value of $\hat{x}$ the two equations

(3) $$F(\hat{x}) = \tfrac{1}{2}; \quad f'(\hat{x}) = 0$$

hold simultaneously.

The second derivative,

$$d^2 \lg [n\sigma^2(\hat{x}_m)]/d^2F = -1/F^2 + 1/(1 - F)^2 - 2f''/f^3 - 4f'^2/f^4,$$

is positive (negative) if

$$f''(\hat{x}) < 0 \; (f''(\hat{x}) > 0).$$

If a symmetrical distribution possesses a mode (a minimum) at the median, the variance of the mth value possesses a minimum (maximum) at the median, and the variances increase (decrease) if we pass from the center of symmetry toward the beginning or the end of the distribution.

If the initial distribution is skewed so that its mode is different from its median, the order statistic with minimum variance defined by (1') differs from both these averages. It is obtained as the solution of

(4) $$1 - 2F = 2f'F(1 - F)f^{-2},$$

provided such a solution exists and is situated in the neighborhood of the central value.

Exercises: 1) Analyze the variance $n\sigma^2(\hat{x}_m)$ for the distributions given in Table 2.1.2 and the distribution $f(x) = xe^{-x}$. 2) Show that the order statistic of minimum variance for the logarithmically normal distribution is obtained from $F = 0.32949$.

Problem: For what distributions is the order statistic with minimum variance conserved under logarithmic transformation?

References: Evans, Eyraud (1934), R. A. Fisher (1922), Westenberg.

2.1.6. Control Band.
If the variate x can be reduced by writing

(1) $$z = \alpha(x - \beta),$$

the standard error $\sqrt{n}\sigma(\hat{z}_m)$ of the reduced mth values,

(2) $$\sqrt{n}\sigma(\hat{z}_m) = \sqrt{F(1-F)}/f,$$

is a pure number independent of the parameters. It is given in columns 2, 3, and 4 of Table 2.1.6 as a function of the probability $F(\hat{z})$, column 1.

Table 2.1.6. *Reduced Standard Errors* $\sigma(\hat{z}_m)\sqrt{n}$ *for Unit Standard Deviation*

Probability	Standard Error of Reduced mth Value			Colorado River
	Normal	Laplacean	Logistic	
$F, 1-F$	$\sigma = 1$	$\sigma = \sqrt{2}$	$\sigma = \pi/\sqrt{3}$	$s(\hat{x}_m)$
.50	1.253	0.707	1.103	1.934
.55	1.257	0.781	1.109	1.940
.60	1.268	0.866	1.126	1.957
.65	1.288	0.964	1.157	1.988
.70	1.318	1.080	1.203	2.035
.75	1.363	1.225	1.273	2.104
.80	1.429	1.414	1.379	2.206
.85	1.532	1.684	1.544	2.365

The standard error of the mth value obtained from (1) and (2) is

(3) $$\sigma(\hat{x}_m) = [\sigma(\hat{z}_m)\sqrt{n}]/(\alpha\sqrt{n})$$

where $1/\alpha$ has the same dimension as x. In most cases, α depends only upon the standard deviation.

If $1/\alpha$ has been estimated, the expression (3) can be used for the construction of control curves. The probability function traced on a probability paper is represented by the straight line

(4) $$x = \hat{\beta} + z/\hat{\alpha}.$$

We now add and subtract from the mth value $\hat{x}_m$ situated on (4), the standard error $\sigma(\hat{x}_m)$. The points $x_m \pm \sigma(\hat{x}_m)$ or $\hat{x}_m \pm 1.96\sigma(\hat{x}_m)$ are joined to form a control curve. Then there is a probability of 2/3 or .95 for each mth observation to be contained within these bands. This gives a graphical criterion for the goodness of fit between theory and observations. An example is shown in Graph 1.2.9(2). The values given in Table 2.1.6, column 2, are multiplied by the factor $1/(\hat{\alpha}\sqrt{n})$ which is 1.544, from 1.2.9(3). All observations in the interval $.15 < F < .85$ are contained in the control band given in Table 2.1.6, column 5. The observations may safely be assumed to be samples from a normal distribution.

The control band for the central value of a normal distribution,

(5) $$\sqrt{n}\sigma_n(\hat{z}) = 1.25331,$$

corresponds to the probability $P = .68269$. Multiplications of $\sigma_n(\hat{z})$ by .6745, 1.960, 2.576, 2.807, and 3.290 lead to bands corresponding to the probabilities $P = .50, .95, .99, .995,$ and $.999$. If we trace $1/\sqrt{n}$ on the ordinate and $\sigma_n(\hat{z})$ on the abscissa, the boundaries of the intervals corresponding to these probabilities become straight lines. In addition we trace in Graph 2.1.6 the cumulative frequencies $F(z)$ and return periods corresponding to the values $\pm\sigma_n(\hat{z})$. The graph shows the probabilities $P(n)$ that the population median appears in a sample of size n as an order statistic within the intervals bounded by the straight lines. For example, for $n = 25$ there is a probability .95 that the population median of a normal distribution is observed as an order statistic with the ranks 8 to 17. This method is parameter free.

The control curves given by (3) cannot be used for the extreme values, since they are no longer asymptotically normally distributed. The question of how to construct control curves covering the whole domain of variation will be settled in 6.1.5.

Exercise: Calculate the control curves for the Cauchy distribution.
References: Eyraud (1935), W. R. Thompson (1935), Wald and Wolfowitz, Wolfowitz, Working and Hotelling.

2.1.7. Joint Distribution of Order Statistics. The joint distribution $\varphi_n(x_m, x_l)$ of the mth and lth among n observations $m < l$, both counted from the bottom, is obtained from the following scheme (1), where F_m stands for $F(x_m)$ and similarly for the other indices

(1) numbers $m - 1$ 1 $l - m - 1$ 1 $n - l$
      probability F_m $f_m\,dx_m$ $F_l - F_m$ $f_l\,dx_l$ $1 - F_l$

as

(2) $$\varphi_n(x_m, x_l) = CF_m^{m-1}(F_l - F_m)^{l-m-1}(1 - F_l)^{n-l}f_m f_l$$

where $C = n!/[(m-1)!(l-m-1)!(n-l)!]$.

M. G. Kendall (1940–41) has shown that the joint distribution of two

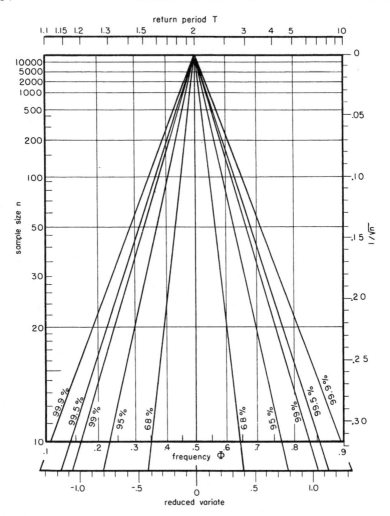

Graph 2.1.6. Control Band for Normal Central Values

central values x_m and x_l converges toward a bivariate normal distribution with variances given by 2.1.4(6) and a covariance

$$(3) \qquad n \operatorname{cov}(x_m, x_l) = \frac{F(\hat{x}_m)[1 - F(\hat{x}_l)]}{f(\hat{x}_m) f(\hat{x}_l)}.$$

Mosteller (1946) has generalized these results for a multivariate distribution. The distribution $\psi_n(w)$ of the non-negative distance

$$w = w_{l,m} = x_l - x_m$$

is, dropping the index m,

(4) $$\psi_n(w) = C \int_{-\infty}^{+\infty} F(x)^{m-1}[F(x+w) - F(x)]^{l-m-1}[1 - F(x+w)]^{n-l} \times f(x+w)f(x)\,dx.$$

A special case will be considered in the next paragraph.

Exercises: 1) Calculate the distribution of w for the rectangular distribution; specialize for $l = m + 1$ and $m = 1$, $l = n$. 2) Calculate the correlation between the mth and lth values (See K. Pearson, 1931b.)

References: Chu, K. Pearson (1920), Siotani, Walsh (1948), Wilks (1948).

2.1.8. Distribution of Distances. In the arrangement of n observations now taken in *decreasing* order of magnitude, the non-negative distances between the mth and the consecutive $(m + 1)$th value from the top is called the mth distance i_m (from the top).

(1) $$i_m = x_m - x_{m+1}, \quad 1 \leq m \leq n - 1.$$

The expressions i_1 and i_{n-1} are the first distances from the top and from the bottom. For a symmetrical distribution, the mth and the $(n - m)$th distances have mutually symmetrical distributions. Let

$$x = x_{m+1}; \quad C = n!/[(n - m + 1)!(m - 1)!],$$

then the distribution $\varphi_n(i)$ of $i = i_m$ is obtained from the joint distribution $\mathfrak{w}_n(x,i)$ of the $(m + 1)$th value and the mth distance,

(2) $$\mathfrak{w}_n(x,i) = CF(x)^{n-m+1} f(x) f(x+i)[1 - F(x+i)]^{m-1},$$

as

(3) $$\varphi_n(i) = \int_{-\infty}^{+\infty} \mathfrak{w}_n(x,i)\,dx.$$

A table (3D) of the corresponding probability function for the first two normal intervals $i_1 = 0.1(0.1)5.0$; $i_2 = 0.1(0.1)4.0$, and $n = 3$, 10(10)100(100)1000 was calculated by I. O. Irwin (K. Pearson's *Tables for Statisticians and Biometricians*. Vol. II, Tables XIX and XX). The probability function of the first interval in a normal sample of size $n + 1$ was derived by St.-Pierre and Zanger on the basis of a formula due to Wilks (1943). The authors give a table (5D) of the probability of $i_1 = 0.0(0.2)2.6$ for $n = 2(1)7$.

The mean mth distance becomes from (3), after introduction of the indices m and $m + 1$,

(4) $$\bar{\iota}_m = C \int_{x_m = -\infty}^{\infty} [1 - F(x_m)]^{m-1} \int_{x_{m+1} = -\infty}^{x_m} [F(x_{m+1})]^{n-m+1}(x_m - x_{m+1}) \times dF(x_{m+1})\,dF(x_m).$$

If we write

(5) $$U_m = \int_{x_{m+1}=-\infty}^{x_m} (x_m - x_{m+1}) \, dF^{n-m}(x_{m+1}),$$

the inner integral in (4) is $U_m/(n-m)$.

The mean mth distance becomes from (4)

$$\bar{i}_m = \frac{C}{(n-m)m} \left[(1 - F(x_m))^m U_m \right]_{-\infty}^{+\infty} + \binom{n}{m} \int_{-\infty}^{+\infty} [1 - F(x)]^m \, dU_m.$$

The first factor vanishes from (5) and we obtain, after dropping the index m, Karl Pearson's (1931) formula,

(6) $$\bar{i}_m = \binom{n}{m} \int_{-\infty}^{+\infty} [1 - F(x)]^m F(x)^{n-m} \, dx.$$

For some distributions the integrations in (3) and (6) can be carried through analytically. Consider the exponential distribution where from (2) for $i = i_m$, after integration of (3),

(7) $$\varphi(i) = m e^{-mi}.$$

The distribution of the mth distance in an exponential distribution is exponential and independent of n. In particular the distribution of the first distance is

$$\varphi(i) = e^{-i}.$$

The mean $\bar{i}_m$ and the standard deviation σ_m of the mth distance,

(8) $$\bar{i}_m = \sigma(i_m) = 1/m,$$

decrease if we pass from the first to the $(n-1)$th distance. The mean first distance is equal to unity, the mean of the distribution.

Galton (1902) studied the ratio R between the difference of the first and the third, and the difference between the second and the third value from the top. This ratio,

$$R = (x_n - x_{n-2})/(x_{n-1} - x_{n-2}),$$

becomes, if the arithmetic means are used as representatives of the extreme values,

(9) $$\check{R} = 1 + \bar{i}_1/\bar{i}_2.$$

For the exponential distribution we obtain from (8)

(10) $$\check{R} = 3,$$

independent of n as stated by Karl Pearson (1902).

As a further example, consider the "logistic" distribution. The variate x expressed by the probability F is

$$x = \lg F - \lg(1 - F).$$

Formula (6) leads from the Beta function to the mean mth distance.

(11) $$\overline{i_m} = \frac{n}{m(n-m)} = \overline{i_{n-m}}.$$

The mean distances diminish if we pass from the extreme toward the middle of the distribution. Galton's ratio,

(12) $$R = 1 + 2(n-2)/(n-1),$$

converges toward 3 for increasing values of n.

Exercises: 1) Prove that the distribution of the mth distance for the uniform distribution is independent of m. 2) Calculate the distribution of the first distance for the logistic distribution.

References: Dixon (1940), Dixon and Mood, Irwin (1925a,b), K. Pearson (1902, 1931b, 1932), Rajalakashma.

2.2. THE DISTRIBUTION OF EXCEEDANCES

"In medio fortissimus ibis."
(*Always choose the middle road.*)

2.2.0. Introduction. In most cases the forecast of extreme values will be based on the knowledge of the initial distribution. However, there are other methods which require only the continuity of the initial variate. Since these procedures are distribution-free, we show them first. Instead of the size of the extremes, we deal only with their frequencies. Such knowledge is sometimes sufficient. If a flood destroys the crop, it does not matter whether the soil is covered by ten centimeters or a meter of water. The same procedure may also be applied to the largest precipitations, extreme temperatures, killing frosts, droughts, etc. We want to forecast the number of cases surpassing a given severity within the next N trials. The methods to be explained lead to a forecast by interpolation, based on the essential and reasonable assumptions that the forthcoming trials are taken from the same population as the prior ones and that the observations are independent.

We study the probability that the mth observation in a sample of size n taken from an unknown distribution of a continuous variate will be exceeded x times in N future trials, and calculate the averages and moments of the number of exceedances. The exact cumulative probabilities can be obtained only after lengthy combinatorial calculations. However, asymptotic approximations are easily calculated.

The original sample size n may be small, or of the same magnitude as the large future sample size N. In both cases, we are interested in the asymptotic distributions of exceedances of values in the neighborhood of the median of the original sample and of the extreme values. If $N = n$ becomes large, m may increase with n so that the quotient m/n remains constant, and the mth values remain near the median. Or, m may remain constant so that $m \ll n$ and the mth values are extremes. This situation is analogous to the transition from Bernoulli's distribution. However, instead of Poisson's law, we find two new asymptotic distributions, one for the largest and another for the smallest observation.

2.2.1. Distribution of the Number of Exceedances. Our starting point is a special case of a distribution studied by S. S. Wilks (1941, 1942), who considered several order statistics, whereas we consider here only one, and later in 3.2.4 two, order statistics. As in 1.2.2, we construct a dichotomy from a continuous variate by choosing a certain mth largest observation $\xi_m = \xi$. In this section, as in the preceding paragraph, *the rank m is counted from the top*. Thus the rank of the largest (smallest) observation is $m = 1$ ($m = n$). The probability for a value less than ξ is $q = F_m$ and the probability of a value equal to or greater than ξ, henceforth called the probability of *exceedance*, is $p = 1 - F_m$. If the probability F_m is known, the probability $w_1(1 - F_m, N, x)$ that x among N *future* trials will exceed the observation ξ is the binomial formula

$$(1) \qquad w_1(1 - F_m, N, x) = \binom{N}{x} (1 - F_m)^x F_m^{N-x}.$$

Thus x is the number of exceedances, a positive integer where $0 \leq x \leq N$. As a rule, the probability p is unknown and the only known data are the n *past* observations. For the elimination of p, consider F_m as a variate and integrate over all its values. The resulting distribution $w(n, m, N, x)$ of exceedances depends on the variate x and the parameters n, m, and N, but not upon the unknown probability p. The convolution of (1) and 2.1.3(1), rewritten for decreasing magnitudes, is

$$w(n,m,N,x) = \binom{n}{m} m \binom{N}{x} \int_0^1 F_m^{N-x+n-m} (1 - F_m)^{m+x-1} \, dF_m.$$

2.2.1 ORDER STATISTICS

This leads from the Beta function to the *distribution of the exceedances*

$$(2) \quad w(n,m,N,x) = \frac{\binom{n}{m} m \binom{N}{x}}{(N+n)\binom{N+n-1}{x+m-1}} ; \quad \sum_{x=0}^{N} w(n,m,N,x) = 1,$$

a fundamental formula first given by H. A. Thomas. If we are interested in the dependency on x only, or on x and m, we write $w(x)$ or $w(m,x)$ respectively. Although no assumption about symmetry of the initial distribution was made, the distribution (2) possesses two symmetries: The probability that *the mth largest among n past observations will be exceeded x times in N future trials is equal to the probability that x among N future trials will fall short of the mth smallest among n past observations and to the probability that the past mth value from the bottom will be exceeded $N - x$ times.*

$$(3) \quad w(n,m,N,x) = w(n, n - m + 1, N, N - x).$$

Formula (2) may also be written

$$(4) \quad \binom{N+n}{n} w(n,m,N,x) = \binom{N+n-m-x}{n-m}\binom{x+m-1}{m-1}.$$

This simplifies the numerical calculation, since, as a rule, n and N are given.

If we write the nN probabilities $w(m,x)$ in the form of a matrix where x is written in the rows and m in the columns, then the $(n - m + 1)$th column is the mth column in reverse order, and the $(N - x)$th row is the xth row in reverse order. In particular the last (next to last) column is equal to the first (second) one in reverse order; the last row is equal to the first one in reverse order, and so on. For $n = N$, the probability in the mth column and the xth row $w(n,m,n,x)$ is related to the probability in the xth column and the mth row $w(n,x,n,m)$ by

$$(5) \quad w(n,m,n,x) = \frac{m}{x} w(n,x,n,m).$$

The matrix for $n = N = 5$ is shown in Table 2.2.1.

The nN probabilities $w(n,m,N,x)$ are linked by two recurrence formulae. For fixed m, i.e., for a determinate column, the probability for $x + 1$ is obtained from the probability for x. For fixed x, i.e., for a determinate row, the probability for $m + 1$ is linked to the probability for m. Starting from $m = 1$, $x = 0$, we obtain all probabilities by recurrent procedures. By virtue of the symmetry, it is sufficient to calculate from $x = 0$ up to

$N/2$, and from $m = 1$ up to $n/2$, and to use (4). In a similar way, the cumulative probability function was calculated by Epstein (1954) for $n = N = 1\ (1)20$.

For particular values of m, n, and N the distribution of the exceedances has many aspects easily worked out. The distribution of the number of exceedances over the largest value $m = 1$ decreases with increasing x,

Table 2.2.1. The Probabilities $w\ (5,m,5,x)$

	$m = 1$	$m = 2$	$m = 3$	$m = 4$	$m = 5$
$x = 0$	.50000	.22222	.08333	.02381	.00397
$x = 1$	.27778	.27778	.17857	.07936	.01984
$x = 2$	.13889	.23810	.23810	.15873	.05952
$x = 3$	.05952	.15873	.23810	.23810	.13889
$x = 4$	.01984	.07936	.17857	.27778	.27778
$x = 5$	.00397	.02381	.08333	.22222	.50000
Mean $\bar{x}$	.83333	1.66667	2.50000	3.33333	4.16667
Median $\tilde{x}$	0	1	2	3	4
Mode $\tilde{x}$	0	1	2,3	4	5
Standard Deviation σ	1.04464	1.32137	1.40153	1.32137	1.04464

and the distribution of the number of exceedances over the smallest value $m = n$ increases with increasing x. If $N = n$ is odd, the distribution of exceedances over the mth value $m = (n + 1)/2$ is symmetrical. Consider the case of no exceedances $x = 0$. If $m = 1$,

(6) $$w(n,1,N,0) = n/(n + N) = w(n,n,N,N),$$

and for $n = N$,

(7) $$w(n,1,n,0) = 1/2 = w(n,n,n,n).$$

The probability that the largest (smallest) of n past observations will never (always) be exceeded in n future trials is equal to 1/2. If $x = 0, m = n = N$,

(8) $$w(n,n,n,0) = \frac{(n!)^2}{(2n)!} = w(n,1,n,n).$$

The probability that the smallest observations will never be exceeded, equal to the probability that the largest value will always be exceeded, is very small, even for moderate sample sizes.

Exercises: 1) Calculate Table 2.2.1 for $n = N = 10$ and trace the distributions. 2) Rewrite the formulae of the sections 2.2 for the mth smallest value. 3) Analyze the modal number of exceedances. (See Gumbel and Schelling.) 4) Show that for $n = N$ the mode is either m or $m - 1$.

References: Gumbel and Schelling, Masuyama, Mosteller (1948), Epstein and Tsao. *Consult also:* K. Sarkadi (1957). On the distribution of the number of exceedances, Ann. Math. Stats., 28: 1021.

2.2.2. Moments.

In the following we calculate the mean number of exceedances and derive the moments from those of the binomial distribution. The mean $\bar{x}(n,m,N)$ is obtained from 2.2.1(4) as the solution of

(1) $\quad \binom{N+n}{n} \bar{x}(n,m,N) = \sum_{x=1}^{N} \binom{N+n-m-x}{n-m}\binom{x+m-1}{m-1} x,$

where from 2.2.1(2)

(2) $\quad \binom{N+n}{n} = \sum_{x=0}^{N} \binom{N+n-m-x}{n-m}\binom{x+m-1}{m-1}.$

After the transformation,

$$x - 1 = z; \; N - 1 = N'; \; m + 1 = m'; \; n + 1 = n',$$

the second member in (1) becomes, from (2),

$(m'-1) \sum_{z=0}^{N'} \binom{N'+n'-m'-z}{n'-m'}\binom{z+m'-1}{m'-1} = (m'-1)\binom{N'+n'}{n'}.$

If we return to N and n, it follows from (1) that

$$\binom{N+n}{n} \bar{x}(n,m,N) = \binom{N+n}{n+1} m.$$

Therefore the mean number of exceedances is simply

(3) $\quad \bar{x}(n,m,N) = mN/(n+1).$

Clearly the mean increases with m. The mean number of exceedances over the mth value from below is such that

(3') $\quad \bar{x}(n, n-m+1, N) + \bar{x}(n,m,N) = N.$

The mean number of exceedances over the smallest value is n times the mean number of exceedances over the largest value. If $N = n+1$, $\bar{x}(n,m,n+1) = m$. If n is odd, and $m = (n+1)/2$, the mean number of exceedances over the median is $N/2$. If n and N are large, the mean number of exceedances over the largest value is unity.

For $N = 1$ formula (3) leads to the plotting position 1.2.7(1).

The mean value (3) may also be obtained from the binomial formula 2.2.1(1) if we replace $p = 1 - F_m$ (where m is counted from the top) by its expectation $m/(n+1)$ derived in 2.1.3(4). This relation leads to all factorial moments. The distribution of the exceedances was derived from the binomial by considering $1 - F_m$ as a variable and integrating over all its values. The same process leads to the moments. The factorial moments of the binomial distribution derived in 1.1.6(19) are

$$\bar{x}_{[k]} = N(N-1)\ldots(N-k+1)p^k .$$

If p^k is replaced by its expectation, given in 2.1.3(3) as

$$\overline{p^k} = \frac{m(m+1)\ldots(m+k-1)}{(n+1)(n+2)\ldots(n+k)} ,$$

we obtain the factorial moments for the number of exceedances:

(4) $\quad \bar{x}_{[k]}(n,m,N) = \dfrac{m(m+1)\ldots(m+k-1)}{(n+1)(n+2)\ldots(n+k)}$
$$\times N(N-1)\ldots(N-k+1); \quad k \leq N .$$

The calculation of the moments (4) is simplified by the recurrence formula

(5) $\quad \bar{x}_{[k]} = \dfrac{(N-k+1)(m+k-1)}{n+k} \bar{x}_{[k-1]} .$

The mean square becomes, from (5) and (3)

$$\overline{x^2} = \bar{x}_{[2]} + \bar{x}$$
$$= \frac{mN}{n+1}\left[1 + \frac{(m+1)(N-1)}{n+2}\right].$$

Therefore the variance of the number of exceedances over the mth value from the top and from the bottom is

(6) $\quad \sigma^2(n,m,N) = \dfrac{Nm(n-m+1)(N+n+1)}{(n+1)^2(n+2)} = \sigma^2(n, n-m+1, N) .$

The variance increases with N and diminishes with n. It reaches a maximum for $m = (n+1)/2$, i.e., for the median, and diminishes if we pass to the largest and smallest initial values, $m = 1$ and $m = n$. The quotient of the variances of the number of exceedances over the median and over the extremes of the initial variate is

(7) $\quad \dfrac{\sigma^2[n, (n+1)/2, N]}{\sigma^2(n,1,N)} = \dfrac{(n+1)^2}{4n} = \dfrac{\sigma^2[n, (n+1)/2, N]}{\sigma^2(n,n,N)} .$

Thus the *variance of the median is about n/4 times the variance of the extremes*. A forecast of the number of times that the largest value will be exceeded is *more* reliable than a forecast of the number of times that the median will be exceeded, a fact which clearly contradicts the opinions held by "practical" people. The remarkable property (7) has its analogue in the distribution of the frequencies 2.1.3(1) and in the binomial distribution, where σ^2 is a maximum for $p = q = \frac{1}{2}$.

The variance of the exceedances is larger than that in the binomial case, because there the probability p is known, while here we know only the position of the corresponding variate. For $N = n + 3$, the variance becomes twice the variance of Bernoulli's distribution. The coefficient of variation obtained from

(8) $$\frac{\sigma^2(n,m,N)}{\bar{x}^2(n,m,N)} = \frac{N+n+1}{N(N+2)} \frac{n-m+1}{m}$$

diminishes if we pass from the largest to the smallest observation.

Exercises: 1) Calculate the first two moments, by summation. 2) Calculate the third and fourth central moments from (4). 3) Show that the generating function is a partial sum of the hypergeometric series.

2.2.3. The Median. If we sum the probabilities $w(x)$ from zero up to a certain x (or from a certain x up to N), we obtain the probabilities $W(x)$ (or $P(x)$) for at most (or at least) x exceedances over the mth past value in N future trials,

(1) $$W(x) = \sum_{z=0}^{x} w(z); \qquad P(x) = \sum_{z=x}^{N} w(z)$$

where
$$W(x) + P(x+1) = 1; \qquad W(x-1) + P(x) = 1.$$

The boundary conditions are

(2) $\quad W(0) = w(0); \qquad W(N) = 1; \qquad P(0) = 1; \qquad P(N) = w(N).$

The summation (1) may lead to a median number of exceedances, i.e., an integer $\check{x}$ such that

(3) $$W(\check{x}) = \sum_{z=0}^{\check{x}} w(n,m,N,z) = \tfrac{1}{2} = \sum_{z=N-\check{x}}^{N} w(n, n-m+1, N, z).$$

Such a number need not exist. Assume, for example, $N = n$. Then $w(n,1,N,0)$ surpasses $1/2$, and the distributions of the number of exceedances over the largest (or smallest) value do not possess a median.

If a median exists, it follows from (3) that

$$\tfrac{1}{2} = 1 - \sum_{z=0}^{N-\check{x}-1} w(n, n-m+1, N, z).$$

If $\check{x}_m$ is the median of the number of exceedances over the mth value from above, then $N - \check{x}_m - 1$ is the median of the number of exceedances over the mth value from below. The relation

(4) $$\check{x}(n,m,N) + \check{x}(n, n-m+1, N) = N - 1$$

differs from the corresponding relation 2.2.2(3′) for the mean. From 2.2.1(6) it follows that the median number of exceedances for $m = 1$ and $n = N$ is zero. Gumbel and Schelling showed that for all m

(5) $$\check{x}(n,m,n) = m - 1.$$

Exercise: Calculate the square matrix $W(n,m,n,x)$ for $n = 2, 3, 4, 5$. (See Epstein, 1954.)

Problem: Prove formula (5) from the properties of the matrix $W(n,m,n,x)$, $1 \leq m \leq n$, $0 \leq x \leq n - 1$. A simple proof was given by K. Sarkadi (1960) Ann. Math. Stat., 31: 225.

2.2.4. The Probability of Exceedances as Tolerance Limit. The probability for the mth largest value to be exceeded at most x times,

(1) $$W(n,m,N,x) = \sum_{z=0}^{x} w(n, n-m+1, N, N-z)$$
$$= \sum_{z=N-x}^{N} w(n, n-m+1, N, z) = P(n, n-m+1, N, N-x)$$
$$= 1 - W(n, n-m+1, N, N-x-1); \quad x \leq N-1,$$

is, from 2.2.1(3) and 2.2.3(1), equal to the probability that the mth smallest value will be exceeded at least $N - x$ times. This relation facilitates the calculation of the probabilities. The probability matrix $W(10, m, 10, x)$ is drawn in Graph 2.2.4 in continuous fashion on normal probability paper.

The probabilities that the mth largest among n past observations will be exceeded *at most once* in $N = n$ future trials are, for $m = 1$ and $m = n$,

(2) $$W(n,1,n,1) = \tfrac{1}{2} \frac{3n-1}{2n-1} = P(n, n, n, n-1) \to \tfrac{3}{4} \text{ and}$$

(3) $$W(n,n,n,1) = \frac{n!\,(n+1)!}{(2n)!} = P(n, 1, n, n-1) \to 0.$$

The probability that *at least* x out of N future trials will exceed the *smallest* value of a previous sample of size n is

(4) $$P(n,n,N,x) = \frac{nN!}{(n+N)!} \sum_{z=x}^{N} \frac{(n+z-1)!}{z!} = W(n, 1, N, N-x).$$

2.2.4 ORDER STATISTICS

This formula, due to S. S. Wilks, leads to an immediate practical application. The smallest number x_α, so that $P(n,n,N,x_\alpha) = \alpha$, called the *lower tolerance limit*, is important in many technical problems. The exact solution presents numerical complications, but the asymptotic

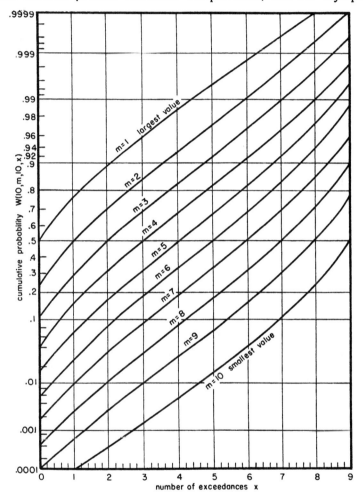

Graph 2.2.4. Probabilities $W(10,m,10,x)$ of Exceedances

solution given by Wilks is easily obtained. If N is large, and x is large and n remains finite, Stirling's formula leads from (4) to

$$\alpha = P(n,n,N,x) = \frac{n}{N^n} \sum_{z=x_\alpha}^{N} z^{n-1}.$$

We replace the sum by an integral whence

(5) $$(x_\alpha/N)^n = 1 - \alpha.$$

For example: If we have made 10 observations, and choose a probability of $\alpha = .99$, then at least 63% of a future large sample will exceed the smallest value in the original sample. Other values of x_α as function of α and n have been calculated by Wilks. Of course the probability that at least x out of N future trials will fall short of the largest observation (upper tolerance limit) is equal to the probability that at least x out of N future trials will exceed the smallest observation.

A related generalized problem is the following: How many trials N have to be made in order to ensure that there is a given probability $\alpha = P$ for the mth largest value of n preceding observations to be exceeded at least once? This probability is

(6) $$P(n,m,N,1) = 1 - \frac{n!}{(n-m)!} \frac{(N+n-m)!}{(N+n)!}$$
$$= W(n, n-m+1, N, N-1).$$

If N and n are large, and m is small, this expression becomes

(7) $$P(n,m,N,1) = 1 - [n/(n+N)]^m = W(n, n-m+1, N, N-1).$$

For $m = 1$, the probability that the largest among n values will be exceeded at least once in N future trials is

$$P(n,1,N,1) = N/(n+N) = W(n, n, N, N-1)$$

and for $n = N$, the probability is $1/2$, independent of n.

For $m = n$, the probability that the smallest among n values will be exceeded at least once is, from (6),

$$P(n,n,N,1) = 1 - \frac{n! \, N!}{(N+n)!} = W(n, 1, N, N-1).$$

Equation (6) leads, if we introduce $\alpha = P$ for the largest value $m = 1$, to

(8) $$\frac{N}{n} = \frac{1}{1-\alpha} - 1.$$

Of course, N/n increases with α. If n is large, and m remains small, equation (7) leads, in first approximation, to

(9) $$N/n = (1-\alpha)^{-1/m} - 1.$$

In a similar way we calculate the probabilities that the largest (and penultimate) among n observations will be exceeded at least twice in N future trials. Let α_2 be this probability. Then we have for the largest value

$$1 - \alpha_2 = w(n,1,N,0) + w(n,1,N,1) = \frac{n}{n+N}\left(1 + \frac{N}{n+N+1}\right).$$

For sufficiently large n the expression simplifies to

(10) $$\frac{1}{1-\alpha_2} \sim \frac{(N/n+1)^2}{2N/n+1}.$$

Finally, for $m = 2$ the probability for the penultimate value to be exceeded at least twice is obtained from

(11) $$\frac{1}{1-\alpha_2} \sim \frac{(N/n+1)^3}{3N/n+1}.$$

If we fix the probabilities α_2, formulae (10) and (11) give the number of future trials for at least 2 exceedances over the largest and the penultimate observation respectively.

References: Birnbaum and Tingey, Birnbaum and Zuckerman, Paulson, Robbins (1944b), Scheffé and Tukey, Simon, Wald (1943), Walsh (1949b), Wilks (1940, 1941, 1942).

2.2.5. Extrapolation from Small Samples. As in the preceding paragraph, let the number of future trials N be large, while the past sample size n remains small. Then m is also small. Instead of x we introduce $q = x/N$, $0 < q \leq 1$, the proportion of future exceedances. Since N is large, consider x and therefore q as continuous variates. Then the distribution $f(n,m,N,q)$ for q becomes, from 2.2.1(2),

$$f(n,m,N,q) = Nm\binom{n}{m}\frac{N!(qN+m-1)!(N-qN+n-m)!}{(N+n)!(qN)!(N-qN)!}.$$

Stirling's formula leads to the asymptotic distribution,

(1) $$f_1(n,m,q) = \binom{n}{m}mq^{m-1}(1-q)^{n-m},$$

which was considered at length in 2.1.3(1). The numerical values of the probability $F_1(n,m,q)$ are obtained from Karl Pearson's *Tables of the Incomplete Beta Function*. The probability corresponding to the smallest value ($m = n$) is

(2) $$F_1(n,n,q) = q^n.$$

In the case $m = 1$ studied by L. D. Harris,

(2') $$F_1(n,1,q) = 1 - (1-q)^n$$

is the probability that in a future large sample, at most a fraction q will exceed the largest or fall short of the smallest value in the original sample. The sample size n being the solution of

$$(3) \qquad n = \frac{\log(1 - F_1)}{\log(1 - q)}$$

is traced in Graph 2.2.5 as function of $\log q$. Since $\dfrac{d \log n}{d \log q}$ is independent

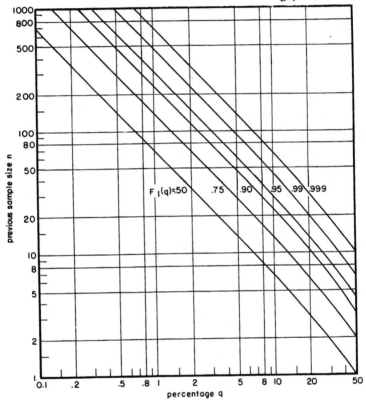

Graph 2.2.5. Sample Size n as Function of Percentage q and Probability $F_1(q)$

of F and fairly constant in the interval chosen, the curves approximate parallel straight lines. The relation (2′) may be used for the design of life tests. If such a test is made simultaneously on n specimens and is discontinued as soon as the first one fails, equation (2′) gives the probability F_1 that, in the future, at most the proportion q will fail in a time shorter than the length of the test, provided that the specimens and the testing procedures are and remain homogeneous. Inversely, Graph 2.2.5 shows the sample size n as function of the percentage q, and the probability F_1, e.g., for $F_1 = .90$, $q = .25$ we obtain $n = 8$. The integral $F_1(n,m,q)$ corresponding to (1) gives the probability that up to q percent of a large future sample will fail in a time shorter than the mth failure in a previous test of n units.

In addition to the extremes $m = 1$ and $m = n$, consider the median of the original variate. If n is large enough (while still small compared to N), the rank of the median is of the order $n/2$. Then the distribution over the median is, from (1),

$$(4) \qquad f_1(n,n/2,q) = C_1 q^{\frac{n}{2}}(1-q)^{\frac{n}{2}}.$$

If we define a new variate z by

$$q = \tfrac{1}{2} + \frac{z}{2\sqrt{n}},$$

its distribution $f_2(z)$ is obtained from

$$(5) \qquad C_2 \left(\frac{1}{4} - \frac{z^2}{4n}\right)^{\frac{n}{2}} \to C_3 e^{-\frac{z^2}{2}}.$$

The expressions C_1, C_2, C_3 are independent of the variates q and z. Therefore, the distribution of the percentage of future exceedances over the initial median converges toward a normal distribution with mean $1/2$ and variance $1/4\,n$.

Exercise: Verify that the moments are conserved in the asymptotic transition.

2.2.6. Normal and Rare Exceedances. In the previous derivation n, although large, was small compared to N. We assume now that $n = N$ becomes large, and consider first the initial median. Let n be odd, then $m = (n+1)/2$ corresponds to the median of the original variate and the symmetry relation 2.2.1(3) becomes

$$w(n, (n+1)/2, N, x) = w(n, (n+1)/2, N, N-x).$$

The distribution of the number of exceedances over the median is symmetrical. Now let $n = N = 2k - 1$ be large, then k is large. To obtain the asymptotic distribution of exceedances over the median we reduce x by writing

$$(1) \qquad x = k + z\sqrt{k},$$

where z remains finite. The same reduction may be applied to mth values in the neighborhood of the median.

The distribution of the number of exceedances over the median is, from 2.2.1(2) and (1),

$$w(2k-1, k, 2k-1, z) = \text{const.} \binom{2k-1}{k+z\sqrt{k}} \bigg/ \binom{4k-3}{2k+z\sqrt{k}-1}.$$

After expansion this is a function of k only multiplied by a function of k and z.

From Stirling's formula it follows, after straight formal calculations, that

(2) $\quad \lim_{k=\infty} w(2k-1, k, 2k-1, z) = Ce^{-\frac{z^2}{2}}; \quad z = (x-k)/\sqrt{k}.$

The number of exceedances over the mth value in the neighborhood of the median, $m \sim N/2$ in a large sample of size N in N future trials, is asymptotically normally distributed with mean and variance equal to $(N+1)/2$. Therefore formula (2) may be called the *distribution of normal exceedances*.

If n and N are large and m remains constant so that $m \ll n$, we deal with mth extreme values. The distribution 2.2.1(2) becomes, from Stirling's formula,

(3) $\quad w(n,m,N,x) \sim \binom{x+m-1}{x} \frac{n^m N^x}{(N+n)^{m+x}} \sim w(n, n-m+1, N, N-x).$

For $x = 0$, the probability that the mth value from above (below) is never (always) exceeded in n future trials is the general term of the geometric series

(3') $\quad w(n,m,N,0) \sim \left(\frac{n}{N+n}\right)^m \sim w(n, n-m+1, N, N).$

For $m = 1$ we obtain, from (3),

(3'') $\quad w(n,1,N,x) \sim \frac{n}{N+n} \left(\frac{N}{N+n}\right)^x \sim w(n, n, N, N-x).$

This is another geometric series decreasing with x.
It follows that for $n = N$ the probability that the largest value will be exceeded at most x times,,

$$W(n,1,n,x) \sim 1 - (\tfrac{1}{2})^{x+1},$$

is independent of n and converges rapidly toward unity. For $N = n$ formula (3) becomes

(4) $\quad w(n,m,n,x) \sim \binom{x+m-1}{x} \left(\frac{1}{2}\right)^{m+x} \sim w(n, n-m+1, n, n-x).$

This *asymptotic* probability that the *m*th largest (smallest) value will be exceeded x times ($n - x$ times) in n future trials is independent of n and contains one single parameter m. Since m is small compared to n, (4) may be called the *law of rare exceedances*. For $m = 1$, the probability that the largest (or smallest) value is exceeded x times (or $n - x$) times is

(5) $$w(n,1,n,x) = (\tfrac{1}{2})^{x+1} = w(n, n, n, n - x).$$

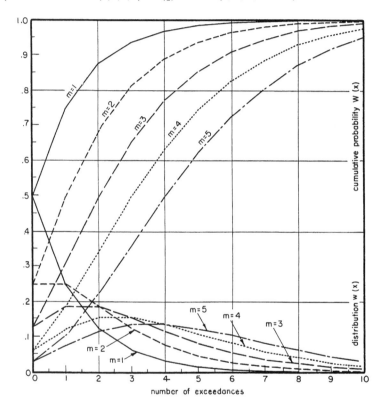

Graph 2.2.6. Distributions and Probabilities of Rare Exceedances

The distribution (4) has two modes,

(6) $$\tilde{x}_1 = m - 2 ; \quad \tilde{x}_2 = m - 1,$$

except for $m = 1$, where the distribution diminishes with x. The probabilities of the modes decrease with m, as shown in Graph 2.2.6, where the probabilities $w(x)$ and the cumulative probabilities $W(x)$ are traced as continuous curves. The distributions are similar to Poisson's distributions for integer m. However, for this distribution the modes are $m-1$ and m.

The first two moments obtained from 2.2.2(3) and 2.2.2(6) are

(7) $$\bar{x} = m\,; \qquad \sigma^2 = 2m\,.$$

The mean number of exceedances over the mth largest value in the distribution of rare exceedances is m itself, as in the Poisson distribution. However, the variance is double the variance in Poisson's distribution, for reasons explained in 2.2.2.

The variance for the normal exceedances was $(N+1)/2$, while the variance for the distribution of rare exceedances $2m$ is much smaller, since m is small compared to N. This is a special case of the relation between the variances stated in 2.2.2. The variance is smallest for $m=1$; i.e., the largest value and the distribution spread with increasing m.

If m is large, the Poisson distribution for the standardized variate $y = (x-m)/\sigma$ converges toward a normal distribution. The same holds for the distribution of rare exceedances. For the proof, consider the generating function $G_x(t)$ obtained from (4) as

(8) $$G_x(t) = \left(\frac{1}{2}\right)^m \sum_{x=0}^{\infty} \binom{x+m-1}{m-1} \left(\frac{e^t}{2}\right)^x,$$

which becomes from the expression for the negative binomial

(9) $$G_x(t) = (2-e^t)^{-m}$$

If we introduce the standardized variate

(10) $$y = (x-m)/\sqrt{2m}\,,$$

the moment generating function $G_y(t)$ becomes, from 1.1.6(3),

$$G_y(t) = (2e^{t/\sqrt{2m}} - e^{2t/\sqrt{2m}})^{-m} = \left[1 - \frac{t^2}{2m} + 0(m^{-3/2})\right]^{-m}.$$

If we neglect the factors $0(m^{-3/2})$, we obtain the normal generating function

(11) $$G_y(t) = e^{t^2/2}.$$

Thus the distribution of rare exceedances converges toward normality in the same way as Poisson's distribution.

2.2.7. Frequent Exceedances. In the previous paragraphs we had $n=N$ large and $m=1$ and $m=(n+1)/2$. It remains to study the number of

2.2.8 ORDER STATISTICS

exceedances over the smallest value where $m = n$. Formula 2.2.1(2) leads to

(1) $\quad w(n,n,n,x) = \dfrac{nn!}{(2n)!} \dfrac{(x+n-1)!}{x!} = w(n, 1, n, n-x).$

For $x = 0$ the probability

(1') $\quad\quad\quad\quad w(n,n,n,0) \sim \sqrt{2\pi n}\, 2^{-2n}$

obtained from Stirling's formula is very small. The larger the number of exceedances, the larger is their probability. Hence the name chosen in the title of this paragraph. The probabilities for $x = n$ and $x = n-1$ are

(2) $\quad w(n,n,n,n) = \tfrac{1}{2}\,; \quad w(n, n, n, n-1) = \dfrac{n}{2(2n-1)}.$

The latter expression converges with increasing n to $1/4$. To obtain an asymptotic expression, consider the variate $z = n - x$ where z is small compared to n. Then the distribution

$$w_1(n,n,n,z) = \dfrac{nn!}{(2n)!} \dfrac{(2n-z-1)!}{(n-z)!}$$

converges to

(3) $\quad\quad\quad\quad w_1(n,n,n,z) \sim (\tfrac{1}{2})^{z+1}.$

Formula (1') shows that this expression becomes invalid for large values of z, i.e., small values of x. The limiting distributions in this and the three preceding paragraphs are summarized in Table 2.2.7.

Table 2.2.7. *The Six Limiting Distributions of Exceedances*

Condition		$N \gg n = O(1)$ $q = x/N$	$N = n \gg 1$
Smallest Value	$m = n$	nq^{n-1}	$(\tfrac{1}{2})^{n-x+1}$; $x \sim n$
Median	$m \sim n/2$	normal, $\bar{q} = \tfrac{1}{2}$ $\sigma^2 = 1/4n$	normal, $\bar{x} = n/2$ $\sigma^2 = n/2$
Largest Value	$m \sim 1$	$n(1-q)^{n-1}$; $m = 1$	$\binom{x+m-1}{x}(\tfrac{1}{2})^{m+x}$

2.2.8. Summary. The probability that the mth largest among n observations will be exceeded x times in N future trials is analogous to Bernoulli's distribution. The mean number of exceedances is the mean in Bernoulli's distribution. However, the variance is larger than in Bernoulli's distribution.

The variance of the number of exceedances is largest for the median of the initial distribution and smallest for its extremes. This superiority of the extremes increases with the sample size.

In 50 percent of all cases, the largest (or smallest) of n past observations will never (always) be exceeded in $N = n$ future trials. The probability for the largest among n observations to be exceeded at most once in n future trials converges toward 3/4. Elementary calculations lead to the setting of sample sizes N corresponding to given probabilities for 1 or 2 exceedances over the past largest and penultimate observation.

If $n = N$ is large, the distribution of the number of exceedances over the median observation is normal with mean and variance $n/2$, while the distribution of the exceedances over the mth extremes (the law of rare exceedances) similar to Poisson's distribution, has mean m and variance $2m$, m being small compared to the sample size.

Chapter Three: EXACT DISTRIBUTION OF EXTREMES

3.1. AVERAGES OF EXTREMES

3.1.0. Problems. The exact distributions of the extremes can be written down immediately as functions of the initial distribution and of the sample size n. The first study on extreme values by Bortkiewicz, who was followed by Von Mises, Neyman, Tippett, Tricomi, and Finetti, started from the normal distribution. However, the analytical results are very complicated, while fundamental theorems can easily be derived from the exponential distribution.

To characterize the distribution of extremes, the usual averages, median, mode, and mean are not sufficient. For this purpose we need a new type of averages introduced by R. Von Mises and studied by E. L. Dodd, who called it the asymptotic value. In a previous French publication (Gumbel, 1934) it was called "la dernière valeur." In 3.1.4 we call it the characteristic extreme. Most of the interesting properties of extreme values are linked to it and to the intensity function introduced in 1.2.1(1).

3.1.1. Exact Distributions. The probability that all of n *independent* observations on a continuous variate are less than x is $F^n(x)$. This may be interpreted as the probability $\Phi_n(x)$ that the largest among n independent observations is less than or equal to x:

(1) $$\Phi_n(x) = F^n(x) .$$

Formula (1) can be used to calculate the probabilities $\Phi_n(x)$ from tables of $F(x)$. If the variate x is multiplied by a positive factor, the largest value is multiplied by the same factor. If a constant is added to the variate, the largest value increases by the same amount. The probability that a given value x is the largest one, diminishes with increasing sample size n. The different functions $\Phi_n(x)$ form a system of curves which shift to the right for increasing n without intersecting. Hence all quantiles, the modes, and the means of largest value increase with n. The Taylor expansion of $\lg F(x)$ leads to the first approximation

(1′) $$\Phi_n(x) \sim e^{-n(1-F(x))}$$

From (1) follows also an important conclusion concerning the sign of the extremes. Consider a variate unlimited in both directions, and let the median be zero. Then the probability P for the largest value x_n to be positive is,

(2) $$P(x_n > 0) = 1 - (\tfrac{1}{2})^n$$

and approaches unity very quickly for increasing n. For $n = 10$, $P > 0.999$; $n = 17$, $P > 0.99999$. Consequently, the probability that x_n is negative may be neglected for all practical purposes, even for moderate sample sizes. For the same reason x_1 may be considered as being negative. However, if we have to integrate over x_1 or x_n for an unlimited initial distribution, the limits of integration are minus infinity to plus infinity.

The probability $1 - {}_1\Phi_n(x)$ that the smallest among n independent observations is less than x is obtained from

(3) $$_1\Phi_n(x) = 1 - (1 - F(x))^n .$$

If the initial variate is unlimited, the extremes are also unlimited.

The distributions $\varphi_n(x)$ and ${}_1\varphi_n(x)$ of the largest and the smallest value obtained from (1) and (3) are

(4) $$\varphi_n(x) = nF^{n-1}(x)f(x) ; \qquad {}_1\varphi_n(x) = n(1 - F(x))^{n-1}f(x) .$$

Siotani derived the formulae corresponding to (3) and (4) for a discontinuous variate and applied them to the binomial distribution.

For increasing sample sizes, the curves representing consecutive distributions of the extremes, have different shapes. It will be shown in 3.1.4 where intersections occur.

If the initial distribution is symmetrical about median zero, the distribution of the largest value is no longer symmetrical since

(5) $$1 - \Phi_n(x) \neq \Phi_n(-x) .$$

The same argument holds for the smallest value. The process of taking extreme values introduces asymmetry. However, the smallest and the largest values x_1 and x_n taken from a symmetrical distribution are mutually symmetrical.

(6) $$\Phi_n(x_n) = 1 - {}_1\Phi_n(-x_1) ; \qquad \varphi_n(x_n) = {}_1\varphi_n(-x_1) ;$$
$$\varphi'_n(x_n) = -{}_1\varphi'_n(-x_1) .$$

If the initial distribution is asymmetrical, the *symmetry principle* (6) means: From a given distribution of the largest value, valid for a variate x, we may obtain a distribution of a smallest value by changing the sign of x. In *two* mutually symmetrical distributions the distribution of the

3.1.1 DISTRIBUTION OF EXTREMES

largest value of the one is the distribution of the smallest value of the other, and vice versa.

Let z be a transformed variate defined by $x = h(z)$; then the distributions of the extremes z_1 and z_n are obtained from the distributions (4) of the extremes x_1 and x_n by the transformation linking z and x. Thus the distributions $v_n(z)$ and $_1v_n(z)$ of the extremes $z = z_n$ and $z = z_1$ are respectively

(7) $\qquad v_n(z) = \varphi_n[h(z)] \left| \dfrac{dx}{dz} \right| \quad : \quad _1v_n(z) = {}_1\varphi_n[h(z)] \left| \dfrac{dx}{dz} \right|.$

Assume a variate x unlimited to the right and a reduced variate $t = (x - \hat{x})/\sigma'$ where $\hat{x}$ stands for an average and σ' for a measure of dispersion. Then the corresponding average x_n of the largest value is

$$\hat{x}_n = \hat{x} + \sigma' \hat{t}_n,$$

where $\hat{t}_n$ increases with n. Assume in addition that the variable is such that the distributions contract with increasing averages. If we compare two populations with averages $\hat{x}_{(1)}$ and $\hat{x}_{(2)}$ and dispersions $\sigma_{(1)}'$ and $\sigma_{(2)}'$, where

$$\hat{x}_{(1)} > \hat{x}_{(2)} \; ; \qquad \sigma_{(1)}' < \sigma_{(2)}',$$

then, there must exist a sample size n so that

(8) $\qquad\qquad\qquad \hat{x}_{n,1} < \hat{x}_{n,2}.$

The average largest value for the population with the smaller average exceeds the average largest value for the other population. This leads to an interesting consequence. A life table may be characterized by the modal age at death $\tilde{x}$, and a measure of dispersion, $e(\tilde{x})$, the expectation of further life, which is the mean deviation of the ages at death counted from the modal age onward. Life tables with large (small) values of $\tilde{x}$ may be called favorable (unfavorable). Many observations have shown that favorable tables have smaller values of $e(\tilde{x})$ than unfavorable tables. If we assume that age at death is an unlimited variate and depends only on two parameters, then the mean largest age at death for a sufficiently large number of years should be larger for the population living under unfavorable conditions. This may be the basis for the claim that extraordinarily long lives have been reached among so-called uncivilized populations. If these statements are true, they do not prove any advantage of barbarism over so-called civilization, since the oldest age at death is not a measure of hygienic progress. The usual explanation is that the claims are false: the alleged ages at death are not exact, since birth certificates do not exist in "uncivilized" countries.

Another application of (8) is to hydrology. A river with a small mean yearly discharge and large dispersion will in the course of years produce a larger flood than a river with a large mean yearly discharge and small dispersion.

If the sample size $n = v$ itself is a statistical variable ($n_1 \leq v \leq n_2$) with a known distribution $g(v)$, then the probability function of the largest value $x = x_n$ becomes

(9) $$\Phi_n(x) = \sum_{v=n_1}^{n_2} F_{(x)}^v g(v)$$

If n_1 and n_2 are both large enough the distribution $g(v)$ may be considered as continuous and the sum is replaced by an integral. The integration may lead to analytical difficulties.

Exercise: Trace the distribution of the largest among n observations for $n = 2, 3, 4, 5$, for the uniform distribution.

References: Bruges, Chung, Chung and Erdös, Darling (1952, 1953), Dixon (1950, 1953), Feller (1950, p. 176), Helmert, Juncosa, Von Mises (1934), Mosteller (1948), Neyman (1923), Olds, Rider (1955), Wald (1947), G. S. Watson (1952).

3.1.2. Return Periods of Largest and Large Values. Consider as largest values the floods of N years and the mean daily average discharges, which surpass a certain basic stage, say the observed minimum of the largest values. They are possible largest values, since large values in certain years may be larger than largest values in other years. More generally, assume N series of n observations, taken from the same population. In each series (year) we take the largest observation (first set). Now we choose the observed minimum ξ of the N largest values, and consider a second set of large values, consisting of the $\bar{k}N$ among the Nn observations which exceed ξ. Here $\bar{k} > 1$ is the mean number of values exceeding ξ in each year.

The mth largest value $x_m(N\bar{k})$ among the $N\bar{k}$ observations in the second set has the observed return period in years T_b. From 1.2.7(2)

(1) $$T_b = (N\bar{k} + 1)/\bar{k}m,$$

which becomes for large N

(2) $$T_b \sim N/m.$$

Now if N is large enough, the value $x_m(N\bar{k})$ or a value close to it will also occur in the first set as the largest observation of some year. We want to find its return period in years within this set. The probability that this value is the largest of the $\bar{k}$ large values of any year, i.e., the probability that this value will occur within the first set is, from the usual combinatorial rules for the number of possible and favorable cases,

$$\binom{N\bar{k}-m-2}{\bar{k}-1} \bigg/ \binom{N\bar{k}-2}{\bar{k}-1} = \frac{(N\bar{k}-m-2)!\,(N\bar{k}-\bar{k}-1)!}{(N\bar{k}-2)!\,(N\bar{k}-\bar{k}-1-m)!}$$

$$\sim \left(\frac{N\bar{k}-\bar{k}-1}{N\bar{k}-2}\right)^m,$$

provided that m is small and $N\bar{k}$ is large. Under the same conditions the probability approaches

$$\left(1-\frac{1}{N}\right)^m \sim e^{-m/N}.$$

The return period T of this largest value is, from (2),

(3) $$T \sim \frac{1}{1-\exp[-1/T_b]}.$$

As soon as $T_b \gg 1$ the return period T converges to

(4) $$T = T_b + \tfrac{1}{2} + 0(T_b^{-1}).$$

The third term can safely be omitted with an error of less than 0.1% from $T = 5$ onward. If we are interested in large return periods, we may safely consider only the set of largest values and need not take into account all values which exceed the basic stage, a procedure proposed by W. Langbein (1949).

Exercise: Trace T from (3) as function of T_b for $T_b \leq 10$.

3.1.3. Quantiles of Extremes. The medians $\check{x}_n$ and $\check{x}_1$ of the extreme values in samples of size n are, from 3.1.1(1), the solutions of

(1) $$F(\check{x}_n) = 1 - F(\check{x}_1) = \exp[-(\lg 2)/n].$$

Obviously, for an unlimited variate, the absolute magnitude of the medians increases with n without limit and the medians converge toward the solution of

(2) $$F(\check{x}_n) = 1 - F(\check{x}_1) \sim 1 - (\lg 2)/n.$$

If 10^{-5} is neglected compared with unity, this approximation may be used for $n \geq 113$. If we plot, for a specific distribution, the probabilities of the medians and other averages of the extremes as functions of the number of observations, it is natural to plot them against $1/n$.

The return period $T(x_n)$ of the median of the largest observation obtained from (1) is asymptotically

(3) $$T(x_n) \sim 1.44269n + \tfrac{1}{2}.$$

Thus, the use of the probability of the median mth value as plotting position (1.2.6) attributes to the largest observation a return period which is 44% larger than the number of observations, and spreads the observations over a domain which is too large compared to the sample size. See Graph 3.1.3(1) which also shows that the linear relation (3) is quickly reached.

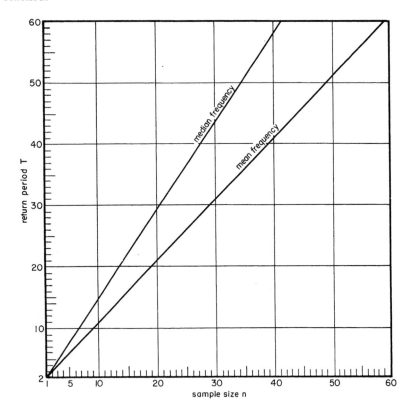

Graph 3.1.3(1). **Return Periods of Largest Values**

For the calculation of the medians of the extremes, we choose certain values of $\check{x}_n$ and $\check{x}_1$ and obtain the corresponding values of n from (1) as

$$n = (\lg 2)/(-\lg F(\check{x}_n)) = (\lg 2)/-\lg(1 - F(\check{x}_1)).$$

From the approximation (2) we find, with the help of a table of the probability function $F(x)$,

(4) $\qquad n \sim (\lg 2)/(1 - F(\check{x}_n)) \sim (\lg 2)/F(\check{x}_1)\,.$

3.1.3   DISTRIBUTION OF EXTREMES   81

The knowledge of the median of the extreme value leads immediately to the knowledge of the quantiles as functions of the number of observations. The first quartile $q_{1,n}$ of the largest value for n observations defined by

(5) $$F(q_{1,n}) = \exp[-(\lg 4)/n]$$

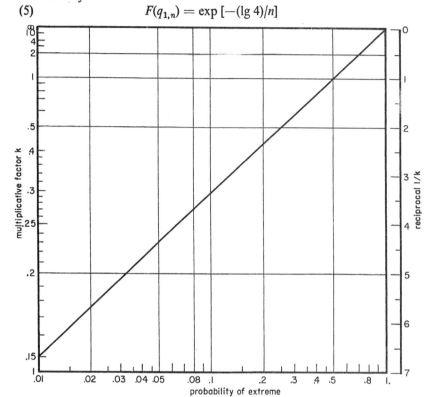

Graph 3.1.3(2). Calculation of Quantiles from the Median Extreme

is from (1) equal to the median of the largest value for $n/2$ observations. In the same way the quantile q_n of the largest value corresponding to the probability

$$F^n(q_n) = q/100$$

for n observations is the median of the largest value for kn observations

(6) $$q_n = \check{x}_{kn},$$

where the multiplicative factor k is obtained from

$$\frac{-\lg(100/q)}{n} = -\frac{\lg 2}{kn}.$$

The relation

(7) $$1/k = 6.63483 - 1.44290 \lg q$$

is traced in Graph 3.1.3(2). E.g., the largest value corresponding to a probability 0.05 for n observations is the median of the largest value corresponding to $0.23n$ observations. Thus the quantiles of the extreme value are known, once a table of the median extreme value as function of n has been calculated.

3.1.4. Characteristic Extremes. Besides the median we introduce a new average of the extremes analogous to the quantiles. Since the probability of a value equal to or larger than x is $1 - F(x)$, we expect $n[1 - F(x)]$ values in a sample of size n to be equal to or larger than x and define the *characteristic largest value* $u_n(n) = u_n$ by writing, for $n \geq 2$,

(1) $$F(u_n) = 1 - 1/n.$$

In n observations the expected number of values equal to or larger than u_n is unity. In the same way, the characteristic smallest value $u_1(n) = u_1$ in a sample of size $n \geq 2$ is defined by

(2) $$nF(u_1) = 1.$$

For $n = 2$ the characteristic largest and smallest values are equal to the initial median, for $n = 4$ to the upper and lower quartiles, for $n = 10$ to the upper and lower deciles of the initial distribution, and so forth. The characteristic extremes are obtained from equations (1) and (2) as functions of n and the parameters of the initial distribution. For large samples, this calculation may be simplified if asymptotic expressions for the probability $F(x)$ exist. The absolute amounts of the characteristic extremes increase with the sample size. If the initial distribution is symmetrical, the absolute amounts of the characteristic extremes coincide for given sample sizes, and the same holds for two mutually symmetrical distributions. The number of observations n corresponding to the characteristic largest and smallest values are the return periods $T(u_n)$ and $T(u_1)$ of a value equal to or larger or smaller than u_n and u_1. From 3.1.3(1) the absolute amounts of the characteristic extremes are smaller than the absolute amounts of the medians of the extremes.

From 3.1.1(1) and 3.1.1(3) the probabilities $\Phi_n(u_n)$ and $_1\Phi_n(u_1)$ converge quickly toward

(3) $$\Phi_n(u_n) \sim \frac{1}{e} \sim 1 - {_1\Phi_n(u_1)}.$$

Hence, if we have N samples, each composed of n observations taken from the same distribution, and if we choose the N largest (smallest) values,

3.1.4 DISTRIBUTION OF EXTREMES

about 36.8% of them will be below (above) the characteristic largest (smallest) value.

The notion of the characteristic extremes leads to an answer to a question raised in 3.1.1. The curves corresponding to two consecutive distributions of extremes for n and $n+1$ observations intersect at a certain value $\dot{x}$ so that

$$\varphi_n(\dot{x}) = \varphi_{n+1}(\dot{x}).$$

From 3.1.1(4)

(4) $$F(\dot{x}) = 1 - 1/(n+1).$$

The distribution of the largest (smallest) among n observations intersects the following (preceding) distribution of the largest (smallest) among $n+1$ observations at the characteristic largest (smallest) value of the latter distribution.

The characteristic extremes are the basis for Chauvenet's criterion for the acceptance or rejection of extreme values. For a given initial distribution, define two theoretical values $\breve{u}_n$ and $\breve{u}_1$ by

(5) $$n[1 - F(\breve{u}_n)] = \tfrac{1}{2}; \qquad nF(\breve{u}_1) = \tfrac{1}{2}.$$

Then the expected numbers of observations above $\breve{u}_n$ and below $\breve{u}_1$ are $\tfrac{1}{2}$. If the expected numbers corresponding to the two observed extremes x_n and x_1 are equal to or larger than $\tfrac{1}{2}$, they are accepted; if not, they are rejected. In the latter case, the parameters in the distribution have to be recalculated without the use of the rejected values. This leads to new values $\breve{u}_n$ and $\breve{u}_1$, which are now used to test the second and penultimate observation. The rejection of an extreme has three possible consequences: we conclude that the extreme is due to a faulty observation, or that a specific cause has given rise to it, or that there is something wrong with the statistical hypothesis.

The notion of characteristic extremes can easily be generalized to cover the mth extremes. We define *the characteristic mth extreme values*

$$u_m(n) = u_m; \qquad {}_m u(n) = {}_m u$$

by

(6) $$F(u_m) = 1 - m/n \qquad F({}_m u) = m/n.$$

The number of observations equal to or larger than u_m (smaller than ${}_m u$) in a sample of size n is a random variable. Its mean value is m. Obviously, the characteristic kth largest value among kn observations is equal to the characteristic largest value among n observations.

Exercise: Trace the characteristic largest value for the Cauchy distribution as function of n.

References: Czuber (p. 211), Dinsmore, Dixon (1950, 1953), Foster and Stuart, Foster and Teichroew, Glaisher, Gould, Grubbs, Jeffreys, King, E. S. Pearson and Sekar, Peirce, Proschan, Rider (1933), Saunder, Schott, Stewart (1920a and b), Stone (1867, 1873, 1874), W. R. Thompson (1935), Walsh (1949b) G. S. Watson (1954), Welch, Winlock.

3.1.5. The Extremal Intensity Function. In connection with the characteristic extreme values we define the *extremal intensity functions* $\alpha_n(n) = \alpha_n$; $\alpha_1(n) = \alpha_1$, in consequence of 1.2.1(1), as

$$(1) \qquad \alpha_n = nf(u_n) \equiv \mu(u_n) ; \qquad \alpha_1 = nf(u_1).$$

Obviously the Alphas are functions of n and have the same dimension as x^{-1}.

For symmetrical initial distributions we have of course $\alpha_n = \alpha_1$. For two mutually symmetrical distributions the α_n for one distribution is the α_1 for the other and vice versa. The densities of the probability $\varphi_n(x)$ and $_1\varphi_n(x)$ of the largest (smallest) values 3.1.1(4) at $x = u_n(x = u_1)$ converge with increasing n to

$$(2) \qquad \varphi_n(u_n) = \alpha_n/e ; \qquad _1\varphi_n(u_1) = \alpha_1/e .$$

It cannot be stated a priori how α_n and α_1 behave with increasing n. If the Alphas increase with (are independent of, or decrease with) n, the same holds for the density of the probability at the corresponding characteristic extreme value. If we trace the distributions of the extremes as functions of n against x, an increase of n shifts the distributions of the largest (smallest) values to the right (left). If, at the same time,

$$(3) \qquad \frac{d\alpha_n}{dn} \gtrless 0 ; \qquad \frac{d\alpha_1}{dn} \gtrless 0,$$

the density of probability at the characteristic values increases (decreases). The operation d/dn may be written

$$(4) \qquad \frac{d}{dn} = \frac{d}{dx}\frac{dx}{dn}.$$

Since the second factor is positive for $x = u_n$, and since any x may be considered as being the characteristic value for an adequate number n, the operation d/dn may be replaced by d/dx. For this purpose, we extend the definitions 3.1.4(1), 3.1.4(2) and (1), of u_n, u_1, α_n, α_1 to all real n, so that the resulting functions are differentiable. Then the inequalities (3) become

$$(5) \qquad \frac{d\alpha_n}{dx} \gtrless 0 ; \qquad \frac{d\alpha_1}{dx} \gtrless 0.$$

The densities of probability at the characteristic values increase (decrease) for increasing numbers of observations, if the extremal intensity function increases (decreases) with the variate for large absolute values of x. Therefore the Alphas play an important role. It will be shown in 5.1 that the asymptotic distributions of the extremes for unlimited variates do not contain any other parameter except α_n and u_n. They will be used for the reduction of the extremes.

3.1.6. The Mode. The modes of the extremes $\tilde{x}_n(n) = \tilde{x}_n$ and $\tilde{x}_1(n) = \tilde{x}_1$ (if they exist) are obtained from the distributions 3.1.1(4) as the solutions of

(1) $\quad \dfrac{n-1}{F(x)} f(x) + \dfrac{f'(x)}{f(x)} = 0 \; ; \quad \dfrac{(n-1)}{1-F(x)} f(x) + \dfrac{f'(x)}{f(x)} = 0 \, .$

These equations may also be written

(2) $\quad n - 1 = -\left(\dfrac{f'(x)}{f^2(x)} F(x)\right)_{x=\tilde{x}_n} ; \quad n - 1 = \left(\dfrac{f'(x)}{f^2(x)} [1 - F(x)]\right)_{x=\tilde{x}_1} .$

For limited initial variates, the equations need not possess solutions. For unlimited variates, the distributions 3.1.1(4) of the extremes vanish with increasing positive and negative values of the variate. Consequently the extremes possess at least one mode. For symmetrical variates with median zero, we have evidently $\tilde{x}_1 = -\tilde{x}_n$, and the same relation holds for two mutually symmetrical distributions.

The solutions of the equations (1) or (2) depend on n and the initial distribution. The equations do not always lead to simple analytical expressions. Numerical values for the modes as functions of n can be calculated, if tables for the probability function and its first two derivatives exist. We choose certain values of $\tilde{x}$ and calculate the corresponding numbers n. This procedure can easily be generalized for the calculation of the modes $\tilde{x}_m(n) = \tilde{x}_m$ of the mth extremes. From 2.1.2(3') they are the solutions of

(2') $\quad m - 1 - (n-1) F(x) = -f'(x) f^{-2}(x) F(x) [1 - F(x)]$

Here m is counted from the bottom. For some distributions, given in Table 2.1.2, equation (2') provides an analytic solution. To find numerical solutions, it is sufficient to calculate the modes of the two extremes proper. The consecutive other modes may then be obtained by the following recurrent procedure. If we put

(3) $\quad 1 - [F(x) + f'(x) f^{-2}(x) F(x) (1 - F(x))]_{x=\tilde{x}_m} = \Delta_m \, ,$

the most probable mth value is the solution of

(3') $\quad\quad\quad F(\tilde{x}_m) = (m - \Delta_m)/n \, .$

The probability $F(\tilde{x}_m)$ is of the order m/n. In particular the most probable smallest value for n observations is the solution of

$$n = (1 - \Delta_1)/F(\tilde{x}_1).$$

If we have solved this equation n, $F(\tilde{x}_1)$ and Δ_1 are known. We now keep the three numerical values $\tilde{x}_1$, Δ_1, and $F(\tilde{x}_1)$, introduce $m = 2$ in (3′), and replace n by $_2n$. Here $_2n$ stands for a number of observations such that $\tilde{x}_1(n) = \tilde{x}_2(_2n)$ is the most probable second among $_2n$ values. This number of observations is obtained from

$$_2n = (2 - \Delta_1)/F(\tilde{x}_1)$$

or

$$_2n = n + 1/F(\tilde{x}_1).$$

In a general way, the number of observations $_mn$ necessary so that $\tilde{x}_1$ is the most probable mth smallest value is obtained from the definition 1.2.3(8) of $_1T(x_1)$ as

(4) $$_mn = n + (m - 1)\,_1T(\tilde{x}_1).$$

This procedure can be carried out for all extreme values from the bottom. If the smallest value has no mode, we have to start from the first rank m so that a mode exists.

A similar algorithm can be established for the modes of the mth largest values. The most probable largest value is, from (2′), the solution of

(5) $$n = \Delta_n/(1 - F(\tilde{x}_n)),$$

which is the first equation in (2). Assume that this equation is solved; then $\tilde{x}_n$, n and Δ_n are known. We keep these three numerical values, introduce into (3′) $m = n_2 - 1$, and replace n by n_2. Thus we obtain the most probable penultimate value,

$$\tilde{x}_{n_2-1}(n_2) = \tilde{x}_n(n),$$

for n_2 observations. The number of observations n_2 is the solution of

$$n_2 = n + 1/T(\tilde{x}_n).$$

In the same way, a number n_m so that $\tilde{x}_n$ is the most probable mth value from the top is obtained by

(6) $$n_m = n + (m - 1)\,T(\tilde{x}_n).$$

Thus, for a given initial distribution, we calculate the most probable smallest (largest) values $\tilde{x}_1$ (and $\tilde{x}_n$) and their occurrence intervals $_1T$ and T_n as a function of n and obtain the modes of the following (preceding) mth values from (4) and (6). This method, based on the recurrent properties of the modes, does not hold for the mean or the median mth values.

If the occurrence intervals $_1T(\tilde{x}_1)$ (and $T(\tilde{x}_n)$) converge toward n, the number of observations $_mn$ (and n_m) so that $\tilde{x}_1$ (and $\tilde{x}_n$) are the modes of the mth extremes, converges toward nm. The mode of the mth extremes for nm observations converges toward the mode of the extremes for n observations. This type of initial distribution will be studied in the next chapter.

Problems: 1) Can the equations (2) have more than one solution? 2) Prove analytically that the mode of the largest value for unlimited distributions increases with the sample size.

3.1.7. Moments. The existence of the moments of the extremes is obviously linked to the existence of the corresponding moments for the initial variate. Both exist (or not) at the same time. If the moments of the extreme values exist, they are

(1) $$\overline{x_n{}^l} = \int_{-\infty}^{\infty} x^l \varphi_n(x)\,dx\,; \qquad \overline{x_1{}^l} = \int_{-\infty}^{\infty} x_1{}^l \varphi_n(x)\,dx\,; \qquad l = 1, 2, \ldots.$$

For most initial distributions, not even the expression for the mean, $l = 1$, can be integrated. The numerical calculations are laborious and complicated.

V. Romanowsky has established a formula linking the successive means of the largest values of a normal distribution for $n+1$, $n+2$, up to $2n+1$ observations. Instead, consider the odd moments of the extremes for any symmetrical distribution with zero mean and unit standard deviation. Then the moments $\mu_{2l+1}(n+1)$ of order $2l+1$ of the largest among $n+1$ observations (provided they exist) are

(2) $$\mu_{2l+1}(n+1) = (n+1) \int_{-\infty}^{+\infty} x^{2l+1}\, F^n(x)\, f(x)\, dx.$$

For the evaluation of (2) consider the integral

$$J = \int_{-\infty}^{+\infty} F^n(x)\,(1 - F(x))^n\, f(x)\, x^{2l+1}\, dx.$$

Since the distribution is symmetrical, the function under the integral is odd and the integral vanishes, whence, by expansion of the binomial,

$$\sum_0^n \binom{n}{v}(-1)^v \int_{-\infty}^{+\infty} F(x)^{n+v} x^{2l+1} f(x)\, dx = 0.$$

We substitute the odd moments (2) in the integral. Then

(3) $$\sum_0^n \binom{n}{v}(-1)^v \frac{\mu_{2l+1}(n+v+1)}{n+v+1} = 0.$$

This linear relation links the odd moments of the largest and of the smallest among $n + 1$, $n + 2$, up to $2n + 1$ observations. For $l = 0$ the relations between successive means of the extremes for n observations are

(4) $$\bar{x}(2n + 1) = (2n + 1) \sum_{0}^{n-1} \binom{n}{v}(-1)^{n+v+1} \frac{\bar{x}(n + v + 1)}{n + v + 1},$$

whence, for $n = 1, 2,$

(5) $$\bar{x}(3) = \frac{3}{2}\bar{x}(2) ; \qquad \bar{x}(5) = 5\left[\frac{\bar{x}(4)}{2} - \frac{\bar{x}(3)}{3}\right].$$

Consequently, it is sufficient for a symmetrical distribution to calculate the means of the extremes for even sample sizes. The means for odd sample sizes are obtained from (4).

In some cases it is possible to construct the generating functions $G_n(t)$ and $_1G_n(t)$ of the extremes. From the distributions 3.1.1(4),

(6)
$$G_n(t) = n \int_0^1 e^{xt} F^{n-1} \, dF = n \int_{-\infty}^{+\infty} e^{xt} F(x)^{n-1} f(x) \, dx$$
$$_1G_n(t) = n \int_0^1 e^{xt}(1 - F)^{n-1} \, dF = n \int_{-n}^{+n} e^{xt}[1 - F(x)]^{n-1} f(x) \, dx \, .$$

In the first formulae the variate x must be expressed as a function of the probability F, a procedure which may turn out to be highly complicated.

For symmetrical initial distributions, the generating functions and the seminvariant generating functions, $L_n(t)$ and $_1L_n(t)$ are linked by the symmetry principle,

(7) $$_1G_n(t) = G_n(-t) ; \qquad _1L_n(t) = L_n(-t) \, .$$

The means are equal in size and opposite in sign. The even (or odd) seminvariants of the smallest value are equal in size and equal (or opposite) in sign to the even (or odd) seminvariants of the largest value, when they exist.

The averages of the largest value increase with n. But we do not yet know how they increase and how the standard deviation of the extreme value depends upon the sample size n. If the standard deviations of the extremes decrease with n, we may increase the precision of the extreme by increasing the sample size from which the extreme observations are taken. This is in accordance with the general rules of statistics. However, the standard deviations of the extremes may converge to a limit or may even increase with the sample size. In the latter case we decrease the precision of the largest value by increasing the corresponding sample size. The larger the sample, the less precise become the extreme values. Of course this possibility has nothing to do with the fact that the precision of the mean of the largest values, taken from N samples, each of size n, increase with the number of samples N from which this mean has been calculated, provided that the population mean and standard deviation exist. If the standard deviation of an extreme is smaller, or becomes smaller than the standard error of the mean, the extreme instead of the mean should be used for the estimation of the parameters in the initial distribution. R. A. Fisher (1935) applied this reasoning to the uniform distribution,

(8) $\quad F(x) = x/\theta \; ; \quad f(x) = 1/\theta \; ; \quad 0 \leq x \leq \theta \; ; \quad \bar{x} = \theta/2 \; ; \quad \sigma^2 = \theta^2/2 \, ,$

where θ is an unknown parameter. The mean and the second moment of the largest value are

(9) $\quad\quad\quad\quad \overline{x_n} = n\theta/(n+1) \; ; \quad \overline{x^2} = n\theta^2/(n+2) \, .$

The parameter θ can be estimated from the sample mean or from the largest observation. Since the variance of the largest observation is

(10) $$\sigma_n^2 = \frac{\theta^2}{(n+1)^2} \frac{n}{n+2} ,$$

the estimate $(n+1) x_n/n$ of the parameter θ from the largest observation x_n is more precise than the estimate $2\bar{x}_0$ from the sample mean, and the advantage of the largest observation increases with the sample size.

Exercises: 1) Prove (4) for the smallest value. 2) Prove that $(x_1 + x_n)/2$ has a smaller sampling variance than any other linear combination of the mth values for a uniform distribution. (The proof was given to me by Mr. John H. Smith.) 3) Verify that (4) does not hold for an asymmetrical distribution.

References: Brookner, Carlton, Egudin.

3.1.8. The Maximum of the Mean Largest Value. The extremes taken from an unlimited variate are again unlimited variates. In this paragraph an upper bound for the mean and a lower bound for the coefficient of variation of the largest value as function of n will be derived. The calculus of variations first applied by Plackett to the variate $x = x(F)$ will lead to those initial distributions (of unusual structure) where the bounds for the mean are reached. As we have done previously, we consider only continuous variates. In addition we require the existence of the first two moments. The initial mean, the initial mean square, and the mean,

mean square and variance of the largest value are, from 3.1.1(4), respectively,

(1) $$\bar{x} = \int_0^1 x \, dF \, ; \qquad \overline{x^2} = \int_0^1 x^2 \, dF$$

and

(2) $$\bar{x}_n = n \int_0^1 x F^{n-1} dF \, ; \qquad \overline{x_n^2} = n \int_0^1 x^2 F^{n-1} dF \, ;$$

$$\sigma_n^2 = n \int_0^1 (x - \bar{x}_n)^2 F^{n-1} \, dF.$$

In order to derive the initial distribution which maximizes the mean largest value $\bar{x}_n$ for given values of the initial mean and standard deviations, we put the first variation of

$$\int_0^1 [nxF^{n-1} - \lambda_1 x^2 - \lambda_2 x] \, dF$$

with respect to x equal to zero. Here the Lagrange multipliers λ_1 and λ_2 will take on the role of parameters. The operation leads to

$$nF^{n-1} - 2\lambda_1 x - \lambda_2 = 0,$$

whence

(3) $$x = (nF^{n-1} - \lambda_2)/2\lambda_1.$$

To obtain the maximum of $\bar{x}_n$, we eliminate λ_1 and λ_2. The mean largest value becomes from (2) after integration of (3)

(4) $$\bar{x}_n = \frac{1}{2\lambda_1} \frac{n^2}{2n-1} - \frac{\lambda_2}{2\lambda_1}.$$

The initial mean obtained for $n = 1$ is

(4′) $$\bar{x} = (1 - \lambda_2)/2\lambda_1.$$

The initial variance obtained from (3) and (4′) is

$$\sigma^2 = \frac{1}{4\lambda_1^2} \int_0^1 (nF^{n-1} - 1)^2 \, dF,$$

whence

(5) $$\sigma = \frac{1}{2\lambda_1} \frac{n-1}{\sqrt{2n-1}}.$$

3.1.8 DISTRIBUTION OF EXTREMES

Combination of the three preceding equations leads to the mean largest value

(6) $$\bar{x}_n = \bar{x} + \frac{\sigma(n-1)}{\sqrt{2n-1}}.$$

If we introduce the standardized variate $z = (x - \bar{x})/\sigma$ with zero mean and unit variance, the mean largest value for the distribution which maximizes this mean is

(7) $$\bar{z}_n = (n-1)/\sqrt{2n-1}.$$

Since this is the maximum, *the mean largest value for any continuous distribution possessing the first two moments increases more slowly than* $\sqrt{n/2}$ *times the initial standard deviation*. This result will be improved in 3.2.5 for symmetrical distributions.

The factors $\bar{x}$, σ, and $\bar{x}_n$ in (6) are, of course, population values. We now establish a similar theorem for the sample values $\bar{x}_0$, s, and x_n. We start from the inequality

$$(x_n - \bar{x}_0)^2 \leq \sum_{m=1}^{n} (x_m - \bar{x}_0)^2,$$

where the equality holds if and only if all observations x_m are identical. Division by $n-1$ and the definition of the sample variance 1.1.4(1) leads to

(8) $$\frac{(x_n - \bar{x}_0)^2}{n-1} \leq s^2,$$

or, excluding the case that all x_m are identical,

(9) $$x_n < \bar{x}_0 + s\sqrt{n-1}.$$

Formula (8) may be used for a rough check on the calculation of the sample standard deviation. The upper boundaries (6) and (9) are traced in Graph 3.1.8.

The initial probability function $F(x,n)$ for which the bound (6) is actually reached is obtained from (1), (4′), and (5) after some calculations as

(10) $$F(x,n) = \left(\frac{n-1}{n} \frac{(x-\bar{x})/\sigma}{\sqrt{2n-1}} + \frac{1}{n} \right)^{1/(n-1)}.$$

The bounds z_0 and z_w for the standardized variate z obtained from $F(x) = 0$ and $F(x) = 1$ are

(11) $$z_0 = -\frac{\sqrt{2n-1}}{n-1} \leq z \leq \sqrt{2n-1} = z_w.$$

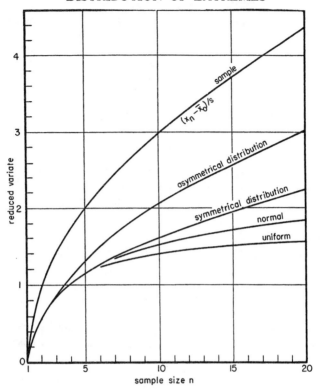

Graph 3.1.8. **Mean Largest Standardized Values**

Therefore, the domain of variation spreads with increasing n and the lower bound z_0 approaches zero in the negative domain.

The probability $\Phi(z,n)$ of the reduced variate z is, from (10),

(12) $\quad \Phi(z,n) = \left(\dfrac{z}{\sqrt{2n-1}} \dfrac{n-1}{n} + \dfrac{1}{n} \right)^{1/(n-1)}$;

$$z = \dfrac{n\sqrt{2n-1}}{n-1} \left(\Phi^{n-1} - \dfrac{1}{n} \right).$$

For $n=2$ the probability function is linear with the domain $-\sqrt{3} \leqq z \leqq \sqrt{3}$. The characteristic largest value u_n becomes from (12), even for moderate values of n,

(13) $\quad u_n = \dfrac{n/e - 1}{n-1} \sqrt{2n-1} < \bar{z}_n$.

It increases asymptotically as $\sqrt{2n}/e$, i.e., more slowly than the mean largest value.

3.1.8 DISTRIBUTION OF EXTREMES

From (12) and (7) the probability function of the largest value becomes

(14) $$\Phi_n(z,n) = [(1 + z\, \bar{z}_n)/n]^{n/(n-1)} .$$

At the mean largest value, the probability converges toward 1/2. Therefore, the mean converges to the median largest value. The probability of the largest value at the characteristic largest value u_n becomes, from (13) and (14),

(15) $$\Phi_n(u_n,n) = \exp[-1 - 1/(n-1)]$$

and converges toward $1/e$ as in 3.1.4(3).

We define a reduced largest value y different from the standardized value z by

(16) $$y = (1 + z \cdot \bar{z}_n)/n = [1 + (x - \bar{x})(x_n - \bar{x})/\sigma^2]/n .$$

Then the probability (14) of the largest value converges to the simple expression

(17) $$\Phi^*(y) = y\ ; \quad 0 \leq y \leq 1 .$$

The rectangular distribution is then the asymptotic distribution of the extreme value of the variate y, defined by (16). It is not surprising that a distribution obtained from a queer condition should show strange properties. Moriguti uses the Schwarz inequality to calculate the minimum of the coefficient of variation of the largest value for a symmetrical distribution with mean zero. The initial variate where the minimum is reached turns out to be

(18) $$x = \text{const}\, \frac{F^{n-1} - (1-F)^{n-1}}{F^{n-1} + (1-F)^{n-1}},$$

and the coefficient of variation of the largest value is obtained asymptotically as

(19) $$\sigma_n/\bar{x}_n > \sqrt{\pi 2^{-n}},$$

a formula which is valid approximately for $n \geq 6$. In a similar way Moriguti derived the minimum for the standard deviation of the largest value. This minimum also converges rapidly toward zero and is therefore not very useful.

To obtain more information we have to assume certain analytical properties of the initial distribution. This will be done in the next chapter.

Exercises: 1) Trace the probability (14) and the corresponding distribution for $n = 2, 3, 4, 5$, and analyze the behavior for large n. 2) Prove that the corresponding standard deviation of the largest value increases asymptotically as $\sqrt{n/6}$. 3) Analyze the distribution (18). 4) Calculate the mean and standard deviation of the largest value and the mean range for this distribution as functions of the sample sizes.

References: Gumbel (1954a), Hartley and David.
Problem: Show that the solution (10) is unique.

3.2. EXTREMAL STATISTICS

3.2.0. Problems. Up to now we have considered the largest and the smallest values separately. Any function depending on both extremes, such as the absolute value of the extremes, the range, the midrange, and the quotient of the extremes, may be called an *extremal statistic*.

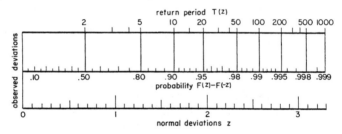

Graph 3.2.1. Probability Paper for Normal Deviations

The absolute extreme is either the largest or the absolute amount of the smallest value, whichever may be larger. The range w is the difference between the largest and the smallest value. The midrange v is their sum. (This definition is analytically simpler than the usual consideration of one half of the sum of the extremes.) The extremal statistics are defined for $n \geq 2$. The range is a non-negative variate, which vanishes if all observations coincide. If the initial variate is limited, the extremal statistics are also limited. The most important case is that of a symmetrical unlimited distribution.

References: Barnard, Burr (1952), Daly, David, Lord, Moshman (1952), Schützenberger, Shone, Walsh (1949a).

3.2.1. Absolute Extreme Values. Bertrand (p. 194) derived the exact distributions for the largest absolute normal value for $n = 2$ and $n = 3$. Charlier gave a probability table (which was not very accurate) for the largest absolute normal value as function of the sample size.

Consider a symmetrical unlimited distribution with median zero. The absolute values $z \geq 0$ of the variate are henceforth called *deviations*.

3.2.1 DISTRIBUTION OF EXTREMES

The largest deviation in a sample of size n is linked to the largest value in a sample of $2n$, since the largest deviation may be either the largest or the smallest value. Therefore, the averages of the largest deviation increase with the sample size and surpass the averages of the largest value. The probability $V(z)$ that a deviation is equal to or less than z, is

(1) $$V(z) = F(z) - F(-z) = 2F(z) - 1.$$

A scheme of a probability paper for normal deviations is shown in Graph 3.2.1.

If we define the *characteristic largest deviation* $\hat{z}_n$ in the same way as the characteristic largest value in 3.1.4(1) by

(2) $$V(\hat{z}_n) = 1 - 1/n,$$

we obtain $\hat{z}_n$ from (1) as

(3) $$\hat{z}_n = u_{2n}.$$

The probability $\Phi_n(z)$ that all deviations are less than z, i.e., that z is the largest among n deviations is, from (1),

(4) $$\Phi_n(z) = (2F(z) - 1)^n.$$

From 1.1.3(3) it follows for symmetrical distributions with zero median, that

(4') $$\Phi_n(z) = \theta^n(x); \quad z = |x|.$$

The median $\check{z}_n$ of the largest deviations is obtained from (4) as the solution of

$$F(\check{z}_n) = [1 + \exp\{-(\lg 2)/n\}]/2.$$

For n large we may write

$$F(\check{z}_n) \sim 1 - (\lg 2)/2n,$$

whence, from 3.1.3(2),

(5) $$\check{z}_n \sim \check{x}_{2n}.$$

The characteristic largest deviation in a sample of size n taken from a symmetrical distribution approaches the characteristic largest value in a sample of size $2n$. The median of the largest deviation for n observations converges toward the median of the largest value for $2n$ observations.

The distribution $\varphi_n(z)$ of the largest deviation obtained from (4) is

$$\varphi_n(z) = 2n(2F(z) - 1)^{n-1} f(z).$$

The first factor in parentheses vanishes for $z = 0$, and the second factor vanishes with increasing z. Therefore the distribution of the largest deviation possesses at least one mode $\tilde{z}_n$, which is the solution of

$$\text{(6)} \qquad \frac{n-1}{2F(z)-1} 2f(z) = -\frac{f'(z)}{f(z)}.$$

The comparison of

$$\frac{1}{2F(\tilde{z}_n)-1} = -\frac{1}{2n-2} \frac{f'(\tilde{z}_n)}{f^2(\tilde{z}_n)}$$

to the corresponding formula 3.1.6(1) for the mode of the largest of $2n$ observations,

$$\frac{1}{F(\tilde{x}_{2n})} = -\frac{1}{2n-1} \frac{f'(\tilde{x}_{2n})}{f^2(\tilde{x}_{2n})},$$

leads to the asymptotic relation,

$$\text{(7)} \qquad V(\tilde{z}_n) = F(\tilde{z}_n) - F(-\tilde{z}_n) \sim F(\tilde{x}_{2n}),$$

between the mode of the largest deviation for n observations and the mode of the largest value for $2n$ observations.

The distribution of the *smallest* deviation is obtained in the same way. Obviously the lower limit of the deviations is zero. The characteristic smallest deviation $\hat{z}_{1,n} = \hat{z}_1$, defined in analogy to the characteristic smallest value 3.1.4(2), is the solution of

$$\text{(8)} \qquad F(\hat{z}_1) = \frac{1}{2} + \frac{1}{2n}$$

and vanishes with increasing sample size.

The probability for a deviation to be larger than z_1 is

$$\text{(9)} \qquad 1 - (F(z_1) - F(-z_1)) = 2 - 2F(z_1).$$

Consequently the median $\check{z}_1$ of the smallest among n deviations is the solution of

$$2(1 - F(\check{z}_1)) = \exp\left[-(\lg 2)/n\right],$$

whence, for a large number of observations,

$$\text{(10)} \qquad F(\check{z}_1) \sim \frac{1}{2} + \frac{\lg 2}{n}.$$

The comparison of (8) and (9) yields

$$\text{(11)} \qquad \check{z}_1 > \hat{z}_1.$$

The distribution $_1\varphi_n(z_1)$ of the smallest deviation decreases monotonically with the deviation, provided that the initial symmetrical distribution has a mode at $x = 0$. For the proof, consider the probability

$$2^n(1 - F(z_1))^n$$

for all deviations to be larger than z_1. The distribution of the smallest deviation,

$$_1\varphi_n(z_1) = 2^n n(1 - F(z_1))^{n-1} f(z_1),$$

is the product of two factors, each of which decreases with increasing values of z. Therefore, the smallest deviation has no mode.

Exercises: 1) Show why the plotting position 1.2.7(1) should be used on the paper 3.2.1. 2) Calculate the averages of the largest deviation for the logistic distribution.

Problems: 1) Under what conditions does the mode of the largest deviation of n observations converge to the mode of the largest value of $2n$ observations? 2) Does the distribution of the smallest deviation converge toward an exponential distribution? (See Smirnov, 1935.)

References: Bendersky, Bortkiewicz (1922b), Marinescu, Smirnov (1935).

3.2.2. Exact Distribution of Range. The range is an evident and simple measure of dispersion. Since it is obtained immediately, and has an intuitive meaning, it is widely used in quality control where a gain in time has to be weighed against a loss in precision. It often turns out that the first choice is preferable. Bortkiewicz derived in 1922 the distribution of the range for the normal distribution and calculated approximations for the mean range. Tippett (1925) calculated the mean, the standard deviation, and the moment quotients for the normal range up to $n = 1000$. His results show that the approximations obtained by Bortkiewicz were quite good. "Student" (1927) reproduced the distribution of the range for small samples, $n = 2, 3, 4, 5, 6, 10$, by Pearson's type I, and for large samples, $n = 20, 60$, by Pearson's type VI, a purely empirical and, therefore, unsatisfactory procedure. This method is misleading, since it tries to put a theoretical distribution into a family where it does not belong. Dederick (1928) calculated the probabilities for the normal range expressed by the probable deviation for $n \leq 12$ observations. A resume of the previous knowledge about the normal range is given in Karl Pearson's Tables, vol. II.

The joint distribution $\mathfrak{w}_n(x_1, x_n)$ of the smallest and largest values x_1 and x_n is such that these two values exist, and that $n - 2$ values are contained between them. Therefore,

(1) $$\mathfrak{w}_n(x_1, x_n) = \frac{n!}{(n-2)!} f(x_1)[F(x_n) - F(x_1)]^{n-2} f(x_n).$$

The distribution $\psi_n(w)$ of the range is obtained by introducing $x_n = x_1 + w$ after dropping the subscript 1 as

$$(2) \quad \psi_n(w) = n(n-1) \int_{-\infty}^{+\infty} [F(x+w) - F(x)]^{n-2} f(x+w) f(x) \, dx.$$

Only in the case $n = 2$, the distribution

$$\psi_2(w) = \int_{-\infty}^{\infty} f(x) f(x+w) \, dx$$

starts with a positive value $\psi_2(0) \neq 0$. For $n > 2$, and for any initial distribution, the density function starts with zero for the range zero, and ends with zero. Consequently, for $n > 2$, the distribution of the range possesses at least one mode. If the variate is limited by

$$(3) \quad a \leq x \leq b,$$

then

$$a \leq x_1 \leq x_1 + w \leq b,$$

and the domain of integration is not (3) but

$$(3') \quad a \leq x \leq b - w.$$

The probability $\Psi_n(w)$ is obtained by integration of (2) over w. If we write z for w as the variable of integration, and change the order of integration, the probability becomes

$$\Psi_n(w) = n(n-1) \int_{x=-\infty}^{+\infty} \int_{z=0}^{w} [F(x+z) - F(x)]^{n-2} f(x+z) f(x) \, dz \, dx.$$

The inner integral can easily be evaluated, whence

$$(4) \quad \Psi_n(w) = n \int_0^1 [F(x+w) - F(x)]^{n-1} \, dF.$$

The beauty of this formula is marred by the fact that, even for simple initial distributions, the probability $F(x+w)$ cannot be expressed by the probability $F(x)$, and that the integration can be achieved only by numerical methods. For the exponential and uniform distributions, however, the integration can be carried through. For such initial distributions it may be simpler to start from (4), and obtain the distributions by differentiation.

The distributions of normal ranges can be reduced to normal functions only for $n = 2$ and $n = 3$. McKay (1935) and McKay and Pearson (1933) obtained

$$(5) \quad \psi_2(w) = 2e^{-w^2}/\sqrt{n} : \quad \psi_3(w) = \frac{6}{\pi\sqrt{2}} e^{-w^2/4} \int_0^{w/\sqrt{6}} e^{z^2/2} \, dz.$$

Sibuya and Toda calculated tables (4D) of the distributions of normal ranges for $n = 3(1)20$ and $w = .05(.05)7.65$ by expanding (2) in a power series.

H. O. Hartley used formula (4) for numerical integration. For a symmetrical variate, the integral (4) may be separated into two parts:

$$n\int_{-\infty}^{-w/2} [F(x+w) - F(x)]^{n-1} f(x) \, dx + n \int_{-w/2}^{\infty} [F(x+w) - F(x)]^{n-1} f(x) \, dx$$
$$= J_1 + J_2.$$

The first integral can be reduced to the second one by introducing

$$y = -x - w.$$

Then

$$J_1 = n \int_{-w/2}^{\infty} [F(-y) - F(-y-w)]^{n-1} f(-y-w) \, dy,$$

and from symmetry

$$J_1 = n \int_{-w/2}^{\infty} [F(y+w) - F(y)]^{n-1} f(y+w) \, dy.$$

Combining the two integrals, and designating the variable of integration by x,

$$\Psi_n(w) = n \int_{-w/2}^{\infty} [F(x+w) - F(x)]^{n-1} [f(x) + f(x+w)] \, dx.$$

This may be written

(6) $\Psi_n(w) = -\int_{-w/2}^{\infty} n[F(x+w) - F(x)]^{n-1} [f(x+w) - f(x)] \, dx$

$$+ 2n \int_{-w/2}^{\infty} [F(x+w) - F(x)]^{n-1} f(x+w) \, dx.$$

The function under the first integral is an exact differential. From the symmetry of $f(x)$ it follows that

$$-\int_{-w/2}^{\infty} d[F(x+w) - F(x)]^n = (2F(w/2) - 1)^n.$$

In the second integral in (6) we let

$$z = x + w;$$

then the probability of the range for a symmetrical distribution is

(7) $\Psi_n(w) = (2F(w/2) - 1)^n + 2n \int_{w/2}^{\infty} [F(z) - F(z-w)]^{n-1} f(z) \, dz.$

The first factor is the probability that all values of the variate lie between $-w/2$ and $w/2$, while the range is less than or equal to w. For a given distribution the first factor can easily be calculated. It becomes increasingly larger than the second one for w increasing. The second term, which is always positive, takes into account all those values which are outside of the previous interval. It can be neglected if w is large. If w is small, the work involved in the calculation is considerably reduced, since the integral starts with $w/2$ instead of w. On this basis the probabilities of the normal ranges were calculated for $n = 4$ up to $n = 20$.

Exercises: 1) Prove formulae (5). 2) Compare for $n = 3$ the mode of the range to the range of the mode. 3) Calculate the moments and the generating function of (5). 4) Calculate the distribution of range for $f(x) = 2(1 - x)$; $0 \leq x \leq 1$, for uniform and exponential distributions. 5) Construct the probability function valid for (1).

References: Baker, Belz and Hooke, Burr (1955), Cadwell (1954a), Conrad, D. R. Cox (1954), Cyffers and Vessereau, Daniels (1952), David (1955a), Dederick, Geary (1943), Gumbel (1944), Hartley (1942), Hartley and Pearson, Hyrenius, N. L. Johnson, Mises (1923), D. Newman, Pearson (1926, 1950, 1952), Pearson and Hartley (1942), Pillai (1950), Rider (1950, 1951a, b, 1953), Robbins (1944a), Schützenberger, G. W. Thomas (1955), W. R. Thompson (1938), Tukey (1955), Walsh (1949a).

3.2.3. The Mean Range. The mode (and the median) of the range, are not the difference of the modes (and medians) of the largest and the smallest values. However, the mean range is the difference between the means of the extremes, provided they exist:

(1) $$\bar{w}_n = \bar{x}_n - \bar{x}_1.$$

No question of dependence enters into this relation. The existence of a population mean range is not evident, since the mean largest value does not exist for distributions where the mean does not exist. Of course, the sample mean range is finite for any number of observations and any distribution.

For most distributions the calculation of the exact values of the mean range is quite intricate. However, L. H. C. Tippett has derived a simple and fundamental formula which may be used for numerical calculations and graphical constructions and does not require the calculation of the means of the extremes. For the derivation consider Pearson's formula for the mean mth distance, 2.1.8(6). The summation of these distances from the first one to the $(n-1)$th leads to the mean distance from the first to the last observation, which is the mean range. Then

(2) $$\bar{w}_n = \sum_{1}^{n-1} \bar{i}_m = \int_{-\infty}^{\infty} \sum_{m=1}^{n-1} \binom{n}{m} (1 - F(x))^m F^{n-m}(x)\, dx.$$

Since, by definition,

$$\sum_{m=0}^{n} \binom{n}{m} F^{n-m}(x)(1-F(x))^m = 1,$$

the mean range becomes

(3) $$\overline{w}_n = \int_{-\infty}^{\infty} [1 - F^n(x) - (1-F(x))^n]\, dx.$$

This is Tippett's Formula, which was found later in an independent way by H. E. Robbins, who applied it to the logistic distribution. By means of a much more complicated procedure, Tippett proved that

(4) $$\overline{w_n^2} = \int_{x_n=-\infty}^{\infty} \int_{z=-\infty}^{x_1} \{1 - F^n(z) - (1-F(x_n))^n\} - [F(z) - F(x_n)]^n\}\, dz\, dx_n.$$

This formula may serve to calculate the standard deviation of the range.

Graph 3.2.3. The Mean Range Interpreted as Area

Siotani gave the formulae for the mean and standard deviation in the discontinuous case and applied them to test the homogeneity of alternative judgments on brands of wine.

Formula (3) can be used for a graphical construction. The minuend $1 - (1-F)^n$ is the probability function, $_1\Phi_n(x)$, 3.1.1(3) of the smallest among n observations; the subtrahend F^n is the probability function, $\Phi_n(x)$, 3.1.1(1), of the largest among n observations. If tables for the two probabilities exist, they may be plotted against x as shown in the schematic Graph 3.2.3 and the mean range is obtained by counting the area, or by using a planimeter. In using formula (3) for an unlimited variate, it is important to resist the temptation of writing the mean range as a sum of integrals of the form

$$\int_{-\infty}^{+\infty} F^v(x)\, dx,$$

since each of these expressions diverges. However, another conclusion may be drawn from (3). The expansions for the mean range $\overline{w}_{2n}$ for an even sample size and for the subsequent odd sample size $2n+1$ both end with $\int_{-\infty}^{\infty} F^{2n}(x)\, dx$. Therefore, $\overline{w}_{2n+1}$ can be expressed by the preceding mean ranges $\overline{w}_v$, $(v = 1, 2, \ldots, 2n)$, while $\overline{w}_{2n}$ cannot be expressed by the preceding ones, since a new factor enters.

The first few ranges obtained from (3) are

(3') $\quad \overline{w}_2 = 2 \int_{-\infty}^{\infty} F(1-F) \, dx, \quad \overline{w}_3 = 3 \int_{-\infty}^{\infty} F(1-F) \, dx,$

$$\overline{w}_4 = 4 \int_{-\infty}^{\infty} F(1-F) \, dx - 2 \int_{-\infty}^{\infty} F(1-F)^2 \, dx.$$

If the variate is symmetrical, the expression (3) is simplified to

(5) $\quad\quad\quad\quad \overline{w}_n = 2 \int_0^{\infty} [1 - (1-F)^n - F^n] \, dx .$

From

(6) $\quad\quad \dfrac{d\overline{w}_n}{dn} = - \int_{-\infty}^{\infty} [F^n \lg F + (1-F)^n \lg (1-F)] \, dx > 0,$

it follows that the mean range increases with the sample size as, of course, is expected. The second derivative, being negative, indicates that this increase diminishes with n increasing.

For symmetrical distributions, the mean range is the double of the mean largest value. Consequently, Romanovsky's linear equations 3.1.7(4) also hold for the mean ranges. However, this system has the disadvantage of linking the mean ranges only for odd sample sizes. P. G. Carlson has shown that Romanovsky's recurrence formula 3.1.7(4) for the mean largest values of a symmetrical distribution also holds for the mean range, for any distribution for which the mean exists. For the proof, he considers the identity

$$(1-z)^{2n} - z^{2n} \equiv (-1)(1-z)^n [-(1-z)^n + (-1)^n z^n]$$
$$+ z^n [-z^n + (-1)^n (1-z)^n]$$

and the relations

$$z^n = [1 - (1-z)]^n = \sum_0^n \binom{n}{v} (-1)^v (1-z)^v ;$$

$$(1-z)^n = \sum_0^n \binom{n}{v} (-1)^v z^v .$$

The first (second) relation introduced into the first (second) bracket leads to

$$(1-z)^{2n} - z^{2n} = (-1)(1-z)^n (-1)^n \sum_0^{n-1} \binom{n}{v} (-1)^v (1-z)^v$$
$$+ z^n (-1)^n \sum_0^{n-1} \binom{n}{v} (-1)^v z^v = \sum_0^{n-1} \binom{n}{v} (-1)^{n+v+1} [(1-z)^{n+v} - z^{n+v}].$$

Integration of this equation over the interval 1, z leads to

$$\frac{1-(1-z)^{2n+1}-z^{2n+1}}{2n+1} = \sum_{0}^{n-1}\binom{n}{v}(-1)^{n+v+1}\frac{1-(1-z)^{n+v+1}-z^{n+v+1}}{n+v+1}.$$

Finally we replace z by $F(x)$ and integrate with respect to x over $-\infty$, $+\infty$. Then Tippett's Formula (3) leads to the recurrence formula for the mean range for odd sample sizes,

(7) $$\bar{w}_{2n+1} = (2n+1)\sum_{0}^{n-1}\binom{n}{v}\frac{(-1)^{n+v+1}}{n+v+1}\bar{w}_{n+v+1},$$

which is of the same structure as 3.1.7(4). The formula expresses the range for samples of size $2n+1$ in terms of those for samples of sizes $n+1$ up to $2n$. For numerical applications it is sufficient to calculate the ranges for even sample sizes. The first expressions are

(8) $$\bar{w}_3 = \frac{3}{2}\bar{w}_2; \quad \bar{w}_5 = 5\left[\frac{-\bar{w}_3}{3}+\frac{\bar{w}_4}{2}\right]; \quad \bar{w}_7 = 7\left[\frac{\bar{w}_4}{4}-\frac{3\bar{w}_5}{5}+\frac{\bar{w}_6}{2}\right].$$

Exercises: 1) Continue the system of equations (8). 2) Calculate the mean range for the uniform and logistic distribution for $n = 2, 3, 4$.

Problems: 1) Is it possible to calculate $\bar{w}_n^2$ by the procedure used for $\bar{w}_n$? 2) Calculate the model normal ranges as a function of n.

References: Barnard, Cohan, D. R. Cox (1954), Craig, Daly, Lord (1947), McKay and Pearson, D. Newman, Patnaik, E. S. Pearson and Haines, E. S. Pearson and Hartley (1935, 1943), Winston.

3.2.4. The Range as Tolerance Limit. Wilks (1941) has used the range for establishing a two-sided tolerance limit. He calculates the probability that a certain proportion of the population falls outside the observed range. The method, based on the probability integral transformation, is distribution free, requires only the continuity of the variate, and is closely related to the procedure used in 2.2.4.

In the joint distribution 3.2.2(1) of the smallest and the largest values,

(1) $$w(x_1, x_n)\, dx_1\, dx_n = n(n-1)(F_n - F_1)^{n-2}\, dF_1\, dF_n,$$

where
$$F_1 = F(x_1); \quad F_n = F(x_n),$$

we introduce a new variate, different from the range,

(2) $$w = F_n - F_1,$$

the proportion of the population contained within the observed range.

Since the Jacobian is equal to unity, the joint distribution (1) becomes

$$n(n-1) w^{n-2} \, dw \, dF_1 \, .$$

To obtain the distribution $h_n(w)$ of w alone, we integrate over F_1. From (2) the domain of integration is

$$0 < F_1 < 1 - w \, .$$

The distribution $h_n(w)$ becomes

(3) $$h_n(w) = n(n-1) w^{n-2}(1-w) \, .$$

The probability P that at least $\beta\%$ of the population is included between the smallest and the largest observation is obtained by integrating (3) from β to unity as

(4) $$P = 1 - n\beta^{n-1} + (n-1)\beta^n \, .$$

There is a probability P that at most $\delta = 1 - \beta$ of the population is outside the observed range of a sample of size n.

If P and δ are known, we may calculate the corresponding sample size n. In practice, we want to calculate the sample size so that the probability P is large, say, .95, and the proportion δ outside the observed range is small, say, less than .01. From (4)

(5) $$1 - P = n(1-\delta)^{n-1} - (n-1)(1-\delta)^n \, .$$

The boundary conditions are $P = 0$, $\delta = 0$, and $P = 1$, $\delta = 1/(n-1)$. For n fixed, δ increases with P. For P fixed, δ decreases with increasing n. Finally, for δ fixed, P increases with n. A first approximation for the right side of (5) valid for $n\delta \ll 1$ is

$$1 - n(n-1) \, \delta^2/2 \, ,$$

and a first approximation for n is

(6) $$n = \tfrac{1}{2} + \frac{\sqrt{2P}}{\delta} \, .$$

Obviously the sample size increases with the probability P, and decreases with the proportion δ. The larger the number of observations the smaller is the proportion outside the observed range.

Formula (6) may be used for a representation of n as a function of P and δ. First choose $P = \tfrac{1}{2}$, whence $n = 1/\delta + \tfrac{1}{2}$. If n is traced as abscissa, (Graph 3.2.4) and δ as ordinate, both in logarithmic scales, a straight line inclined under 45° is obtained. Then we choose

$P =$	.02;	.045;	.08;	.125;	.18;	.245;	.32;	.405;	.605;	.72;	.845;	.98
$\sqrt{2P} =$	.20;	.30;	.40;	.50;	.60;	.70;	.80;	.90;	1.1;	1.2;	1.3;	1.4

3.2.4 DISTRIBUTION OF EXTREMES

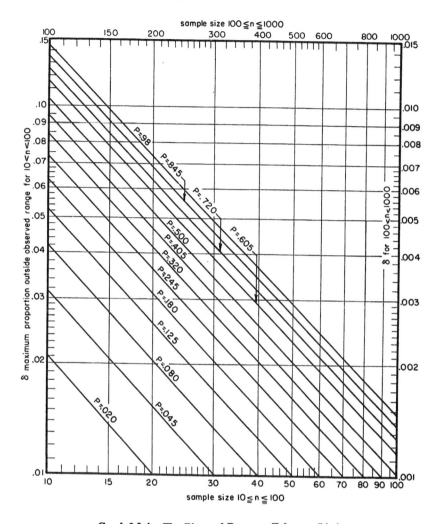

Graph 3.2.4. The Observed Range as Tolerance Limit

and obtain parallels to the first line. The lower (upper) horizontal scale in Graph 3.2.4 applies to sample sizes from 10 to 100 (100 to 1000). The left (right) vertical scale gives the maximum proportions of the population lying outside of the observed range. In a sample of size 100, the probabilities are

P:	.5 ;	.605 ;	.72 ;	.845 ;	.98
$1 - \beta$:	1 ;	1.15 ;	1.25 ;	1.36 ;	1.47

that not more than $1 - \beta$ percent of the population falls outside of the sample range. This allows a forecast about future observations based on the observed range alone. The calculation cannot be continued beyond $\sqrt{2P} = 1$, since for $P = 1$ equation (5) leads to negative values for δ.

The previous considerations are distribution free. S. K. Mitra calculated tolerance limits based on sample mean and range or mean range for the normal distribution. The author gives the values of k_1, which ensure a probability β that $\bar{\bar{x}}_0 + k_1 w$ in a sample of size $n = 2(1)20$ include at least a proportion p of the population. Let $\bar{\bar{x}}_0$ be the grand mean and $\bar{w}$ the mean range in $N = 4(1)20(10)50, 75, 100, 125, \infty$ samples of size 4, then a second table gives the values of k_2 which ensure a probability β that $\bar{\bar{x}}_0 + k_2 \bar{w}$ includes at least a proportion p of the population. In both tables $\beta = .75, .90, .95, .99$, and $p = .75, .90, .95, .99, .999$.

Exercises: 1) Calculate the second approximation of equation (5). 2) Construct a similar graph for the exponential distribution.

Problem: Construct a nomogram based on equation (5).

References: Birnbaum and Zuckerman, Bliss et al., May, Noether (1955), Purcell, Scheffé and Tukey, Shimada, Tukey and Ran, Wald (1943), Wilks (1941, 1942), Wolfowitz.

3.2.5. The Maximum of the Mean Range.

R. L. Plackett derived the maximum of the ratio of mean range to standard deviation as function of the sample size, and gave the initial distribution for which this maximum is actually reached. His method is based on the calculus of variations. Moriguti (1951) used the Schwarz inequality to obtain the maximum for the mean largest value of a symmetrical distribution, which turned out to be one half the value given by Plackett.

As in 3.1.8, we look for the initial probability F_2 so that the mean range obtained from 3.1.8(2),

$$(1) \qquad \bar{w}_n = n \int_0^1 x\{F_2^{n-1} - (1 - F_2)^{n-1}\} \, dF,$$

is a maximum for given values of the initial mean and standard deviations. To find the unknown probability function F_2, the first variation of

$$\int_0^1 [nx\{F_2^{n-1} - (1 - F_2)^{n-1}\} - k_1 x^2 - k_2 x] \, dF_2$$

with respect to x is put equal to zero. The Lagrange multipliers k_1 and k_2 are parameters to be determined later. The operation leads to

$$n\{F_2^{n-1} - (1 - F_2)^{n-1}\} - 2k_1 x - k_2 = 0,$$

whence

$$(2) \qquad x = [n\{F_2^{n-1} - (1 - F_2)^{n-1}\} - k_2]/2k_1.$$

3.2.5 DISTRIBUTION OF EXTREMES

Thus the variate is given here as a function of the differences of probabilities. The initial mean is, from (2),

(2') $$\bar{x} = -k_2/2k_1.$$

The initial variance σ^2 is obtained from the difference

(3) $$x - \bar{x} = n\{F_2^{n-1} - (1 - F_2)^{n-1}\}/2k_1$$

as

(4) $$\sigma^2 = \frac{n^2(1 - \varepsilon_n)}{2k_1(2n - 1)},$$

where the factor

(5) $$\varepsilon_n = (n - 1)!^2/(2n - 2)!$$

tends to zero with increasing n, as 2^{-2n}. The standardized variate z becomes, from (2') and (3),

(6) $$z = \sqrt{\frac{2n - 1}{2(1 - \varepsilon_n)}} \{F_2^{n-1} - (1 - F_2)^{n-1}\}.$$

It follows from (1) that the mean range is

(7) $$\bar{w}_n = n\sqrt{\frac{2(1 - \varepsilon_n)}{2n - 1}},$$

which is Plackett's formula for the maximum of the mean range. From (6) it follows that the variate z for which this maximum is reached has a symmetrical distribution. Therefore the mean of the largest value which is $1/2$ of the value given by (7), is

(8) $$\bar{x}_n = n\sqrt{\frac{1 - \varepsilon_n}{2(2n - 1)}}.$$

This is the maximum which this mean can reach for any symmetrical distribution. This result, traced in Graph 3.1.8, was derived by Moriguti by a different approach. The maximum of the mean largest value for an asymmetrical distribution increases as $\sqrt{n/2}$, while for symmetrical distributions it increases as $\sqrt{n}/2$. For large values of n, the maximum of the mean largest value for an asymmetrical distribution is 41% larger than for a symmetrical one.

Moriguti (1954) also studied by the same method and for symmetrical distributions the least upper bound for $\overline{w_n^2}$ and the greatest lower bounds for the coefficient of variation and for the variance of the range. They can be obtained as characteristic values of certain complicated integral equations. Numerical solutions for initial unit standard deviation are given only for $n = 2, 3, 4, 6, 8$.

Exercises: 1) What is the relation between Tippett's formula 3.2.3(3) and (1)? 1) Trace the probability and distribution function corresponding to (6) for $n = 2, 3, 4, 5$. (See Moriguti, 1951.) 3) Calculate the standard deviation of the largest value for the distribution (6). 4) Prove that the asymmetrical distribution 3.1.8(10), for which the mean largest value is a maximum, yields a mean range which is only 71% of the maximum of the mean range.

References: Gumbel (1954a), Hartley and David, Hoeffding, Thomson.

3.2.6. Exact Distribution of the Midrange.

The midrange v is defined by $v = x_n + x_1$. For symmetrical distributions, one half of the population midrange coincides with the mean, median, and mode (or anti-mode), and the sample value estimates the center of symmetry. For asymmetrical distributions, the sample midrange depends on the sample size, and does not correspond to any fixed value of the population. Therefore what it estimates cannot be stated. Owing to this deficiency, not much theoretical work has been done on the midrange.

The joint distribution $w_n(x_1, v)$ of the smallest value and the midrange is, from 3.2.2(1),

$$(1) \quad w(x_1, v) = n(n-1) f(x_1) [F(v - x_1) - F(x_1)]^{n-2} f(v - x_1) .$$

The distribution $h_n(v)$ of the midrange is obtained by integration over x_1. Since

$$x_n = v - x_1 \geq x_1 ,$$

the upper limit of integration is

$$(2) \quad x_1 \leq \frac{v}{2} .$$

Therefore,

$$(3) \quad h_n(v) = \int_{-\infty}^{v/2} w(x_1, v) \, dx_1 .$$

The probability function $H_n(v)$ is obtained by integration over v. Substitution of y for v as the variable of integration leads, from (2) and (3), after dropping the index 1, to

$$(4) \quad H_n(v) = \int_{y=2x}^{v} \int_{x=-\infty}^{v/2} n(n-1) f(x) [F(y-x) - F(x)]^{n-2} f(y-x) \, dx \, dy .$$

3.2.6 DISTRIBUTION OF EXTREMES

The introduction of a variate s defined by

(5) $$y - 2x = s$$

leads to

(6) $$H_n(v) = \int_{s=0}^{v-2x} \int_{x=-\infty}^{v/2} n(n-1) f(x) [F(s+x) - F(x)]^{n-2} f(s+x) \, dx \, ds,$$

whence

(7) $$H_n(v) = n \int_{-\infty}^{v/2} [F(v-x) - F(x)]^{n-1} f(x) \, dx,$$

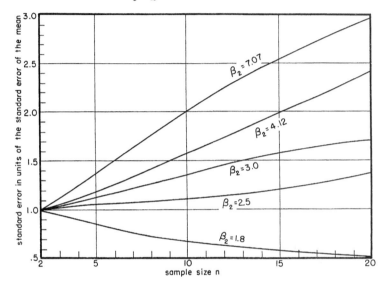

Graph 3.2.6. Standard Error of the Center as Function of Sample Size

which is analogous to the probability 3.2.2(4) of the range. Unfortunately, the integral cannot in general be expressed as a function of v.

E. S. Pearson and N. K. Adyanthaya call $v/2$ the "center" of a distribution. They calculate the standard error of the center for certain symmetrical distributions belonging to Pearson's system. The figures for the rectangular and the normal distribution follow from the theory. The rest of the data shown in Graph 3.2.6 was obtained from experimental sampling. The standard error expressed in units of the standard error of the mean is traced in this graph as a function of the sample size for different values of β_2. The estimate of the mean from the center increases in reliability as β_2 decreases. The authors state: "It is only for the uniform distribution that the center becomes more reliable than the mean."

Fréchet (1954) has shown that no probability function with a limited first and a continuous second derivative exists so that the range and the midrange are independent. In other words, under the given conditions the joint probability function of range and midrange cannot be split into a product of the separate probability functions of the range and of the midrange.

Exercises: 1) Prove that the distribution of the normal midrange is normal for $n = 2$. 2) Calculate the distribution of the midrange for the uniform distribution. (See Neyman and Pearson.)

Problems: 1) Prove that the distribution of the midrange taken from a symmetrical unimodal unlimited distribution is symmetrical, unimodal, and unlimited. 2) Why does the estimate of the mean from the "center" increase in reliability as β_2 increases?

References: Gumbel (1944), Homma (1952), Murty, Neyman and Pearson, Pillai (1950), Rider (1956), Walsh (1949a). *Consult also:* P. R. Rider (1957). The midrange of a sample as an estimator of the population midrange, J. Am. Stat. Assn., 52: 537.

3.2.7. Asymptotic Independence of Extremes. Tippett and other authors assumed in their numerical calculations that the joint distribution of the extremes split up for sufficiently large samples as the product of the distributions of the smallest and of the largest value. This procedure is legitimate for all probability functions which are monotonically increasing.

Let $_m x$ be the mth smallest observation; let x_l be the lth largest observation where m and l are small compared to n, n being large. Then the joint distribution $w_n(_m x, x_l)$ is

(1) $\quad w_n(_m x, x_l) = C F^{m-1}(_m x) [F(x_l) - F(_m x)]^{n-m-l} [1 - F(x_l)]^{l-1} f(_m x) f(x_l)$

Cramér (p. 370) introduces the transformation

(2) $\qquad n[(1 - F(x_l)] = \eta \, ; \qquad n F(_m x) = \xi \, ,$

which is a generalization of 3.1.4(1). Then the joint distribution $v_n(\xi, \eta)$ of the variates ξ and η becomes

$$v_n(\xi, \eta) = \frac{n!}{n^2 (m-1)!(l-1)!(n-m-l)!} \left(\frac{\xi}{n}\right)^{m-1} \left(1 - \frac{\xi + \eta}{n}\right)^{n-m-l} \left(\frac{\eta}{n}\right)^{l-1},$$

where $m + l$ is assumed to be small compared to n. As n increases, $v_n(\xi, \eta)$ passes to the limit, obtained from 1.1.8(7),

(3) $\qquad\qquad v(\xi, \eta) = \dfrac{\xi^{m-1} e^{-\xi}}{\Gamma(m)} \dfrac{\eta^{l-1} e^{-\eta}}{\Gamma(l)} \, ,$

so that ξ and η become independent. If we now impose the mild restriction that $F(x)$ be monotonically increasing, then (2) defines a one to one transformation, and therefore there must exist an inverse function uniquely defining $_m x$ as a function of ξ and x_l as a function of η. From the limiting independence of ξ and η follows then the limiting independence of the extremes $_m x$ and x_l. In particular the two extremes x_n and x_1 are asymptotically independent.

Reference: Homma (1951).

3.2.8. The Extremal Quotient. The asymptotic independence of the extremes leads to interesting asymptotic properties of

(1) $$q = x_n/(-x_1),$$

which may be called the *extremal quotient*. For the analysis we assume zero median, symmetry of the initial distribution, and the existence of an asymptotic distribution of the extremes (this is to be proven in Chapter 5),

(2) $$\varphi(x_n) = \varphi(-x_1).$$

Since, from 3.1.1(2), the largest (smallest) value of an unlimited variate may be assumed to be positive (negative) for a sufficiently large sample size, and since x_n and x_1 are independent, the distribution $h(q)$ of the extremal quotient is for $x_n = x$

(3) $$h(q) = \int_0^\infty x\varphi(qx)\varphi(x)\,dx.$$

It is easily seen that this distribution is invariant under a reciprocal transformation. In addition we obtain from (3)

(4) $$h(1/q) = q^2 h(q) > h(q); \quad \text{for } q > 1.$$

If the distribution (2) is continuous, the derivative of (3) with respect to q leads for $q = 1$ to

$$h'(1) = -h(1) < 0.$$

If the distribution possesses a unique mode, and this is possible if and only if $f(x)$ possesses a unique mode, it must be less than unity. The probability function $H(q)$ becomes from (3), if $\Phi(x)$ is the probability function corresponding to (2),

(5) $$H(q) = \int_0^\infty \Phi(qx)\varphi(x)\,dx,$$

and after integration by parts

(6) $$H(q) = 1 - q\int_0^\infty \Phi(x)\varphi(qx)\,dx.$$

The probability at the value $1/q$ leads to the symmetry relation

(7) $$H(1/q) = 1 - H(q).$$

The first quartile is the reciprocal of the third quartile, etc. In particular, *the median of the extremal quotient is equal to unity.* This result is intuitively clear, since the distribution (3) is invariant under a reciprocal transformation and since the variates from which the quotient is derived are mutually symmetrical. Finally, equation (5) leads through the transformation $q = \lg z$ to a probability function $H^*(z)$,

$$(8) \qquad H^*(z) = \int_0^\infty \Phi(x \cdot \lg z)\varphi(x)\, dx,$$

which is symmetrical about median zero.

The geometrical mean of q exists and is equal to unity. Since the moments of a product of two independent variates are the products of their moments, the moments of order l are

$$(9) \qquad \overline{q^l} = \overline{x^l} \cdot \overline{x^{-l}} = \overline{q^{-l}}.$$

These moments exist if and only if the moments and reciprocal moments for the initial variate exist simultaneously. Even if the initial variate possesses all moments, the mean $\bar{q}$ need not exist.

The properties of the extremal quotient also hold for the quotient of two positive variates, x and y, which have the same distribution. Therefore, the usual procedure in economic and meteorological statistics of calculating the quotients of two series of independent positive variates in order to test whether the ratio is constant may be misleading. If the theoretical mean does not exist, the sample mean does not characterize the relation between the two series.

References: Broadbent, Dixon (1951), Geary (1930), Gumbel and Herbach, Gumbel and Keeney.

Chapter Four: ANALYTICAL STUDY OF EXTREMES

"Sunt certi denique fines quos ultra citraque nequit existere verum."
(All variates are limited in both directions.)

4.1. THE EXPONENTIAL TYPE

4.1.0. Problems. We want to know the interrelations among the median, mode, mean, characteristic largest value, and extremal intensities. How do the averages increase with sample size and how do the extremal intensities depend on sample size? Analytic answers can be given immediately for the exponential distribution. All statements which are exact for the exponential distribution are asymptotically valid for the exponential type (to be defined in 4.1.4). Thus, the theory of extreme values will be centered about the exponential distribution.

4.1.1. Largest Value for the Exponential Distribution. The exponential distribution 1.2.2(8) of a reduced variate x is asymmetrical and limited toward the left. It is basic in many phenomena, especially in modern physics, since it governs the decay of radioactive substances. Using the notations 1.2.1(1) and 1.2.2(4),

(1) $F(x) = 1 - e^{-x}$; $\quad f(x) = -f'(x) = 1 - F(x) = e^{-x}$;
$$T(x) = e^x; \quad \mu(x) = 1; \quad x \geq 0.$$

All moments exist. The population mean and the standard deviation are both unity. Therefore, x may be considered as being expressed in one of these units. A probability paper is obtained by tracing $\lg T(x)$ on the abscissa and x on the ordinate. The usual semi-logarithmic paper can serve for this purpose.

We first study the largest value. The result also covers the properties of the smallest value for the distribution,

$$f_1(x) = e^x; \quad x \leq 0,$$

which is symmetrical to $f(x)$. The characteristic largest value defined in 3.1.4(1) is simply

(2) $$u_n = \lg n.$$

The probability function $\Phi_n(x)$ for the largest value is

(3) $$\Phi_n(x) = (1 - e^{-x})^n .$$

The median $\check{x}_n$ of the largest value converges toward

(4) $$\check{x}_n \sim \lg n - \lg \lg 2 ,$$

and exceeds the characteristic largest value. See Graph 4.1.1(1).

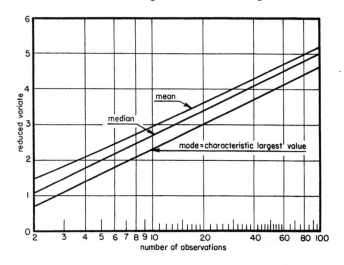

Graph 4.1.1(1). Averages of Largest Exponential Values

The distribution of the largest value obtained from (3) is

(5) $$\varphi_n(x) = n(1 - e^{-x})^{n-1} e^{-x} .$$

Consequently, the mode of the largest value is

(6) $$\tilde{x}_n = \lg n = u_n ,$$

a simple and important formula. *The most probable and the characteristic largest value are equal.* The occurrence interval (2.1.2) $T(\tilde{x}_n)$ for n observations is n itself. *To double the most probable largest value, we have to square the number of observations.* If we double the number of observations, the most probable largest value increases only by 59.3% of the initial standard deviation.

The distributions (5) are traced for $n = 1, 2, 3, 4, 5$, in Graph 4.1.1(2). The density of probability at the mode $\varphi_n(\tilde{x}_n)$ converges by virtue of (5) and (6) toward $1/e$, independent of n. The distributions of the largest values shift to the right with increasing n and without change of shape.

4.1.1 ANALYTICAL STUDY OF EXTREMES 115

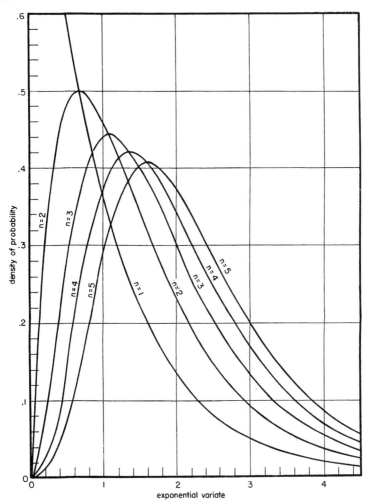

Graph 4.1.1(2). Distributions of Largest Exponential Values

This asymptotic behavior is not yet visible for the small value of n used in Graph 4.1.1(2).

The generating function $G_n(t)$, 3.1.7(6), of the largest value is

$$G_n(t) = n \int_0^1 F^{n-1}(1-F)^{-t}\,dF = \Gamma(n+1)\Gamma(1-t)/\Gamma(n+1-t),$$

whence, from 1.1.8(6),

(7) $$\lg G_n(t) = -\sum_1^n \lg(1 - t/v).$$

Consequently the mean of the largest value,

$$\bar{x}_n = \sum_1^n 1/v, \tag{8}$$

is asymptotically equal to

$$\bar{x}_n \sim \lg n + \gamma. \tag{8'}$$

Its distance from the mode (6) tends to $\gamma \sim .57722$.

The asymptotic value (8') is very near the exact value (8), even for moderate values of n. For example: for $n = 50$, the exact and asymptotic values are 4.49920, and 4.48924 respectively. The distance from the mean (8) to the mode (6) of the largest value decreases with increasing numbers n. For $n = 100$ the difference shrinks to .58221, while the asymptotic value is $\gamma = .57722$. From (8') and (1) the return period of the mean largest value is asymptotically

$$T(\bar{x}_n) = n e^\gamma = 1.78107 n. \tag{9}$$

The averages of the largest value increase as the natural logarithm of n. However, the standard deviation behaves differently. From (7) the variance σ_n^2 of the largest value is

$$\sigma_n^2 = \sum_1^n 1/v^2, \tag{10}$$

an expression converging toward

$$\pi^2/6 = 1.6449340668. \tag{11}$$

For $n = 100$, the variance $\sigma_{100}^2 = 1.63498$ differs from the asymptotic value only by 0.6%. With increasing numbers of observations the standard deviation no longer changes. An increase of the sample size does not increase the precision of the largest value.

Exercises: 1) Draw the quantiles of the largest values and the modes of the mth extremes by the procedures described in 2.1.2 and 3.1.6. 2) Calculate the averages of the mth extremes. 3) By what factor must the sample size be multiplied in order to double the mean (and median) largest value? 4) Calculate the two Betas from (7) for $n = 10, 20, 50, 100$. 5) Calculate the mean and standard deviation of the largest value for the Laplacean distribution 1.2.5(3).

References: Bartholomew (1957), Deemer and Votaw, Epstein and Sobel (1952, 1953, 1954), Gilbert, Goodman (1953), Gumbel (1937b), Sarhan and Greenberg (1957).

4.1.2. Order Statistics for the Exponential Distribution. Since the exponential distribution will be the basis for the study of extremes, we consider its order statistics. The distribution $\varphi_n(x_m)$ of the mth smallest value $x_m = x$ among n observations is

(1) $$\varphi_n(x_m) = \binom{n}{m} m(1 - e^{-x})^{m-1} e^{-x(n-m+1)}.$$

The generating function

$$G_m(t) = \binom{n}{m} m \int_0^\infty (1 - e^{-x})^{m-1} e^{-x(n-m+1-t)} \, dx$$

becomes by the transformation $\exp(-x) = z$

$$G_m(t) = \frac{n!}{(n-m)!} \frac{(n-m-t)!}{(n-t)!}.$$

Hence

(2) $$\lg G_m(t) = \lg n! - \lg(n-m)! - \sum_{v=0}^{m-1} \lg(n-t-v).$$

Consequently, the mean and variance of the mth smallest value is

(3) $$\bar{x}_m = \sum_{n-m+1}^n 1/v ; \qquad \sigma_m^2 = \sum_{n-m+1}^n 1/v^2.$$

The distribution $_1\varphi_n(x)$ of the smallest value, $m = 1$, obtained from (1),

(4) $$_1\varphi_n(x) = ne^{-nx} = nf(nx),$$

is again an exponential function. If the distribution of an extreme is equal to the initial distribution except for a linear transformation of the variate, the initial distribution is called *stable with respect to this extreme*. Thus the exponential distribution is stable with respect to the smallest value.

The median of the smallest value is

$$\check{x}_1 = (\lg 2)/n.$$

The characteristic smallest value u_1 converges toward

(5) $$u_1 = 1/n^2.$$

The generating function $G_1(t)$ of the smallest value is from (4)

(6) $$G_1(t) = (1 - t/n)^{-1}.$$

The moments of order k for the smallest value are equal to the corresponding moments of the initial variate divided by the kth power of n. The averages decrease toward zero as n increases, and the distribution contracts with increasing numbers of observations. The standard deviation $\sigma_1 = 1/n$ of the smallest value is smaller than the standard error of the mean, which is $1/\sqrt{n}$.

The distribution $\psi_n(w)$ of the distance $w_{l,m} = w$ from the mth to the lth observation is obtained from 2.1.7(4). We write, from 4.1.1(1),

$$F(x_m + w) = 1 - (1-F_m)e^{-w}\,; \quad f(x_m+w) = (1-F_m)e^{-w},$$

whence

$$\psi_n(w) = C \int_0^1 F_m^{m-1}[1-(1-F_m)e^{-w}-F_m]^{l-m-1}(1-F_m)^{n-l+1}e^{-w(n-l+1)}\, dF_m$$

$$= Ce^{-w(n-l+1)}(1-e^{-w})^{l-m-1} \int_0^1 F_m^{m-1}(1-F_m)^{l-m-1+n-l+1}\, dF_m\,.$$

The constant factor is

$$C = \frac{n!\,(m-1)!\,(n-m)!}{(m-1)!\,(l-m-1)!\,(n-l)!\,n!} = \binom{n-m}{l-m}(l-m)\,.$$

Integration leads from (1) to

(7) $$\psi_n(w_{l,m}) = \varphi_{n-m}(x_{l-m})\,.$$

The distance from the lth to the mth among n observations is distributed as the $(l-m)$th among $n-m$ observations. This is a generalization of the relation 2.1.8(7). In particular for $l=n$, $m=1$, the distribution of the range of n observations is the distribution of the largest among $n-1$ observations.

Exercises: 1) Prove that the smallest observation and the range are independently distributed. 2) Calculate the first two Betas from (2).

References: Epstein and Sobel (1953), Epstein and Tsao, Gumbel (1937b), Shimada, Sukhatme.

4.1.3. L'Hôpital's Rule. The properties of the largest value taken from an exponential distribution are now used to construct a type of initial distributions. The rule from which we start is the equality 4.1.1(6) of the characteristic and the modal largest values. For any continuous distribution, the modal largest values are the solutions of 3.1.6(1). Asymptotic solutions of these equations can be given for a wide category of unlimited distributions by the following device:

For large positive (negative) values of the variate, the density of probability $f(x)$ and the probabilities $1-F(x)$ (and $F(x)$) are small. The derivatives $f'(x)$ are also small and negative (positive), and the quotients $f/(1-F)$ and f/F become indeterminate for large and small values of the variate respectively. If this holds, it is natural to apply L'Hôpital's Rule,

(1) $$\lim_{x=\infty} f(x)/(1-F(x)) = -\lim_{x=\infty} f'(x)/f(x)\,;$$
$$\lim_{x=-\infty} f(x)/F(x) = \lim_{x=-\infty} f'(x)/f(x)\,.$$

4.1.3 ANALYTICAL STUDY OF EXTREMES

In general, the asymptotic equations,

(2) $\quad f(x)/(1 - F(x)) = -f'(x)/f(x) \; ; \quad f(x)/F(x) = f'(x)/f(x)$,

will be approximately valid for very large (or very small) values of the variate. If we use the intensity function, 1.2.1(1), the first equation (2) can be written asymptotically

(2') $\qquad\qquad -\mu(x) \sim \dfrac{d \lg f(x)}{dx}$.

Thus we suppose the existence of the four limits given in equation (1) for unlimited variates, or the existence of the first (second) pair for distributions which are limited to the left (right). Furthermore, it is assumed that the quotients of the left and right sides of each equation (1) approach unity. The introduction of this assumption, which for our purposes is henceforth called L'Hôpital's Rule, means that we deal only with a specific type of unlimited distributions.

The first equation (2) is fulfilled for all positive values of the exponential distribution $f(x)$ in the positive domain,

$$f_1(x) = e^{-x} \; ; \quad x \geqq 0 \; ; \quad f_1(x) = 0 \; ; \quad x < 0 ,$$

and the second equation in (2) is fulfilled for all negative values of the exponential distribution,

$$f_2(x) = e^{x} \; ; \quad x \leqq 0 \; ; \quad f_2(x) = 0 \; ; \quad x > 0 ,$$

in the negative domain. The same property holds for the composition of these two distributions, the Laplacean distribution,

$$f_3(x) = \tfrac{1}{2} e^{-|x|} .$$

For this distribution, when $x \neq 0$,

$$\mu(x) = -f'(x)/f(x) = 1 .$$

For unlimited distributions we define the *critical quotient* $Q(x)$ by

(3) $\qquad\qquad Q(x) \equiv \dfrac{\mu(x)}{-f'(x)/f(x)} = \dfrac{-f^2(x)}{f'(x)\,[1 - F(x)]} > 0$.

Its values for large positive x may be

(4)
(5) $\qquad\qquad Q(x) = \begin{cases} 1 + |\varepsilon(x)| \\ 1 \\ 1 - |\varepsilon(x)| \end{cases}$
(6)

with

(7) $\qquad\qquad \lim_{x = \infty} |\varepsilon(x)| = 0$.

The distinctions among these three values will be studied in the next paragraphs. If L'Hôpital's rule holds, the return period becomes asymptotically

$$(8) \qquad T(x) = -\frac{1}{f(x)} \frac{d \lg f(x)}{dx},$$

which is the converse to 1.2.3(3).

4.1.4. Definition of the Exponential Type. Those unlimited distributions for which the equations 4.1.3(2) hold are said to be of the *exponential type*. It will be shown in Chapter 5 that the probability densities converge towards zero for $|x| \to \infty$ in a way similar to that of an exponential distribution. Therefore, all moments exist for these distributions.

The notion of "exponential type" is defined only for unlimited variates. If the variate is limited toward the left (right), it may be of exponential type for large (small) values of the variate. A distribution of the exponential type is said to belong to the first, second, or third class according to whether one of the equations 4.1.3(4), (5) or (6) holds. For example, the probability function

$$F(x) = 1 - \exp[-x^k]; \quad x \geq 0$$

belongs to the first (second or third) class with respect to the largest value if $k > 1 (= 1; < 1)$.

For distributions of the exponential type, asymptotic expressions for the modes $\tilde{x}_n$ and $\tilde{x}_1$ are obtained from 3.1.6(2) and 4.1.3(1) as the solutions of

$$(1) \qquad n - 1 \sim \left(\frac{F(x)}{1 - F(x)}\right)_{x=\tilde{x}_n}; \quad n - 1 \sim \left(\frac{1 - F(x)}{F(x)}\right)_{x=\tilde{x}_1}.$$

These equations simplify readily into

$$(2) \qquad F(\tilde{x}_n) \sim 1 - 1/n; \quad F(\tilde{x}_1) \sim 1/n,$$

which are the definitions 3.1.4(1) of the characteristic largest values. *For distributions of the exponential type, the most probable extremes converge toward the characteristic extremes.* The equations

$$(3) \qquad \tilde{x}_n \sim u_n(n); \quad \tilde{x}_1 \sim u_1(n)$$

hold asymptotically. For given small n, and a given initial distribution, the modes may be larger or smaller than the characteristic values.

It is not possible to decide a priori the number of observations beyond which the asymptotic equations (2) may be used. This number depends upon the initial distribution, and upon the precision which is chosen as sufficient.

From equation 3.1.3(1) for the median, and from (2) it follows that, for sufficiently large n, the medians of the largest (smallest) value are larger (smaller) than the modes. Furthermore, for the distribution of the largest (smallest) value, the area on the right side of the mode is larger (smaller) than the area on the left side. *The distribution of the largest (smallest) value taken from a distribution of the exponential type is positively (negatively) skewed.*

Since the modes of the extremes converge toward the characteristic values, the probabilities at the modes converge by virtue of (2) toward

$$\Phi_n(u_n) \sim \frac{1}{e} \sim 1 - {}_1\Phi_n(u_1) \,. \tag{3'}$$

The mode $\tilde{x}_m$ of the mth largest value (counted from the top) is linked to the characteristic largest value u_m defined in 3.1.4(6). The mode is the solution of the second equation 2.1.2(3),

$$\frac{n-m}{F_m} f_m - \frac{m-1}{1-F_m} f_m + \frac{f'_m}{f_m} = 0 \,, \tag{4}$$

where f_m stands for $f(\tilde{x}_m)$. To find $\tilde{x}_m$ we use the asymptotic equations 4.1.3(2), which become for the mth largest values

$$\frac{-f'_m}{f_m} \sim \frac{f_m}{1-F_m} \,. \tag{5}$$

This approximation requires that n be sufficiently large, and m sufficiently small. Then the mode is obtained from

$$\frac{n-m}{F_m} \sim \frac{m}{1-F_m} \,.$$

The most probable mth largest value $\tilde{x}_m$ converges toward u_m, the characteristic largest value, i.e.,

$$\tilde{x}_m \to u_m \,. \tag{6}$$

Finally, consider the largest deviation defined in 3.2.1. If n is large enough, and if the initial distribution is of the exponential type, the equation 3.2.1(6) for the modal largest deviation $\tilde{z}_n$ becomes, from 4.1.3(2),

$$\frac{2n-2}{2F(z)-1} \sim \frac{1}{1-F(z)} \,,$$

whence

$$F(\tilde{z}_n) \sim 1 - 1/(2n) \,. \tag{7}$$

The mode of the largest deviation in a sample of size n converges toward the mode of the largest value in a sample of size $2n$. From the comparison of 3.2.1(5) and (7), it follows that the median of the largest deviation exceeds asymptotically the mode of the largest deviation. The distribution of the largest deviation is positively skewed.

Exercise: Prove that the mode of the kmth largest among kn observations converges, under the condition (5), to the mode of the mth observation.

Problem: L'Hôpital's Rule is sufficient for the convergence of the characteristic to the modal largest value. Is it also necessary?

4.1.5. The Three Classes. The convergence of the mode toward the characteristic extreme values, and all consequences of this, hold only for distributions of the exponential type. The convergence leads to two questions for a finite number of observations: 1) How is the occurrence interval, (2.1.2), related to the number of observations which define the characteristic largest value? 2) How is the mode related to it?

The occurrence intervals were defined by

(1) $$T(\tilde{x}_n) = 1/(1 - F(\tilde{x}_n)) \; ; \quad {}_1T(\tilde{x}_1) = 1/F(\tilde{x}_1) \,,$$

while the characteristic values u_n and u_1 were defined by

(2) $$n = 1/(1 - F(u_n)) \; ; \quad n = 1/F(u_1) \,.$$

The right sides of equations (1) are functions of the modal extremes, and thus of the sample size. The occurrence intervals converge toward n for distributions of the exponential type in the same way as the probabilities $1 - F(x_n)$ and $F(x_1)$ converge toward $1/n$. For a sufficiently large number of observations, the occurrence interval approaches the number of observations. If we have made n observations we may expect, for a sufficiently large n, that the largest value of this sample will be equalled or exceeded in n future observations. The convergence may also be interpreted in the following way: *As soon as n is large enough, the variate to which a value of T corresponds becomes the most probable largest value for T observations.* A graph showing the variate as a function of the return period also shows, for large n, the most probable value as a function of the number of observations. The convergence of the occurrence interval toward the sample size may be fast or slow. This depends upon the approach of the initial distribution toward zero. Furthermore, the convergence need not be monotonic.

To obtain the relation between the occurrence interval $T(\tilde{x}_n)$ and the sample size, the return period is written in the form

$$T(\tilde{x}_n) - 1 = F(\tilde{x}_n)/(1 - F(\tilde{x}_n)) \,.$$

On the other hand, we have for the mode, from 3.1.6(2),

$$n - 1 = -\left(\frac{f'(x) F(x)}{f^2(x)}\right)_{(x=\tilde{x}_n)}.$$

Division of the two equations leads to another expression for the critical quotient 4.1.3(3).

(3) $$\frac{T(\tilde{x}_n) - 1}{n - 1} = Q(\tilde{x}_n).$$

The conditions 4.1.3(4) to (6) determine whether the occurrence interval is less than, equals, or exceeds the number of observations and whether, with increasing n, the occurrence interval increases or decreases toward the number of observations. If $T = n$, the mode coincides with the expected largest value, and the distribution belongs to the second class of distributions of the exponential type. If $T > n$ $(T < n)$, the distribution belongs to the first (or third) class. In other words: If a sample of size n has led to a largest observation x_n, we may expect that, in a second sample, this value, or a larger one, will happen after more (or less) than n observations. If $T(\tilde{x}_n) < n$ and if a sample of size n has led to a largest observation x_n, we may expect a larger one in a second sample of the same size.

From (1) and (2), the inequalities 4.1.3(4) to (6) may be written

$$F(\tilde{x}_n) \gtreqless F(u_n),$$

whence

(4) $$\tilde{x}_n \gtreqless u_n.$$

For distributions of the second class the mode is equal to the characteristic largest value. For distributions of the first (third) class the mode is greater (less) than the characteristic largest value. The study of the mode, median, and characteristic extreme is important since, in general, the expectation of the extremes cannot be stated analytically.

4.1.6. The Logarithmic Trend. The mode of the largest reduced value taken from an exponential distribution 4.1.1(6) is the logarithm of the sample size. A similar relation holds for all distributions of the exponential type. Assume n to be so large that the probabilities of the characteristic extremes,

(1) $$F(u_n) = 1 - 1/n; \quad F(u_1) = 1/n,$$

may be considered as continuous functions of n. Differentiation with respect to n leads from the definition 3.1.5(1) of the extremal intensity to

(2) $$\frac{du_n}{d\lg n} = \frac{1}{\alpha_n}; \qquad \frac{du_1}{d\lg n} = -\frac{1}{\alpha_1}$$

or

(3) $$du_n = \frac{d\lg n}{\alpha_n}; \qquad du_1 = -\frac{d\lg n}{\alpha_1}.$$

Since the characteristic value for $n = 2$ is the median of the initial distribution, integration from 2 to n leads to

(4) $$u_n = \check{x} + \int_2^n \frac{d\lg z}{\alpha_n(z)}; \qquad u_1 = \check{x} - \int_2^n \frac{d\lg z}{\alpha_1(z)},$$

where z stands for the variable of integration. Of course, this procedure has only a heuristic value since, in principle, n is discrete. If Alpha increases (decreases) with n, the absolute amount of the characteristic extremes increases more slowly (quickly) than the logarithm of n. Therefore, the logarithm of n is a yardstick to measure the velocity of the absolute increase of the characteristic extremes. The same relation holds asymptotically for the modes. This convergence may be called the *logarithmic trend of the increase of the extremes*, valid for the exponential type. It is natural to plot the averages of the extremes against the logarithm of the sample size.

In the following, we determine the curvature of the characteristic largest value traced as a function of $\lg n$. If we write $\mu(u)$ instead of α_n, the sign of the second derivative leads from (2) to

(5) $$\text{sign} \frac{d^2(u_n)}{(d\lg n)^2} \gtrless 0 \quad \text{if} \quad \mu'(u) \lessgtr 0.$$

The characteristic largest value traced as a function of $\lg n$ turns its convex (concave) side toward the horizontal axis if, for large values of the variate, the derivative of the intensity function is negative (positive). See Graph 4.1.6. This holds also for the mode, provided that the number of observations is sufficiently large. The sign of $\mu'(x)$ may depend upon the numerical values of the parameters existing in this function. Finally a distribution may be such that the sign of $\mu'(x)$ changes if we pass from small to large values of x.

We are mainly interested in the specific condition which is fulfilled for large values of x. The derivative $\mu'(x)$ of the intensity function is

(6) $$\mu'(x) = \mu(x)\left(\frac{f'(x)}{f(x)} + \mu(x)\right).$$

4.1.7 ANALYTICAL STUDY OF EXTREMES

Consequently, for finite values of x,

(7)
$$\mu'(x) > 0 \quad \text{if} \quad \mu(x) > -f'(x)/f(x)$$
$$\mu'(x) = 0 \quad \text{if} \quad \mu(x) = -f'(x)/f(x)$$
$$\mu'(x) < 0 \quad \text{if} \quad \mu(x) < -f'(x)/f(x) \,.$$

These are the conditions for the critical quotient stated in 4.1.3(4) to (6). Consequently, the classification of a distribution is not only a statement about its analytic nature or the orders of the parameters, but also about

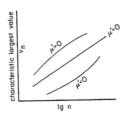

Graph 4.1.6. The Three Classes of the Exponential Type

the sample size. A distribution of the first or third class enters into the second class if $|\mu'(x)|$ becomes very small from a certain size onward. From (4) the characteristic largest value increases for the first (second, third) class more slowly than (as fast as, more quickly than) the natural logarithm of the sample size. The properties of the three classes are summarized in Table 4.1.6.

Table 4.1.6. The Three Classes of Distributions of the Exponential Type

Notion	Symbol	First Class	Second Class	Third Class
Intensity	$\mu(u_n) = \alpha_n$	Increases with n	Invariant	Decreases with n
Critical Quotient	Q	>1	$=1$	<1
Characteristic Value	u_n	Increases slower than lg n	Increases as lg n	Increases quicker than lg n
Curvature	$u''(\lg n)$	Concave	Straight	Convex
Mode	$\tilde{x}_n$	$>u_n$	$=u_n$	$<u_n$
Occurrence Interval	$T(\tilde{x}_n)$	$>n$	$=n$	$<n$
Density	$\varphi_n(u_n)$	Increases	Invariant	Decreases

4.1.7. The Characteristic Product. It remains to establish the *asymptotic* behavior of α_n and of $\alpha_n u_n$, henceforth called the *characteristic product*, for distributions of the exponential type. To this end, consider the equation

(1)
$$\lg(\alpha_n u_n) = \lg n + \lg f(u_n) + \lg u_n \,,$$

obtained from the definition of α_n in 3.1.5.(1). Again assuming n to be continuous, the derivative with respect to n is, from L'Hôpital's rule, 4.1.3(2) and 3.1.5(1),

$$\frac{d \lg (\alpha_n u_n)}{dn} \sim 1/n - nf(u_n)u'_n + u'_n/u_n.$$

From the definition of u_n and α_n in 3.1.4(1) and 3.1.5(1), $f(u_n)u'_n = n^{-2}$. Therefore,

(2) $$\frac{d\alpha_n u_n}{dn} \sim \alpha_n u'_n > 0.$$

It follows that $\alpha_n u_n$ increases asymptotically with $\lg n$, while α'_n converges to zero, for distributions of the exponential type.

Exercise: Prove that $\alpha_n u_n$ increases slower than $\sqrt{n}$.

4.2. EXTREMES OF THE EXPONENTIAL TYPE

4.2.0. Problems. To show the importance of the exponential type, consider four interesting distributions. The first is the logistic, which is widely used in studies on dosage. It will turn out later, in 8.1.4, to be the asymptotic distribution of the midrange for symmetrical distribution of the exponential type. In addition, we study the normal, the Chi square, and the logarithmic normal distributions.

4.2.1. The Logistic Distribution. As an example of the first class of the exponential type and its quick transition to the second class, consider the symmetrical distribution of a reduced variate x:

(1) $F(x) = 1/(1 + e^{-x})$; $f(x) = F(x)(1 - F(x))$; $f'(x) = f(x)(1 - 2F(x))$
$T(x) = e^x + 1$; $\mu(x) = F(x)$.

This function called logistic—for no reason whatsoever—was derived by Verhulst in 1845. He assumed that the increase of the logarithm of the population size $P(t)$ for a given country as a function of the time t is a constant minus a function which increases with the population. One of the solutions of his system of equations is

$$P(t) = \frac{P_\infty}{1 + ce^{-\alpha t}},$$

where P_∞ is the asymptotic size of the population alleged to be independent of technical and social conditions, and c and α are positive constants. Verhulst's Formula has been widely used, especially by Pearl, and much abused as a general law of growth.

4.2.1 ANALYTICAL STUDY OF EXTREMES

The shape of the distribution is similar to the normal one, and some authors, e.g., Berkson, prefer it to the normal distribution. It has the analytic advantage that the density of probability and the variate can be expressed by the probability, since

(1') $$e^x = F/(1 - F).$$

For large x the probability converges toward the exponential probability. The moment generating function $G_x(t)$ becomes, from (1'),

$$\int_0^1 e^{xt} dF = \int_0^1 F^t (1 - F)^{-t} dF,$$

which is the Beta function. Therefore,

(2) $$G_x(t) = \Gamma(1 + t)\,\Gamma(1 - t),$$

and the variance is

(2') $$\sigma^2 = 2S_2 = \pi^2/3.$$

From (1) it follows that the critical quotient $Q(x)$ defined in 4.1.3(3) is

$$\frac{F(x)}{2F(x) - 1} = 1 + \frac{1 - F(x)}{2F(x) - 1}.$$

It decreases and converges toward unity. Thus, the distribution belongs to the first class of initial distributions of the exponential type. The characteristic largest value u_n, 3.1.4(1), is

(3) $$u_n = \lg(n - 1).$$

The median of the largest value converges toward

(4) $$\check{x}_n \sim \lg n - \lg \lg 2 - (\lg 2)/2n.$$

The mode of the largest value is obtained from 3.1.6(2) as

(5) $$\tilde{x}_n = \lg n.$$

Thus, the characteristic largest value is slightly smaller than the mode towards which it converges rapidly. The occurrence interval is

(6) $$T(\tilde{x}_n) = n + 1.$$

The formulae (3) to (6) resemble closely the corresponding formulae 4.1.1(2) to (6) for the exponential distribution.

The moment generating function $G_n(t)$ of the largest value becomes, from (1), (1') and 3.1.7(6),

(7) $$G_n(t) = n \int_0^1 F^{n-1+t} (1-F)^{-t} \, dF,$$

whence

(8) $$G_n(t) = \Gamma(n+t) \, \Gamma(1-t)/\Gamma(n),$$

In consequence of the procedure 1.1.8(6),

(9) $$G_n(t) = \prod_1^{n-1} (1 + t/v) \cdot \Gamma(1+t) \, \Gamma(1-t).$$

The moment generating function for the largest value differs from the moment generating function (2) of the variate by a multiplicative factor which causes the distribution of the largest values to become asymmetrical. The mean of the largest value,

(10) $$\bar{x}_n = \sum_1^{n-1} 1/v,$$

differs only by $1/n$ from the mean of the largest value of the exponential distribution 4.1.1(8), and converges toward

(11) $$\bar{x}_n = \gamma + \lg n.$$

The variance σ_n^2 of the largest value obtained from the seminvariant generating function is

(12) $$\sigma_n^2 = 2S_2 - \sum_1^{n-1} 1/v^2$$

and converges toward

(13) $$\sigma_n^2 \sim S_2 = \pi^2/6.$$

The variance of the largest value is for large samples equal to one half the variance of the variate and has the same value as for the exponential distribution.

Exercises: 1) Plot the normal, the logistic, and the Cauchy distributions for zero median and unit interquartile range on logistic paper. 2) Calculate the generating function of the mth value. 3) Derive the distribution from the generating function. 4) Calculate the variance of the mth values. 5) Verify the formula 2.1.8(6) for the mean mth distance. 6) Use the transformation $x = \alpha(v - \beta)$ and prove that the estimate of β from the mean is asymptotically more precise than from the median and

4.2.2 ANALYTICAL STUDY OF EXTREMES

that the estimate of $1/\alpha$ from the mean deviation is asymptotically more precise than the estimate from the standard deviation. (See Tukey.)
7) Estimate the parameters by the method of maximum likelihood and successive approximation. (See Berkson.)

References: Anscombe, Gumbel (1944).

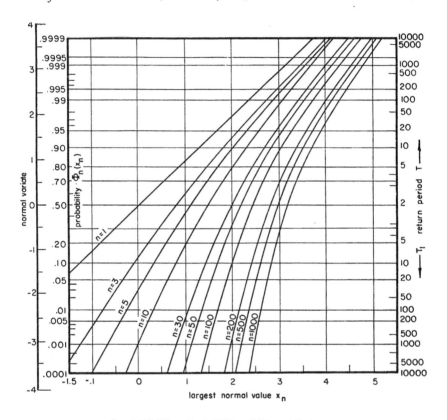

Graph 4.2.2(1). Probabilities of Normal Extremes

4.2.2. Normal Extremes, Numerical Values. The study of the normal distribution is necessary, since it plays a central role in statistical theory and practice. However, the formulae for the extremes are complicated. Bachelier gave a formula based on an expansion of the normal distribution. His line of approach has not been followed up. Tricomi reduced the asymptotic mean of the largest normal value to the evaluation of a certain integral. Tippett calculated the numerical value of the probability function $\Phi_n(x)$, 3.1.1(1), for $n = 3, 5, 10, 20, 30, 50, 100(100)1000$. The results, given in K. Pearson's Tables, vol. II, are plotted in Graph 4.2.2(1)

Table 4.2.2(I). Normal Extremes

1	2	3	4	5	6	7	8	9
F	$(\tilde{x}_n); u_n$	$T(\tilde{x}_n); n(u)$	$n(\tilde{x})$	n_2	n_3	n_4	n_5	$1/\alpha_n$
.50	0	2	1	3	5	7	9	1.25331
.60	0.25334	2.5	1.3935	3.894	6.394	8.894	11.39	1.03535
.70	0.52440	3.333	2.0558	5.389	8.722	12.06	15.39	.86283
.80	0.89162	5.0	3.4050	8.405	13.41	18.41	23.41	.71438
.90	1.2816	10	7.5721	17.57	27.57	37.57	47.57	.56981
.95	1.6449	20	16.151	36.15	56.15	76.15	96.15	.48480
.96	1.7507	25	20.503	45.50	70.50	95.50	120.5	.46418
.97	1.8808	33.33	27.812	61.15	94.48	127.8	161.1	.44090
.98	2.0538	50	42.569	92.57	142.6	192.6	242.6	.41307
.985	2.1701	66.67	56.441	123.1	189.8	256.4	323.1	.39609
.990	2.3264	100	87.413	187.4	287.4	387.4	487.4	.37520
.992	2.4089	125	110.02	235.0	360.0	485.0	610.0	.36497
.994	2.5121	166.7	147.86	314.5	481.2	647.9	814.5	.35288
.995	2.5758	200	178.25	378.3	578.3	778.3	978.3	.34579
.996	2.6521	250	223.96	474.0	724.0	974.0	1224.0	.33764
.997	2.7478	333.3	300.43	633.8	967.0	1300	1633	.32790
.998	2.8782	500	454.05	954.1	1454	1954	2654	.31545
.999	3.0902	1000	917.86	1918	2918	3918	4918	.29699

4.2.2 ANALYTICAL STUDY OF EXTREMES

on normal paper. The chart shows how the probability functions of the largest value shift toward the right, becoming more and more asymmetrical and concentrated with increasing n. Not much numerical knowledge existed before and not much has been gained since the publication of Tippett's paper. The probability points calculated on this basis by

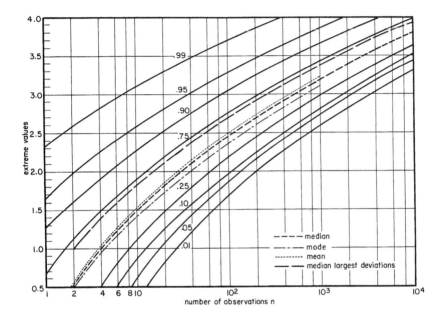

Graph 4.2.2(2). Quantiles of Normal Extremes

E. S. Pearson (K. Pearson's Tables II) traced in Graph 4.2.2(2) against lg n approach straight lines, a fact which will become important in the next paragraph.

The medians of normal extremes calculated by Finetti from 3.1.3(1) and the means of normal extremes calculated by Tippett for $2 \leq n \leq 1000$ are also traced in Graph 4.2.2(2). The distributions $\varphi_n(x)$ of normal extremes calculated by Finetti from 3.1.1(4) are traced in Graph 4.2.2(3) for $n = 1, 2, 3, 5, 10$. For $n = 10, 100, 1000$, see Graph 6.2.1(4).

The characteristic extremes defined by 3.1.4(1) are obtained in the following way: in Table 4.2.2(1) we choose certain normal probabilities F (first column), take the corresponding values x (column 2) from a normal table, and calculate $n = 1/[1 - F(x)]$ (column 3). Then the values x are the characteristic largest normal values, u_n corresponding to the sample size n given in column 3. The relation between u_n and n is traced in

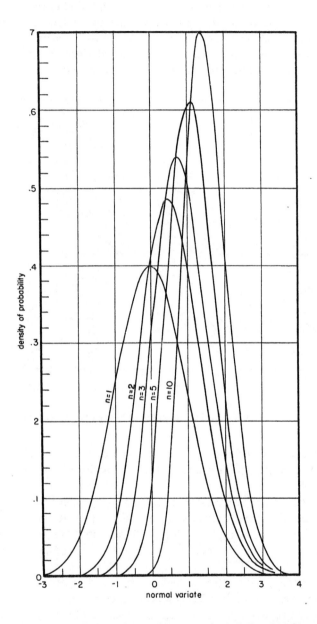

Graph 4.2.2(3). Distributions of Normal Extremes for 2, 3, 5, and 10 Observations

Graph 4.2.2(4). The reason for the scale used will be explained in the next paragraph.

Table 4.2.2(2). Characteristic Normal Extremes u_n for Large Samples

n	1×10^n	2×10^n	5×10^n	8×10^n
10^3	3.091	3.291	3.540	3.662
10^4	3.719	3.891	4.107	4.215
10^5	4.265	4.417	4.611	4.708
10^6	4.754	4.892	5.069	5.158
10^7	5.199	5.327	5.491	5.573
10^8	5.612	5.731	5.884	5.962
10^9	5.998	6.110	6.254	6.327
10^{10}	6.361	6.467	6.604	6.673

The modes of the normal extremes become, from 3.1.6(1),

(1) $$n - 1 = [xF(x)/f(x)]_{x=\tilde{x}_n}.$$

For chosen values of x, the sample sizes n are obtained from the quotients $F(x)/f(x)$ (K. Pearson's *Tables*, vol. II). This leads to the modes, Table 4.2.2(1) (column 2), as function of the corresponding sample sizes n (column 4),

Table 4.2.2(3). Normal Extremes and Intensity

1	2	3	4	5
Characteristic Value	Largest	Normal Intensity	Sample Size	Parameter
u_n	$u_n + 1/u_n$	α_n	n	$1/k_n = (u_n^2 - 1)/(u_n^2 + 1)^2$
3.0	3.3333	3.2831	$7.4080 \cdot 10^2$	0.08000
3.5	3.7857	3.7514	$4.2987 \cdot 10^3$	0.06408
4.0	4.2500	4.2257	$3.1574 \cdot 10^4$	0.05190
4.5	4.7222	4.7043	$2.9432 \cdot 10^5$	0.04263
5.0	5.2000	5.1864	$3.4886 \cdot 10^6$	0.03550
5.5	5.6818	5.6715	$5.2661 \cdot 10^7$	0.02995
6.0	6.1667	6.1584	$1.0136 \cdot 10^9$	0.02557
6.5	6.6539	6.6472	$2.4900 \cdot 10^{10}$	0.02205

traced in Graphs 4.2.2(2) and 4.2.2(4). They turn out to be smaller than the medians. The convergence of the occurrence interval $T(\tilde{x}_n)$ (column 3) to the sample size n is shown in graph 4.2.2(5). The sample sizes n_m (columns 5 to 8) corresponding to the modes of the mth extremes (column 2) are obtained from the procedure outlined in 3.1.6. The coefficients α_n defined in 3.1.5(1) taken from K. Pearson's *Tables*, which are given in column 9 and traced in Graph 4.2.2(4), increase with n and turn out to be greater than the characteristic values. Table 4.2.2(1) holds for values of n up to 1000. The characteristic values u_n and the coefficient α_n for larger values of n are given in Tables 4.2.2(2) and (3). Column 5 of Table 4.2.2(3) will be explained in 7.3.3.

The standard deviations of the largest normal values obtained from Tippett's tables are plotted, together with the standard error of the mean

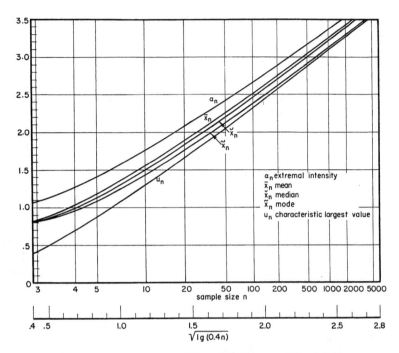

Graph 4.2.2(4). Averages of Normal Extremes in Natural Scale

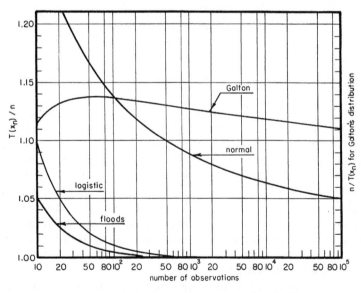

Graph 4.2.2(5). Occurrence Interval and Sample Size

4.2.2　ANALYTICAL STUDY OF EXTREMES

against $1/\sqrt{n}$, in Graph 4.2.2(6). The moment quotients $\sqrt{\beta_1}$ and β_2 taken from the same source are traced in semi-logarithmic scale in Graph 4.2.2(7) as functions of n. Finally, they are traced one against the other in Graph 4.2.2(8). The other values traced in these graphs will be explained in 4.2.4.

Exercise: Calculate tables similar to 4.2.2(1), (2), and (3) for the exponential and logistic distributions.

References: Daniels (1941) David (1957), Graf and Wartmann, Grubbs and Weaver, Gumbel (1944). Haag, Haag and Borel, Hald (p. 331), Howell

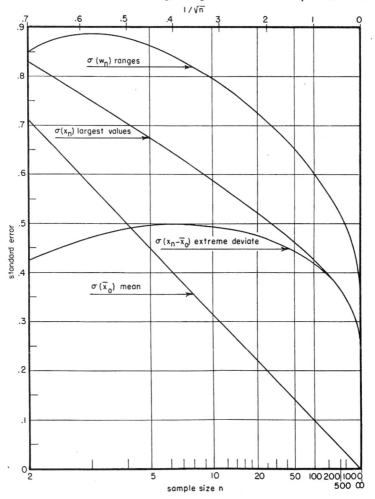

Graph 4.2.2(6).　Standard Deviations for Normal Extremes, Deviates, and Ranges

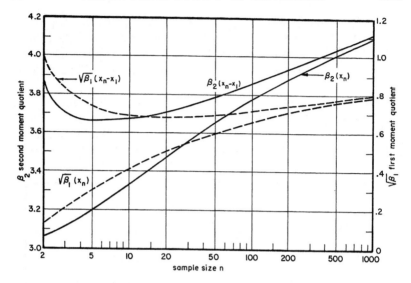

Graph 4.2.2(7). Moment Quotients for Largest Normal Values and Deviates as Functions of Sample Size

(1949, 1950), Kozelka, Kudô, McKay, Ogawara (1951), Robbins (1944a), Smirnov (1933).

4.2.3. Analysis of Normal Extremes. In the following, we give the analytic reasons for some of the numerical relations stated before. The results were first obtained by R. A. Fisher and L. H. C. Tippett (1928) in their fundamental paper. However, our derivation differs from theirs.

For large values of x, the normal probability function $F(x)$ and the distribution $f(x)$ are related by the well-known formula (Kendall vol. I, p. 129):

(1) $\quad 1 - F(x) = f(x)(1 - x^{-2} + 3x^{-4} - 15x^{-6} + \ldots)/x.$

The intensity is, therefore,

(2) $\quad 1/\mu(x) = (1 - x^{-2} + 3x^{-4} - 15x^{-6} + \ldots)/x,$

while the logarithmic derivative is

(3) $$\frac{-d \lg f(x)}{dx} = x.$$

Consequently, the critical quotient 4.1.3(7),

(4) $$Q(x) = \mu(x)/x,$$

4.2.3 ANALYTICAL STUDY OF EXTREMES

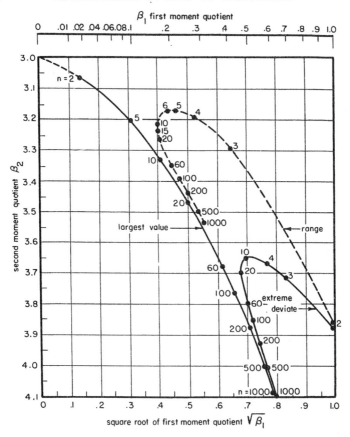

Graph 4.2.2(8). Beta Diagram for Normal Extremes, Deviates, and Ranges

decreases very slowly toward unity. The normal distribution belongs to the first class of distributions of the exponential type. From the critical quotient we obtain, after substituting u_n for x and α_n for $\mu(x)$,

(4') $$1/\alpha_n = u_n^{-1}(1 - u_n^{-2} + 3u_n^{-4} - 15u_n^{-6} + \ldots).$$

Since u_n increases with n, the same holds for α_n. The density of probability at the most probable largest value increases with the sample size as shown in Graph 4.2.2(3). Inversion of the power series (4') leads to

(5) $$\alpha_n = u_n + u_n^{-1} - 2u_n^{-3} + 10u_n^{-5} \ldots$$

The first approximation valid for $u_n^4 \gg 1$,

(6) $$\alpha_n = u_n + 1/u_n,$$

was given by Fisher and Tippett in the form

(6')
$$\frac{1}{\alpha_n} = \frac{u_n}{u_n^2 + 1}.$$

With increasing n, the extremal intensity converges very slowly toward the characteristic largest value.

The characteristic largest value $u = u_n$ is, from (1) the solution of

(7)
$$\frac{1}{n} = \frac{e^{-u^2/2}}{u\sqrt{2\pi}} (1 - u^{-2} + 3u^{-4} - 15u^{-6} + \ldots),$$

where the index n is dropped. For example, for $u = 3$, we obtain from the first four members of the power series $n = 744.6$, while Table 4.2.2(3) shows $n = 740.8$. Therefore (7) may be used for the calculation of u_n for $n > 750$.

For reasons which will appear in Chapter 7, we are interested in $d\lg \alpha / d\lg n$ as a function of u. From (6') and (7) it follows that

$$\lg \alpha = \lg(u^2 + 1) - \lg u; \qquad \frac{d\lg \alpha}{du} = \frac{u^2 - 1}{u(u^2 + 1)}$$

$$\lg n = u^2/2 + \lg u + \lg \sqrt{2\pi}; \qquad \frac{d\lg n}{du} = \frac{u^2 + 1}{u},$$

whence, after division,

(8)
$$\frac{d\lg \alpha}{d\lg n} = \frac{u^2 - 1}{(u^2 + 1)^2},$$

a result which is also due to Fisher and Tippett.

If we assume only that $u^6 \gg 1$, the corresponding formulae are, from (5) and (7),

$$\frac{d\lg \alpha}{du} = u^{-1} - 2u^{-3} + 13u^{-5} - 92u^{-7}$$

$$\frac{d\lg n}{du} = u + u^{-1} - 2u^{-3} + 13u^{-5} - 92\,u^{-7}.$$

Hence,

(9)
$$\frac{d\lg \alpha}{d\lg n} = u^{-2} - 3u^{-4} + 18u^{-6} - 129u^{-8},$$

which differs from (8) only by a factor of the order $10u^{-6}$.

The modal largest value $\tilde{x}_n$ obtained from 3.1.6(1) and (1) becomes asymptotically the solution of

$$n - 1 = \frac{x}{f(x)} - (1 - x^{-2} + 3x^{-4} - 15x^{-6}),$$

4.2.3 ANALYTICAL STUDY OF EXTREMES

whence, neglecting the third and subsequent members,

(10) $$n = \sqrt{2\pi}\, \tilde{x}\, e^{\tilde{x}^2/2} + \tilde{x}^{-2}.$$

For $\tilde{x} = 3$, the calculation leads to $n = 677.9$, whereas interpolation from the graph 4.2.2(4) gives $n = 680$. Therefore the approximation (10) may be used from $x \geq 3$ onward. From 4.1.5(4) the mode of the largest value exceeds and converges toward the characteristic largest value. As a first approximation, the modal largest value obtained from (10) is, for $\tilde{x}_n^2/2 \gg \lg \tilde{x}_n$,

(11) $$\tilde{x}_{n,1} \sim \sqrt{2 \lg (n/\sqrt{2\pi})} \sim \sqrt{2 \lg (0.4n)},$$

The most probable largest normal value increases as the square root of the logarithm of the number of observations. Formula (11) was first used by Philipp Frank to calculate the velocity of the quickest gas molecule. From (11) it is natural to trace the averages of extreme values against $\sqrt{\lg (0.4n)}$, as is done in Graph 4.2.2(4).

The asymptotic formula (11) gives the answer to the following questions:
1) What is the most probable largest value for $2n$ observations? The triangular relation,

(12) $$\tilde{x}_{2n}^2 = \tilde{x}_n^2 + 2 \lg 2,$$

is shown in the schematic Graph 4.2.3. 2) By what factor should the sample size be multiplied in order to obtain the double of a given modal largest value? Let k be this factor; then it follows from

(13) $$\tilde{x}_{kn}^2 = 4\tilde{x}_n^2$$

Graph 4.2.3. Modal Largest Normal Value for $2n$ Observations

that $k = 0.06349 \cdot n^3$. To double the most probable largest value, we must take 6% of the fourth power of the original sample size. Of course, these asymptotic relations have to be modified if the initial distribution has a mean different from zero and a standard deviation different from unity. For numerical purposes, the approximation (11) is insufficient for the characteristic largest value $u = u_n$. A second approximation is obtained by a procedure due to Cramér (p. 374). From the expansion (1), an approximate value of u is obtained as the solution of

(14) $$u^2/2 + \lg u = \lg (0.4n).$$

Let u_1 be the value given by (11). To obtain a second approximation, we introduce a small correction ϵ defined by

$$u_2 = \sqrt{u_1^2 - \epsilon}; \qquad \epsilon^2 \ll u_1^4.$$

Then ϵ becomes from (14) the solution of

$$\epsilon = \lg (u_1^2 - \epsilon),$$

whence, after the usual expansion,

$$\epsilon = (u_1^2 \lg u_1^2)/(u_1^2 + 1).$$

The resulting formula,

(15) $\qquad u_2 = u_1 [1 - (\lg u_1^2)/(u_1^2 + 1)]^{1/2},$

is sufficiently accurate for n equal to or larger than 200. However, in general, no analytic formula for the characteristic largest value is needed, since n as a function of u_n can easily be obtained from the usual tables. Their inversion is given in Tables 4.2.2(1) and (2).

Exercises: 1) Show how the occurrence interval converges toward n. 2) Prove that the mode exceeds the characteristic largest value. 3) Compare u_2 to the numerical values obtained from tables of the normal probability.

Problems: 1) Calculate the power series for the mean, median, and characteristic largest normal value. 2) Reconsider problem 1.2.9(2) in the light of the asymptotic increase of the extremes.

References: Anis, Cramér, Cochran (1941), Darling and Erdös, Eberhard, Edgeworth, Grubbs, Howell (1949), Irwin (1925a), Jeffreys, Kudô, McKay, Mises (1923), Nair (1948a), E. S. Pearson and Sekar, Ruben, Smirnov (1935).

4.2.4. Normal Extreme Deviates. The difference between the observed extreme x_n and the sample mean $\bar{x}_0$, henceforth called the *extreme deviate*, may be used as criterion for the acceptance or rejection of the first or last normal value. A. T. McKay derived the probability function for the normal extreme deviate,

(1) $\qquad \xi_n = x_n - \bar{x}_0,$

for unit population standard deviation and studied especially its generating function. The distribution $\varphi_n{}^*(\xi_n)$ can be written down explicitly only for $n = 2$, where

(2) $\qquad \varphi_2{}^*(\xi_2) = 2e^{-\xi_2{}^2}/\sqrt{\pi}$

becomes the right half of the normal distribution. The moments are

$$\overline{\xi_2{}^l} = \Gamma[(l + 1)/2]/\sqrt{\pi}.$$

For $n = 3$, McKay obtained

(3) $\qquad \varphi_3{}^*(\xi_3) = 3\sqrt{3/\pi}\, e^{-3\xi^2/4}\, F(3\xi/\sqrt{2}),$

where F stands for the normal probability. For larger sample sizes, the probability $\Phi_n^*(\xi_n)$ can be reduced to the probability $\Phi_{n-1}^*(\xi_{n-1})$ by

(4) $\quad \Phi_n^*(\xi_n) = \dfrac{n^{3/2}}{\sqrt{2\pi(n-1)}} \displaystyle\int_0^{\xi_n} \exp\left[-\dfrac{nx^2}{2(n-1)}\right] \Phi_{n-1}^*\left(\dfrac{nx}{n-1}\right) dx.$

The derivation of this formula was considerably simplified by Grubbs. McKay calculated the numerical values of the standard deviation of ξ_n for selected values of n from 2 to 1000. In addition to this, he showed that the correlation between x_n and $\bar{x}_0$ diminishes from 0.856 for $n = 2$ to 0.090 for $n = 1000$. Nair calculated numerical values to six decimals of the probability (4) for the extreme deviate 0.00 (0.01) 4.70 and sample sizes $n = 3$ (1) 9 and two decimals for the upper and lower percentage points .0001, .005, .01, 025, .05, .10 for $n = 3(1)9$. Grubbs calculated the probability function by the methods used by Nair for the extreme deviates 0.00 (0.05) 4.60, and $n = 2(1)25$ to 5 decimals. For $n = 2, 3, 4$, the standard deviation of the extreme deviate is smaller than the standard error of the mean. The standard deviation of the extreme deviate first increases, later decreases, with the sample size n and converges toward the standard deviation of the largest value. See graph 4.2.2(6). The reason is as follows: With increasing sample size the correlation between the largest value and the mean decreases, the influence of the mean diminishes, and the standard error of the mean decreases more quickly than the standard deviation of the largest value. Therefore, the contribution of the standard error of the mean to the standard deviation of the extreme variate diminishes.

Grubbs also gave the numerical values of the two moment quotients of the extreme deviate. For small n, they differ widely from the corresponding moment quotient of the extreme value, and the convergence is slow. See Graph 4.2.2(7). The initial difference between the convergence of the two types of moment quotients is shown in Graph 4.2.2(8).

The use of the extreme deviate for testing outlying observations requires an estimation of the population standard deviation. This is not required for the "studentised" extreme deviate,

(5) $\quad\quad\quad\quad\quad\quad\quad t_n = (x_n - \bar{x}_0)/s,$

first proposed by Thompson (1935). Its distribution is quite intricate, and related to the distribution of $r\sqrt{n-1}$, where r is the sample coefficient of correlation of n pairs of normal independent observations. E. S. Pearson and Chandra Sekar derived probability points for t_n. Grubbs derived the exact distribution, following methods due to McKay and Nair, and calculated the probability points traced in Graph 4.2.4. If the probabilities $\Phi_n(t_n)$ of extreme normal deviates are considered as normal probabilities $F(x)$, a linear relation seems to hold between $x(F)$ and $t_n(\Phi)$ for the

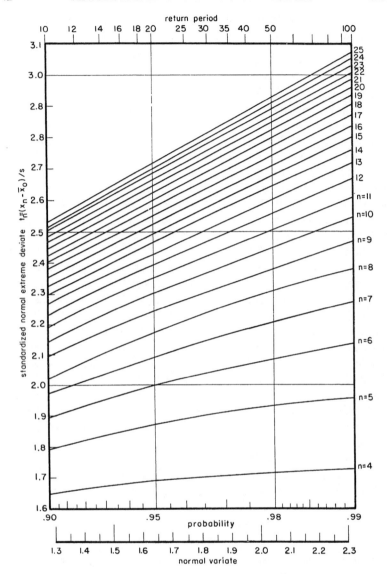

Graph 4.2.4. Probabilities of Studentized Normal Extreme Deviate

same numerical values for F and Φ and for large n. This relation, obtained from Grubbs' table IA, is quite unexpected.

As an alternative for the statistic t_n, Grubbs proposed to use S_n^2/s^2, the quotient of the sample variance obtained by omission of the first or last observations, by the sample variance of all observations. The two statistics are related by

(6) $$S_n^2/s^2 = 1 - t_n^2/(n-1).$$

This leads to a table of the percentage points for S_n^2/s^2 given to four decimals, for the probabilities .90, .95, .975, .99 and for $n = 3(1)25$. In itself, the complicated statistic S_n^2/s^2 has no advantage over the simple statistic t_n. However, it permits a generalization: A test for the second extreme is obtained by the probability of the ratio of the sample variance of the first or last $n-2$ observations and the total sample variance. Grubbs gave the percentage points for this statistic to four decimals for .90, .95, .975, .99 and for $n = 4(1)20$.

Problems: 1) Calculate the characteristic largest value of the sample correlation coefficient r for n observations in N repetitions for the value zero of the population coefficient. (See Thompson, 1935.) 2) What is the reason for the linear relation shown in Graph 4.2.4? 3) Calculate probability points for the extreme deviate $\bar{x}_0 - x_1$ for the exponential distribution.

References: David (1956b), Dixon (1953), Halperin, King, Moshman (1952), Nair (1948b), Nair and David.

4.2.5. Gamma Distribution. In the following we calculate the characteristic largest value for the reduced Gamma distribution (Pearson's type III),

(1) $$f(x) = x^{v-1}e^{-x}/\Gamma(v), \quad x \geq 0; \quad v \geq 1.$$

which is related to the χ^2 distribution, and constitutes a generalization of the exponential distribution for which $v = 1$. The probability function $F(x)$ becomes, after repeated partial integrations, for integer values of v,

(2) $$1 - F(x) = f(x)\left(1 + \frac{v-1}{x} + \frac{(v-1)(v-2)}{x^2} + \ldots + \frac{(v-1)!}{x^{v-1}}\right)$$

or

(3) $$1 - F(x) = e^{-x}\sum_{k=0}^{v-1}\frac{x^k}{k!}.$$

The generating function is

(4) $$G_x(t) = (1-t)^{-v}.$$

Hence, the mean is v itself, and the density function at the mean is the probability of the mean in the Poisson distribution. A mode exists for $v > 1$, and its value is $v - 1$. The variance is also v. Consequently the distribution spreads with the parameter v.

To prove that the distribution (1) is of the exponential type for large values of x, consider the reciprocal of the intensity function obtained from (2) as

(5) $$\frac{1}{\mu(x)} = 1 + \frac{v-1}{x} + \frac{(v-1)(v-2)}{x^2} + \ldots + \frac{(v-1)!}{x^{v-1}}.$$

The negative reciprocal of the logarithmic derivative is, from (1),

$$-\frac{f(x)}{f'(x)} = \frac{1}{1-(v-1)/x},$$

whence

(6) $$-\frac{f(x)}{f'(x)} = 1 + \frac{v-1}{x} + \frac{(v-1)^2}{x^2} + \ldots.$$

Division of (6) by (5) leads to the critical quotient

(7) $$\lim_{x=\infty} \frac{\mu(x)}{-f'(x)/f(x)} = 1 + |\epsilon(x)|; \quad |\epsilon(x)| \to 0,$$

which, from 4.1.3(4) is the criterion for the first class of the exponential type. The characteristic largest value for $v=1$, i.e., for the exponential distribution is, of course,

(8) $$u_n = \lg n.$$

For $v > 1$ the value u_v as function of n and v is obtained from (3) as the solution of

$$e^{u_v} = n \sum_{k=0}^{v-1} u_v^k/k! \qquad u_v = u_n(v).$$

This may be written

(9) $$u_v = \lg n + \lg\left(1 + u_v + \frac{u_v^2}{2!} + \ldots + \frac{u_v^{v-1}}{(v-1)!}\right).$$

To obtain numerical values of u_v, we use an algorithm. For $v=2$ ($v=3$), the sample size $n_2(n_3)$ corresponding to the characteristic largest value u_v is obtained from (9) as

(10) $$\lg n_2 = u_2 - \lg(1 + u_2); \quad \lg n_3 = u_3 - \lg(1 + u_3 + u_3^2/2),$$

and so on. We choose $u_v = 1, 2, 3, 4, 5$ and obtain the corresponding consecutive sample sizes n_2, n_3, n_4, n_5, from (10) and the consecutive equations. We plot n on a logarithmic scale as abscissa, u_v on a linear scale as ordinate, and read u_v as function of n and v from Graph 4.2.5.

This approach holds for small n. If n is large, $u_n(v)$ is large too and (2) leads approximately to the characteristic largest value $u_n(v) = u$ as the solution of

(11) $$u - (v-1)(\lg u + 1/u) = \lg[n/\Gamma(v)].$$

4.2.5 ANALYTICAL STUDY OF EXTREMES

The first approximation is
$$u_1 = \lg(n/\Gamma(v)).$$
The second approximation is obtained by writing
$$u_2 = u_1 + \varepsilon,$$

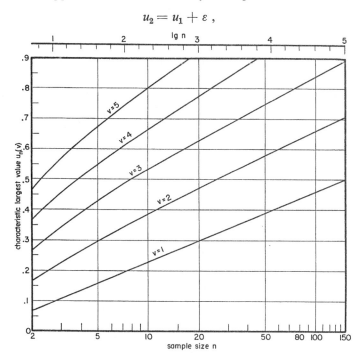

Graph 4.2.5. Characteristic Largest Value for Gamma Distribution

where ε is obtained from (11) after the usual expansions as the solution of
$$\varepsilon = (v-1)(\lg u_1 + \varepsilon/u_1 + 1/u_1 - \varepsilon/u_1^2).$$
Again in first approximation,
$$\varepsilon = \frac{(v-1)u_1 \lg u_1}{u_1 - v + 1}.$$
If v is small and n sufficiently large, the characteristic largest value for the Gamma distribution is thus

(12) $$u_2 = \lg[n/\Gamma(v)] + (v-1)\lg\lg[n/\Gamma(v)],$$

a formula which is analogous to 4.2.3(15). The treatment may serve as an example for distributions where the probability can only be written as a series.

Exercise: Calculate u_n and α_n for $F(x) = 1 - \exp(-x^k)$, $k > 0$.

Problems: 1) Calculate α_n for the Gamma distribution. 2) Study the generalized Gamma distribution $f(x) = k^{-v} x^{v-1} e^{-x/k} \Gamma(v)^{-1}$; $v > 0$; $k > 0$.

References: Cadwell (1953b), Cochran (1941), Doornbos, Eisenhart, Hartley (1950), C. Thompson (1941), Vora.

4.2.6. Logarithmic Normal Distribution. The logistic, the Gamma, and the normal distributions belong to the first class of the exponential type. The logarithmic normal distribution, which is unlimited to the right, belongs to the third class, if we study the largest values. For this aim, it is sufficient to consider the reduced form obtained from 1.1.9(6) for $\beta = \delta = 1$,

(1) $$f_g(z) = \frac{1}{z\sqrt{2\pi}} \exp[-(\lg z)^2/2]; \qquad F_g(z) = F(\lg z),$$

where F stands for the normal probability function. The logarithmic derivative of (1) is

(2) $$-\frac{f'_g(z)}{f_g(z)} = \frac{1 + \lg z}{z}.$$

To obtain the asymptotic value of the intensity function $\mu(z)$, we neglect the higher terms in 4.2.3(1), whence

$$1 - F(\lg z) \sim \frac{\exp[-(\lg z)^2/2]}{\sqrt{2\pi} \lg z}.$$

Consequently, from (1),

(3) $$\mu(z) \sim (\lg z)/z.$$

The critical quotient 4.1.3.(3) obtained from (2) and (3),

(4) $$Q(z) \sim \frac{\lg z}{\lg z + 1} = 1 - |\epsilon(z)|,$$

increases toward unity and the distribution belongs to the third class of distributions of the exponential type.

The characteristic largest value $u_{n,g}$ is obtained from (1) as

(5) $$u_{n,g} = e^{u_n},$$

where u_n stands for the characteristic largest normal value. The coefficient $\alpha_{n,g}$ is obtained from

$$\frac{1}{\alpha_{n,g}} = \frac{1 - F(\lg z)}{f(\lg z)/z},$$

4.2.7 ANALYTICAL STUDY OF EXTREMES

and the asymptotic expression 4.2.3(1) and 4.2.3(4') as

$$\frac{1}{\alpha_{n,g}} = \frac{e^{u_n}}{\alpha_n},$$

where α_n is the normal intensity function. Hence

(6) $$\alpha_{n,g} = \alpha_n e^{-u_n}$$

and

(7) $$\alpha_{n,g} u_{n,g} = \alpha_n.$$

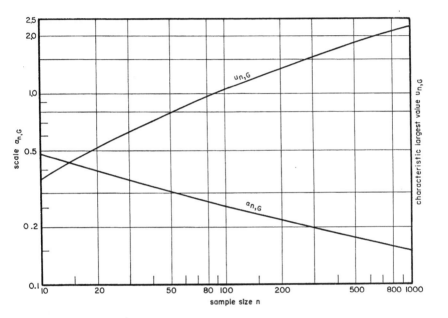

Graph 4.2.6. Asymptotic Parameters for the Logarithmic Normal Distribution

These relations are traced in Graph 4.2.6. Therefore, the characteristic largest value and the Alpha for the logarithmic normal distribution are obtained immediately from the corresponding values of the normal distribution. Graph 4.2.6 shows how $\alpha_{n,g}$ diminishes with n. The decrease of the hazard rate $\alpha_{n,g} = \mu(u_{n,g})$ contradicts the behavior expected in life testing. The distribution of the largest value spreads with increasing sample size. (The scale for $u_{n,g}$ is ten times the scale for $\alpha_{n,g}$.)

Exercise: Prove that $\tilde{x}_n$ increases faster than $\lg n$.

4.2.7. The Normal Distribution as a Distribution of Extremes. Two cases are known where the normal distribution plays the role of an extremal distribution.

In his study on the linear Brownian movement, Paul Lévy has shown the astonishing fact that the normal law holds at the same time for the variable position of a particle and its extreme values. He writes $X(t)$ for the position of a particle at time t, $M(t)$ and $m(t)$ for the maximum and minimum, respectively, of $X(t)$ in the interval $0 < \tau < t$, and proves by beautiful symmetry considerations that, under the assumption $X(0) = 0$, the five values $X(t)$, $M(t)$, $-m(t)$, $M(t) - X(t)$, and $X(t) - m(t)$ have the same probability function, namely

$$(1) \qquad F(x) = \sqrt{\frac{2}{\pi t}} \int_0^x \exp[-\xi^2/(2t)] \, d\xi; \quad x > 0.$$

Thus the normal distribution holds for the absolute value of the variable, the extremes, and the differences of the variable from the extremes. The position $X(t)$ will exceed any bound with probability as close to 1 as desired if the time considered is large enough. The particle may be expected to drift indefinitely far from the origin. However it is also known that the particle will return with probability 1 to the origin no matter how far out it drifts.

Paul Lévy also proved that the joint distribution $w(\xi,x)$ of the maximum $M(t) = \xi$ and the variable $X(t) = x$ is

$$(2) \qquad w(\xi, x) = \sqrt{\frac{2}{\pi t}} \frac{2\xi - x}{t} \exp\left[-\frac{(2\xi - x)^2}{2t}\right]; \quad \xi > 0; \; x < \xi.$$

It follows for $x = \xi - y$ that the joint distribution $w_1(\xi,y)$ of the maximum ξ and the difference $y(t) = M(t) - X(t)$,

$$(3) \qquad w_1(\xi, y) = \sqrt{\frac{2}{\pi t}} \frac{\xi + y}{t} \exp\left[-\frac{(\xi + y)^2}{2t}\right]; \quad \xi > 0; \; y > 0,$$

is symmetrical with respect to the maximum and the difference of the variable from it.

A normal limit to an asymptotic distribution of breaking strengths of bundles of n parallel threads was developed by Daniels (1945). He assumes that the breaking strengths x have a continuous probability function $F(x)$ so that

$$\lim_{x=\infty} x(1 - F(x)) = 0.$$

His physical model of rupture is as follows: After rupture of one thread the load is supported by the remaining threads until the next one breaks, and so on. Let S be the load and let $x_i (i = 1, 2, 3 \ldots n)$ be the breaking strengths of the individual threads arranged in decreasing order of magnitude, then the bundle breaks if

$$(5) \qquad 0 < x_i < S/n; \; x_{i-1} < x_i < S/(n - i + 1).$$

4.3.1 ANALYTICAL STUDY OF EXTREMES

Let $\acute{S}$ be the largest value of S for which these inequalities are not all true. Then $\acute{S}$ is a random variate representing the breaking load of the bundles of n threads. Finally, let $\dot{x}$ be the maximum of $x(1 - F(x))$, then Daniels proves that for large n the value $\acute{S}$ is normally distributed with mean

(6) $$\overline{S} = n\dot{x}(1 - F(\dot{x}))$$

and variance

(7) $$\sigma^2(\acute{S}) = n\dot{x}^2 F(\dot{x})(1 - F(\dot{x})).$$

Of course, the validity of this interesting result is linked to the existence of the model which does not hold, say, for metals or capacitors. Here the opposite theory that no chain is stronger than its weakest link may be more appropriate.

4.3. THE CAUCHY TYPE

4.3.0. Problems. The two preceding sections dealt with the exact distribution and the averages of extreme values for the exponential type. However, there are unlimited distributions, some possessing all moments which do not belong to this type. To obtain a definition of another type of unlimited distribution we start from Pareto's distribution. Its generalization leads to the Pareto type, to be defined by the asymptotic value of the characteristic product 4.1.7. Distributions which belong to the Pareto type in both directions are said to be of the Cauchy type. Here either *no moments or only a finite number of moments* exist, and this holds also for the distribution of extreme values. The mode of the extremes increases more quickly with n than the mode of the exponential type. The Cauchy type is less important than the exponential type. Finally, it will be shown that there are unlimited distributions which belong to neither category.

4.3.1. The Exponential Type and the Existence of Moments. The results obtained in (4.2) are based on the convergence of the critical quotient toward unity. However, as will be shown, this convergence does not hold for all unlimited distributions, not even for all those for which all moments exist.

The distributions of the exponential type possess all moments. The converse is not true: a distribution may possess all moments without being of the exponential type. For the proof, R. von Mises (1936) considers a non-negative, unlimited variate x with the probability

(1) $$F(x) = 1 - \exp\left(-x - \frac{\sin x}{2}\right) \qquad 0 \leq x \leq \infty.$$

This is a continuous function increasing with x possessing all derivatives. The distribution

(2) $$f(x) = \left(1 + \frac{\cos x}{2}\right) \exp\left[-x - (\sin x)/2\right]$$

starts with $f(0) = 3/2$ and decreases with increasing x. It shows a certain periodicity, since, for integral values k,

$$f(2k\pi) = f(0)\, e^{-2k\pi}.$$

The distribution (2) and the exponential distribution intersect at the values $x_0 + k$; $k = 0, 1, 2\ldots$, where $x_0 = 0.7028$ is the first solution of

$$\exp\left[\frac{\sin x}{2}\right] = 1 + \frac{\cos x}{2}.$$

The characteristic largest value is the solution of

(3) $$u_n + \frac{\sin u_n}{2} = \lg n\,; \qquad \lg n - \tfrac{1}{2} \leq u_n \leq \lg n + \tfrac{1}{2}.$$

Therefore, the differences $|1 - u_n/\lg n|$ approach zero. In this sense, the characteristic largest value approaches $\lg n$, the value for the exponential distribution.

The lth moment $\overline{x^l}$ about origin is

$$\int_0^\infty \left\{\exp\left[-x - \frac{\sin x}{2}\right]\right\}$$
$$\times \left(1 + \frac{\cos x}{2}\right) x^l\, dx < \frac{3}{2}\int_0^\infty \left\{\exp\left[-x - \frac{\sin x}{2}\right]\right\} x^l\, dx.$$

Since

(4) $$\overline{x^l} < \frac{3}{2}\sqrt{e}\, l!,$$

all moments exist. The logarithmic derivative is

(5) $$-\frac{f'(x)}{f(x)} = \mu(x) + \frac{\sin x}{2 + \cos x}.$$

Therefore, the critical quotient does not converge toward unity. Although all moments exist, the distribution (2) does not belong to the exponential type, and the modal largest value does not converge to the characteristic largest value.

Exercise: Trace the distribution (2) and the exponential one on the same scale.

4.3.2 ANALYTICAL STUDY OF EXTREMES

4.3.2. Pareto's Distribution. In the previous distribution all moments exist yet L'Hôpital's Rule does not hold. Now consider a distribution lacking certain moments where this rule also does not hold. One form of Pareto's distribution is

(1) $\quad F(z) = 1 - z^{-k}; \quad f(z) = kz^{-k-1}; \quad \mu(z) = k/z;$
$\quad -f'(z)/f(z) = (k+1)/z; \quad T(z) = z^k; \quad z \geq 1; \quad k > 0.$

The variate is written z to distinguish the distribution from the exponential type. The distribution (1) may be obtained from the exponential distribution,

$$f(x) = ke^{-kx},$$

by the transformation $x = \lg z$. The critical quotient 4.1.3(3) is

(2) $\quad Q(z) = k/(k+1).$

Since k is a fixed value the distribution is not of the exponential type.

Formula (1) is used in economic statistics to reproduce income distributions. The variable z is the income expressed in units of its minimum, and has no dimension. The logarithm of the relative number of persons with an income greater than z, traced against the logarithm of z, is a straight line with slope k. The essence of this "law" is that there are more poor than rich people. This is not a natural law, but a social statement approximately valid for a capitalist society. It often gives a good fit, although the observed number of persons with small incomes generally falls short of the theoretical number. Pareto's distribution has no mode and the moments of order $l \geq k$ do not exist. The maximum likelihood estimate for $1/k$ is

(3) $\quad 1/\hat{k} = \overline{\lg z}.$

The sum of all incomes is not invariant in this estimate. Observations show that $\hat{k}$ surpasses unity. For $k > 1$ the income concentration decreases if k increases.

The probability and the distribution of the largest value,

(4) $\quad \Phi_n(z) = (1 - z^{-k})^n; \quad \varphi_n(z) = nk(1 - z^{-k})^{n-1} z^{-k-1},$

may also be obtained from the corresponding expressions for the exponential distribution by the logarithmic transformation. The distribution of the largest value spreads with increasing n. The median $\check{z}_n$, the characteristic largest value v_n, and the modal largest value $\tilde{z}_n$,

(5) $\quad \check{z}_n^k \sim n/\lg 2 > v_n^k = n > \tilde{z}_n^k = (nk+1)/(k+1),$

all decrease with increasing k.

In contrast to the exponential type, the most probable value $\tilde{z}_n$ does not converge toward the characteristic value v_n, and the occurrence interval is always smaller than the sample size.

Exercises: 1) Trace the distributions of the largest value for $n = 2, 3, 4, 5$ and different values of $k \geq 1$. 2) Calculate the modes and show that the consecutive distributions intersect for $k = 1$ at $n = 2, 3, 4, 5 \ldots$. 3) Restate the results for $k < 1$.

References: Muniruzzaman. *See also* E. J. Gumbel. Ein Masz der Konzentration, etc. Archiv f. Socialwissenschaft, 58: 113, 1927.

4.3.3. Definition of the Pareto and the Cauchy Types. In the Pareto distribution 4.3.2(1), the product of the intensity and the variate is constant. This property is taken as the starting point for the definition of a new type Let v_n be the characteristic largest value and let μ_n be the intensity function at $x = v_n$, then the distributions with which we are concerned are such that the limit of the characteristic product defined in 4.1.7 exists and

(1) $$\lim_{n=\infty} \mu_n v_n = k,$$

is a positive constant independent of n. The probability obtained after integration from 1.2.1(2) is such that

(2) $$\lim_{z=\infty} (1 - F(z))z^k = A \quad \text{where } A > 0, \quad \text{whence}$$

(3) $$\lim_{z=\infty} f(z) \cdot z^{k+1} = A \cdot k.$$

In the same way we impose upon the smallest value the condition

(4) $$\lim_{z=-\infty} F(z)(-z)^{k_1} = A_1.$$

If the initial distribution is symmetrical, $k_1 = k$ and $A_1 = A$. Limited distributions, for which one or the other condition holds, are said to be of the Pareto type, since equation (1) holds for all values z for Pareto's distribution, 4.3.2(1). Distributions which are unlimited in both directions and for which conditions (2) and (3) hold, are said to be of the Cauchy type. For the justification of the name, consider the Cauchy distribution

(5) $$f(z) = \frac{1}{\pi(1+z^2)}; \quad F(z) = \tfrac{1}{2} + \frac{\arctan z}{\pi},$$

which is symmetrical about the median zero. No moments exist. For large absolute values of z, the usual expansion leads to

(6) $$F(z) = \begin{cases} 1 - 1/\pi z + 0(z^{-2}); & z > 0 \\ -1/\pi z - 0(z^{-2}); & z < 0. \end{cases}$$

If the factors $0(z^{-2})$ are neglected, the parameters in (2) and (3) are

$$A_1 = A = \pi^{-1} \; ; \quad k_1 = k = 1 \; .$$

The conditions (2) and (3) are fulfilled. The characteristic largest value v_n, which is asymptotically

(7) $$v_n \sim n/\pi \; ,$$

is quickly approached. The distribution 4.3.1(2) does not belong to the Pareto type, since the characteristic product does not converge to a fixed limit. Therefore, there exist distributions unlimited toward the right which belong neither to the Pareto nor to the exponential type.

Exercise: Calculate the median and the mode of the largest value for Cauchy's distribution.

4.3.4. Extremal Properties. Since distributions of the Cauchy type are defined by an asymptotic property, we can derive only asymptotic statements about the extreme value. The most interesting difference between the exponential and the Cauchy types is the behavior of the characteristic extreme values as a function n. Equation 4.1.6(2), rewritten as

$$\frac{dv_n}{d \lg n} = \frac{1}{\mu_n} ,$$

leads asymptotically from the characteristic product 4.3.3(1) to

(1) $$\frac{d \lg v_n}{d \lg n} = 1/k \; .$$

Then, after integrating from 2 to n, with the same limitation as in 4.1.6,

(2) $$v_n \sim \check{z}(n/2)^{1/k} \; .$$

The characteristic largest value v_n increases asymptotically as a power of n. For distributions of the Cauchy type, a yardstick for the increase of the characteristic largest value is the number of observations itself, while, for the exponential type, the yardstick is the logarithm of n. Distributions without moments have longer tails than distributions with moments. Therefore, for the same number of observations the distributions of the Cauchy type must show larger extremes.

The density of probability at the characteristic largest value is

(3) $$\varphi_n(v_n) = nF^{n-1}(v_n)f(v_n) \; .$$

From the definition of the characteristic largest values and of the characteristic product 4.3.3(1) we obtain asymptotically

(4) $$\varphi(v_n) \sim k/ev_n \; .$$

Since the characteristic largest value increases with n, its density of probability diminishes with n. The distributions of the largest values are shifted to the right and expand at the same time with n. The larger the number of observations, the less precise is the largest value. In this respect all distributions of the Pareto type are analogous to the third class of the exponential type. However, the three classes have an analogy within the Pareto type in the following sense: If $k = 1$, the expected largest value is proportional to n. This corresponds to the exponential distribution. If $k > 1$ ($k < 1$), the expected largest value traced as a function of n turns its concave (convex) side toward the horizontal axis. From (1) it follows that

$$\frac{d^2 v_n}{dn^2} = \frac{1}{k} \frac{v_n}{n^2} \left(\frac{1}{k} - 1\right).$$

Consequently,

(5) $$\frac{d^2 v_n}{dn^2} \gtrless 0 \quad \text{if} \quad k \gtrless 1.$$

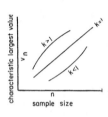

Graph 4.3.4. The Three Classes for the Cauchy Type

In this sense the cases $k > 1$ ($k < 1$) correspond to the first (third) class of the exponential type. See the schematic Graph 4.3.4.

However, the distributions of the first and third class of the exponential type tend to the second class for $n \to \infty$, while the classification in the Pareto type is independent of the sample size.

4.3.5. Other Distributions without Moments.

It was shown in 4.3.1 that not all distributions possessing all moments are of the exponential type. We now show that not all distributions without moments are of Pareto's type. For the proof, consider

$$F(z) = 1 - 1/\lg z\,; \quad f(z) = 1/(z \lg^2 z)\,; \quad \mu(z) = 1/(z \lg z)\,; \quad z \geqq e$$

The intensity function decreases monotonically toward zero. The distribution does not belong to the exponential type. The logarithmic derivative is

$$-\frac{d \lg f(z)}{dz} = \frac{1}{z} + \frac{2}{z \lg z}.$$

Therefore, the critical quotient converges toward zero instead of unity. No moment exists; the mean diverges. The characteristic largest value v_n is

$$v_n = e^n.$$

The characteristic product,

$$\mu_n v_n = 1/n,$$

converges toward zero. According to 4.3.3(1) the distribution is not of Pareto's type.

4.3.6. Summary. Since the distribution of the smallest value is related to the distribution of the largest value, it is, in general, sufficient to consider the latter. Those unlimited distributions for which the intensity function converges, for x increasing, toward the logarithmic derivative with negative sign, are said to be of the exponential type. All moments exist. However, if the moments exist, the distribution need not be of the exponential type. Those unlimited distributions for which the product of the variate by the intensity function converges toward a finite value are said to be of Pareto's type. Then the higher moments diverge. However, not all distributions for which the higher moments diverge are of Pareto's type. Within the two types, there exist three classes, and outside the two types there exist other unlimited distributions.

To characterize an extreme value of an unlimited variate, we choose the characteristic extreme. The mode has several advantages over the mean, since it is obtained by differentiation, whereas the mean has to be calculated from an integration which cannot always be evaluated in closed form. For distributions of the exponential type, the mode converges toward the characteristic largest value, and the return period of the modal largest value, called the occurrence interval, converges toward the number of observations. This does not hold for return periods corresponding to the mean or the median largest value.

For distributions of the Pareto type, the mode does not converge to the characteristic largest value, and the occurrence interval does not converge to the sample size. For distributions of the exponential type, the yardstick for the increase of the largest value is the logarithm of the sample size. For distributions of the Cauchy type, it is a power of the sample size. Both types are unlimited with respect to the variate in which we are interested.

A third type consists of those initial variates which are limited either towards the right, if we are interested in the largest value, or towards the left, if we are interested in the smallest value. In this case the extremes are also limited. In the next chapters we will prove the existence, and show the general properties, of the asymptotic distributions of the extreme values for the three types.

Chapter Five: THE FIRST ASYMPTOTIC DISTRIBUTION

"Sint ut sunt aut non sint."
(Accept them as they are or
deny their existence.)

5.1. THE THREE ASYMPTOTES

5.1.0. Introduction. In the preceding chapter, we studied the exact distribution of the extremes as functions of the sample size for two types of initial distributions. Now it will be shown that three asymptotic distributions exist. Each assumes a specific behavior for absolute large values of the variate. The main part of the initial distribution plays a minor role; only the parameters in the asymptotic distributions depend on it. The three asymptotic distributions of largest values are linked by the symmetry principle 3.1.1(6) to asymptotic distributions of smallest values.

An asymptotic distribution is henceforth called an *asymptote*. The first asymptote holds for initial distributions of the exponential type (4.1.4); the second holds for Cauchy's type (4.3.3). In both cases the initial variates are unlimited, either to the right (for the largest value), or to the left (for the smallest value), or in both directions. The third asymptote of largest (smallest) values holds for distributions which are bounded toward the right (left). As shown in 4.3.1 and 4.3.5, the three types of initial distributions are not exhaustive. In addition, an initial distribution belonging to one type for one extreme may belong to another type, or to no type, for the other extreme.

5.1.1. Preliminary Derivation. Table 5.1.1. shows how the asymptotic distributions of the largest value are obtained for the two prototypes, namely, the exponential and Pareto's distributions (4.1.1) and (4.3.2) and a limited distribution defined in the third column.

The characteristic largest value, Table 5.1.1, line 4, is introduced into the probability of the largest value, line 5. This leads to the asymptotic probabilities in line 6. These derivations serve only to introduce the three asymptotes in a preliminary and simple way, and do not assert mathematical rigor. The first asymptote contains only two parameters, α and u_n,

5.1.2 FIRST ASYMPTOTIC DISTRIBUTION

Table 5.1.1. *The Three Asymptotic Probabilities of Largest Values*

	Exponential	Pareto	Limited
1) Initial Distribution			
2) Conditions	$x \geq 0; \alpha > 0$	$z > 1 + \varepsilon; k \geq 1$	$\omega - 1 < x \leq \omega; k > 1$
Bound	0	$\varepsilon \geq 0$	$\omega > 0$
3) Initial Probability	$1 - e^{-\alpha x}$	$1 - (z - \varepsilon)^{-k}$	$1 - (\omega - x)^k$
4) Characteristic Largest Value	$1/n\ e^{\alpha u_n} = 1$	$n(v_n - \varepsilon)^{-k} = 1$	$n(\omega - u_n)^k = 1$
5) Probability of Largest Value	$(1 - e^{-\alpha x})^n$	$[1 - (z-\varepsilon)^{-k}]^n$	$[1 - (\omega - x)^k]^n$
$x = x_n; z = z_n$	$[1 - 1/n\ e^{-\alpha(x-u_n)}]^n$	$\left[1 - 1/n\left(\dfrac{z-\varepsilon}{v_n - \varepsilon}\right)^{-k}\right]^n$	$\left[1 - 1/n\left(\dfrac{\omega - x}{\omega - u_n}\right)^k\right]^n$
6) Asymptotic Probability $n = \infty$	$\exp[-e^{-\alpha(x-u_n)}]$	$\exp\left[-\left(\dfrac{z-\varepsilon}{v_n-\varepsilon}\right)^{-k}\right]$	$\exp\left[-\left(\dfrac{\omega-x}{\omega-u_n}\right)^k\right]$

studied in 3.1.4, while the second (third) asymptote contains, besides k and v_n (or u_n), the lower (upper) bound of the initial variate.

The first two initial probabilities (line 3) and the two asymptotic probabilities (line 6) of the largest value are linked by the common transformation,

(1) $x = \lg(z - \varepsilon)$; $u_n = \lg(v_n - \varepsilon)$; $\alpha = k$.

The second and third initial probabilities and the corresponding two asymptotic probabilities of the largest value are linked by the common transformation

(2) $z = -x$; $\varepsilon = -\omega$; $v_n = -u_n$

and a change in the sign of the parameter k. These relations simplify considerably the analysis of the asymptotes.

In the preceding shortcut based on *specific* initial distributions, α and k are parameters of the initial distribution, and therefore independent of n. In the general derivations to be shown later for three *types* of initial distributions, α and k may depend upon n.

Exercise: Use Cramér's transformation 3.2.4(2) to obtain the asymptotic distribution of the largest value for the exponential and Laplace's distribution. (See Cramér (1948), p. 373.)

References: Epstein (1948), Finetti, Leme, Smirnov (1933, 1935, 1952).

5.1.2. The Stability Postulate. We now show the Fisher and Tippett derivation of the three asymptotes. Consider N samples, each of size n, taken from the same population. In each sample there is a largest value and the largest value in the Nn observations is the largest of the N largest values taken from samples of size n. The distribution of the largest value in Nn observations will tend to the same asymptotic expression as the distribution of the largest value in samples of size n, provided that such an asymptote exists. Consequently, the asymptote must be such that the largest value of a sample of size n taken from it must have the same asymptotic distribution.

Since a linear transformation does not change the form of the distribution, the probability that the largest value is below x should be equal to the probability of a linear function of x,

(1) $$F^n(x) = F(a_n x + b_n),$$

the two parameters a_n and b_n being functions of n. Equation (1) is called the Stability Postulate. It was first used by Fréchet (1927), who derived the second asymptote. Since the fundamental paper of Fisher and Tippett is difficult to read, their deviation will be explained in detail.

1) If a_n is unity, the selection of the largest value merely shifts the initial probability by the amount b_n. The curve $F^n(x)$ traced as a function of x is parallel to the initial curve $F(x)$. The equation

(2) $$F^n(x) = F(x + b_n)$$

contains two unknown functions, namely, the parameter b_n considered as a function of n, and the probability $F(x)$. First we determine b_n. Taking logarithms twice,

(3) $$\lg n + \lg(-\lg F(x)) = \lg(-\lg F(x + b_n)).$$

Thus $\lg(-\lg F(x))$ increases by $\lg n$ if x increases by b_n. In other words,

(3′) $$\lg(-\lg F(x)) - \frac{x \lg n}{b_n} = \text{Constant},$$

where the constant, say k, is independent of x. Equation (3′) is immediately verified from (3).

We are interested only in nonperiodic solutions of (3). To determine b_n as a function of n, we write, from (2),

$$F^{nm}(x) = F(x + b_{nm}).$$

The same result must be obtained by taking the mth power of (2). Since this procedure is the same as the addition of a factor b_m,

$$[F^n(x)]^m = F(x + b_n + b_m) = F^{nm}(x).$$

The two preceding equations combined give a functional equation for b_n alone, namely,

(4) $$b_{nm} = b_n + b_m.$$

Its solution is $$b_n = c \lg n + d,$$

where c and d are constants. Substitution into (4) leads to $d = 0$. Consequently, the factor b_n as a function of n is

(4') $$b_n = c \lg n .$$

This solves one of the functional equations contained in (3). We have still to determine the probability function $F(x)$. Introduction of (4') into (3') leads to

$$\lg [-\lg F(x)] = x/c + k .$$

Now, for x increasing, $-\lg F(x)$ decreases toward zero. Hence the factor c must be negative. We write

(5) $$c = -1/\alpha ; \quad \alpha > 0 ; \quad ck = -u ,$$

and obtain

$$-\lg F(x) = e^{-\alpha(x-u)} .$$

The *first initial probability* which is stable with respect to the largest value is thus equal to the *first asymptote*,

(I) $$F(x) = \exp [-e^{-\alpha(x-u)}] ,$$

where α and u are positive parameters independent of n. It is immediately verified that the functional equation (2) is satisfied, provided that

$$b_n = -(\lg n)/\alpha,$$

which is the combination of equations (4') and (5).

The probability (I) of the largest value is equal to the initial probability shifted toward the right by the amount $(\lg n)/\alpha$ without change in shape. In a later derivation (5.2.1) α and u are interpreted as parameters in the distribution of the largest value, become functions of n, and depend also upon the initial distribution. This leads to a change in shape, if we pass from the initial distribution to the distribution of the largest value.

2) If the factor a_n in (1) differs from unity, the two curves $F^n(x)$ and $F(x)$ traced against x are no longer parallel. There is a value x' where the two probabilities are equal. This value is

$$x_n' = a_n x' + b_n ; \quad x' = b_n/(1 - a_n) .$$

Now the equation

$$F^n(x') = F(x')$$

can be satisfied if and only if

(6) $$F(x') = 0 \quad \text{or} \quad F(x') = 1 .$$

In the first (second) case, the distribution starts (ends) with x' and is bounded to the left (right). This leads to the second (third) asymptote, where the variate is bounded by $x \geq x'$ ($x \leq x'$). If we shift the origin of the variate, we may put $x' = 0$, hence $b_n = 0$.

The second and third asymptotic distributions of the largest values are now derived from the assumption $b_n = 0$, i.e., from the functional equation

(7) $$F^n(x) = F(a_n x)$$

and one of the two conditions (6) respectively. Here the scale of the variate is changed if we pass from the initial distribution to the distribution of the largest value. As before, we have to find two functions, namely, the parameter a_n as a function of n, and the initial probability $F(x)$. We start again by determining the parameter. Raising F to the mth power is the same as multiplying x by a_m. Therefore,

$$[F^n(x)]^m = F^{nm}(x) = F(a_n a_m x) .$$

On the other hand, from (7),

$$F^{nm}(x) = F(a_{nm} x) .$$

Therefore, the parameter a_n is subject to the functional equation,

(8) $$a_{n \cdot m} = a_n \cdot a_m .$$

The solution,

(9) $$a_n = n^l ,$$

where l is a constant, is immediately verified. The sign of l will differ for the two asymptotes. To obtain the probability function $F(x)$, we take the logarithm of equation (7) and introduce (9), which leads to

$$\lg n + \lg [-\lg F(x)] = \lg [-\lg F(n^l x)] .$$

This equation may be interpreted from (9) in the following way: if $\lg x$ is increased by $\lg a_n$, then $\lg (-\lg F(x))$ is increased by $\lg n$ so that, excluding, as before, periodic solutions,

(10) $$\lg [-\lg F(x)] - (\lg x)/l = \text{constant} .$$

This equation is easily verified. Using (8), equation (10) can be changed into

$$\lg [-\lg F(x)] = (\lg x - \lg v)/l ,$$

where v is a transformation of the constant in (10). Consequently,

(11) $$-\lg F(x) = (x/v)^{1/l} ; \quad F(x) = \exp [-(x/v)^{1/l}] .$$

5.1.2 FIRST ASYMPTOTIC DISTRIBUTION

To obtain the two asymptotes from (11), we introduce the boundary conditions. If $F(0) = 0$, i.e., if the variates x is non-negative, the factor l is negative, and

(12) $$1/l = -k \; ; \quad k > 0 .$$

The *second* initial *probability* which is stable with respect to the largest value is thus equal to the *second asymptote*,

(II) $\quad F(x) = \exp\left[-\left(\dfrac{v}{x}\right)^k\right] ; \quad x \geqq 0 ; \quad v > 0 ; \quad k > 0 .$

The functional equation (7) leads to

$$F^n(x) = \exp\left[-n\left(\frac{v}{x}\right)^k\right]$$
$$= \exp\left[-\left(\frac{v}{xn^{-1/k}}\right)^k\right].$$

The effect of taking the largest value is to *increase* the scale of the variate by the multiplicatice factor $n^{1/k}$ while the origin zero remains unchanged. The interpretation

(13) $$a_n = n^{-1/k}$$

is obtained from equations (9) and (10).

3) If $F(0) = 1$, i.e., if the variate is non-positive, the factor in (11) is positive, and

$$1/l = k > 0 .$$

Then the *third initial probability* function which is stable with respect to the largest value is the *third asymptote*

(III) $\quad F(x) = \exp\left[-(x/v)^k\right] ; \quad x \leqq 0 ; \quad v < 0 ; \quad k > 0 .$

The functional equation (7) leads to

(14) $$F^n(x) = \exp\left[-\left(\frac{xn^{1/k}}{v}\right)^k\right],$$

or to the interpretation

(15) $$a_n = n^{1/k} .$$

The effect of taking the largest value is to *decrease* the scale of the variate by the multiplicative factor $n^{-1/k}$, while the end $x = 0$ remains unchanged. The second and the third types differ only by the signs of the variate and of the parameter k. If the reduced initial variate is written $-z$, the probability of the largest value becomes

(III′) $\quad F(z) = \exp -[-z^k] ; \quad z \leqq 0 ; \quad k > 0 .$

From the deviations of I, II, and III, it follows that a linear function of the variate in the three extremal distributions is also an extreme value.

Problems: 1) Do other nonperiodic functions $F(x)$ exist which satisfy the functional equation (1)? (For solution see 7.1.6.) 2) What are the asymptotic distributions of the largest values taken from bivariate (and multivariate) initial distributions? (See Finkelstein.)

References: Cramér (p. 373), Gnedenko (1941, 1943), Jenkinson, Kendall (1946, Vol. 1, p. 218).

5.1.3. Outline of Other Derivations. In the stability postulate 5.1.2(1) no assumption is made about the sample size. It may be as small as $n = 2$, and we look for an *initial* distribution defined by a functional equation. In the following, we start from known classes of initial distributions, calculate the distribution of the largest values, and prove that, for *large samples*, the distributions of the extremes are given by certain formulae.

In Table 5.1.3(1), which is arranged in the same way as Table 5.1.1, we state three asymptotic probabilities valid for large and small values of the variate and interpret the parameters contained in the formulae. The formulae 3.1.1(1) and 3.1.1(3) lead to the corresponding asymptotic probabilities of the extremes given in Table 5.1.3.(2), where the three asymptotic probabilities of largest values are designated by upper indices (1), (2), (3), and the corresponding probabilities of smallest values by a lower index (1).

Table 5.1.3(1). *Asymptotic Probabilities of* Initial *Variates*

Initial Type	Largest Value	Smallest Value	Symmetry
1) Exponential	$F(x_n) = 1 - \dfrac{e^{-\alpha_n(x_n - u_n)}}{n}$	$F(x_1) = \dfrac{1}{n} e^{\alpha_1(x_1 - u_1)}$	$\alpha_n = \alpha_1 > 0$
Interpretation of Parameters	$F(u_n) = 1 - 1/n$ $F(u_n - 1/\alpha_n) = 1 - e/n$	$F(u_1) = 1/n$ $F(u_1 + 1/\alpha_1) = e/n$	$u_n = -u_1 > 0$
2) Cauchy	$F(z_n) = 1 - \dfrac{1}{n}\left(\dfrac{v_n - \varepsilon}{z_n - \varepsilon}\right)^{k_n}$	$F(z_1) = \dfrac{1}{n}\left(\dfrac{\omega - v_1}{\omega - z_1}\right)^{k_1}$	$k_n = k_1 > 0$
Interpretation of Parameters	$\varepsilon < z_n$ $F(v_n) = 1 - 1/n$	$\omega > z_1$ $F(v_1) = 1/n$	$\omega = -\varepsilon > 0$ $v_n = -v_1 > 0$
3) Limited	$F(x_n) = 1 - \dfrac{1}{n}\left(\dfrac{\omega - x_n}{\omega - v_n}\right)^{k_n}$	$F(x_1) = \dfrac{1}{n}\left(\dfrac{x_1 - \varepsilon}{v_1 - \varepsilon}\right)^{k_1}$	$k_n = k_1 > 0$
Interpretation of Parameters	$\omega \geq x_n$ $F(v_n) = 1 - 1/n$ $F(\omega) = 1$	$\varepsilon \geq x_1$ $F(v_1) = 1/n$ $F(\varepsilon) = 0$	$\omega = -\varepsilon$ $v_n = -v_1 > 0$

5.1.3 FIRST ASYMPTOTIC DISTRIBUTION

Table 5.1.3(2) summarizes the asymptotic theory of extreme values. The first line is derived in 5.2 and analyzed in Chapters 5 and 6; the second and third are derived and analyzed in Chapter 7. The parameters α are positive, and the formulae are normalized in such a way that all probabilities are equal to $1/e$ for $x = u_n$ or v_n respectively.

The initial probabilities given in Table 5.1.3(1) are sufficient conditions for the existence of these asymptotic distributions. B. Gnedenko (1943) has shown the necessary and sufficient conditions for the existence of the first, second, and third asymptotic distributions of largest values. They are in our notations for reduced variates y and x:

(1)   for I: $\lim_{n=\infty} n[1 - F(u_n + y/\alpha_n)] = e^{-y}$,

where $F(u_n) = 1 - 1/n$; $F(u_n + 1/\alpha_n) = 1 - 1/ne$; $\alpha_n > 0$.

(2)   for II: $\lim_{x=0} \dfrac{1 - F(x)}{1 - F(cx)} = c^k$; $c > 0$; $k > 0$.

(3)   for III: $\lim_{x=-0} \dfrac{1 - F(cx + \omega)}{1 - F(x + \omega)} = c^k$,

where $F(\omega) = 1$; $F(\omega - \epsilon) < 1$; $\varepsilon > 0, c > 0, k > 0$.

Gnedenko designates the asymptotic probability (II) corresponding to the Cauchy type by $\Phi(x)$ and calls it No. 1. The asymptotic probability (III) corresponding to the limited type is designated by $\Psi(x)$ and called No. 2. Finally, the asymptotic probability (I) corresponding to the exponential type designated by $\Lambda(x)$ is called No. 3. In addition, Gnedenko has shown that the three asymptotic distributions are the only ones which fulfill the stability postulate. Lack of space prevents us from reproducing his important proof, which is easily available.

Jenkinson writes the three asymptotes in a common form

(4) $\Phi(x) = \exp[-(1 - k\alpha(x - u))^{1/k}]$; $\alpha > 0$,

which fulfills the stability postulate. If $k > 0$ ($k < 0$) there is an upper limit ω (lower limit ϵ), with $\omega, \epsilon = u + 1/k\alpha$ and we reach the third (second) asymptote, while $k \to 0$ leads to the first one. In general, the transformation from x to y is

(5) $x = u + (1 - e^{-yk})/k\alpha$.

Exercises: 1) Prove that Gnedenko's condition (I) holds a) for the exponential distribution b) for the logistic distribution c) for the third asymptotic distribution of smallest values d) for the distribution 1.1.9(17). 2) Determine α_n and u_n and check the boundary conditions.

Table 5.1.3(2). Asymptotic Probabilities

Initial Type	Largest Value	Conditions
1) Exponential	$\Phi^{(1)}(x) = \exp[-e^{-\alpha_n(x-u_n)}]$	$\alpha_n > 0$
Interpretation of Parameters	$\Phi^{(1)}(u_n) = 1/e$ $\Phi^{(1)}(u_n - 1/\alpha_n)$	$= 0.36788$ $= 0.06599$
2) Cauchy	$\Phi^{(2)}(x) = \exp\left[-\left(\dfrac{v_n - \varepsilon}{x - \varepsilon}\right)^{k_n}\right]$	$k_n > 0; x \geqq \varepsilon$ $v_n > \varepsilon \geqq 0$
Interpretation of Parameters	$\Phi^{(2)}(\varepsilon)$ $\Phi^{(2)}(v_n)$	$= 0$ $= 1/e = 0.36788$
3) bounded	$\Phi^{(3)}(x) = \exp\left[-\left(\dfrac{\omega - x}{\omega - v_n}\right)^{k_n}\right]$	$x \leqq \omega; v_n < \omega$ $k_n > 0$
Interpretation of Parameters	$\Phi^{(3)}(\omega)$ $\Phi^{(3)}(v_n)$	$= 1$ $= 1/e = 0.36788$

Problems: 1) Prove that Gnedenko's condition (I) holds a) for the normal, b) for the Gamma distribution.

References: Gnedenko (1953), Meizler (1949, 1955), Smirnov (1935, 1941), Wilks (1948). *Consult also:* J. Geffroy (1957). Etude des diverses majorations asymptotiques des valeurs extrèmes d'un échantillon, Compt. Rend. Ac. Sc., 244: 1712; Sur la stabilité en probabilité des valeurs extrèmes d'un échantillon, Compt. Rend. Ac. Sc., 245: 1215.

5.1.4. Interdependence. It was pointed out in 3.1.1 that the exact distribution of extreme values holds only for independent observations. For the asymptotic distribution of extremes, the initial distribution to be used for interdependent observations is very complicated. However, the distribution of extreme values depends only on the properties of the initial distribution for large values of the variate where the influence of interdependence may vanish. Therefore, the asymptotic distribution of extremes may still be valid for interdependent observations. This idea has been put in a mathematical form by G. S. Watson (1954).

A sequence of random variable x_i is called m-dependent if $|i - j| > m$ implies that x_i and x_j are independent. If the variables have a finite upper bound, the largest among n observations tends with probability one to this bound. Watson shows that if the variables are unlimited, the asymptotic distribution of the largest value is the same as in the case of independence. For the proof, he assumes that the probability $P(x > c)$ of a value larger than c is such that

5.1.4 FIRST ASYMPTOTIC DISTRIBUTION

of Extreme Values

Smallest Value	Conditions	Symmetry
$1 - {}_1\Phi^{(1)}(x) = \exp[-e^{\alpha_1(x-u_1)}]$	$\alpha_1 > 0$	$\alpha_1 = \alpha_n;\ u_1 = -u_n$
$\quad = 1 - {}_1\Phi^{(1)}(u_1)$		
$\quad = 1 - {}_1\Phi^{(1)}(u_1 + 1/\alpha_1)$		
$1 - {}_1\Phi^{(2)}(x) = \exp\left[-\left(\dfrac{\omega - v_1}{\omega - x}\right)^{k_1}\right]$	$k_1 > 0;\ x \leqq \omega$ $v_1 < \omega$	$v_1 = -v_n$ $k_1 = k_n$ $\omega = -\varepsilon$
$\quad = 1 - {}_1\Phi^{(2)}(\omega)$		
$\quad = 1 - {}_1\Phi^{(2)}(v_1)$		
$1 - {}_1\Phi^{(3)}(x) = \exp\left[-\left(\dfrac{x - \varepsilon}{v_1 - \varepsilon}\right)^{k_1}\right]$	$x \geqq \varepsilon, k_1 > 0$ $v_1 > \varepsilon$	$v_1 = -v_n$ $k_1 = k_n$ $\omega = -\varepsilon$
$\quad = 1 - {}_1\Phi^{(3)}(\varepsilon)$		
$\quad = 1 - {}_1\Phi^{(3)}(v_1)$		

(1) $$\lim_{c=\infty} \frac{\max\limits_{|i-j|\leqq m} P(x_i > c, x_j > c)}{P(x_i > c)} = 0,$$

and writes for ξ fixed

(2) $$\xi = nP[x_i > c_n(\xi)].$$

Then the formula for the probabilities of the joint occurrence of a set of events in terms of probabilities of occurrence of their contraries leads to

$$\sum_{q=0}^{l-1} \frac{(-\xi)^q}{q!} \leqq \lim_{n=\infty} P[x_i \leqq c_n(\xi)] \leqq \sum_{q=0}^{l} \frac{(-\xi)^q}{q!}$$
$$i = 1, 2, \ldots, n;\quad l \leqq n.$$

Thus we reach the approximation 3.1.1(1') in the form

(3) $$\lim_{n=\infty} P[x_i \leqq c_n(\xi)] = e^{-\xi};\quad i = 1, 2, \ldots, n.$$

Watson shows then that the condition (1) in the form

(4) $$\lim_{c=\infty} \frac{P(x > c,\ y > c)}{P(x > c)} = 0$$

holds for a bivariate normal distribution with mean zero, variance unity, and covariance ρ.

The Gamma distribution (4.2.5) is of the exponential type. Consequently, the distribution of its largest value converges to the first asymptote. Since the mean of a Gamma distribution is again subject to a Gamma distribution, the first asymptotic distribution holds also for the largest means, provided the observations are independent. However, Gurland (1955) has shown that this remains valid for the largest means of uncorrelated and, what is more, for positively correlated observations taken from a multidimensional Gamma distribution. Again the independence is less important for the theory of extreme values than it seemed at first sight.

Exercises: 1) Calculate the distribution of the largest value of the sum of n independent observations taken from an exponential and a Gamma distribution.

Problems: 1) Calculate the distribution of the largest value of the sum of n independent observations taken from a logarithmic normal distribution. 2) What is the relation between Watson's condition and Gnedenko's conditions 5.1.3(1) for the existence of the first two asymptotic distributions of largest values? 3) How should the three types of initial distributions be defined for bivariate and multivariate distributions?

References: Darling (1956), Wald (1947).

5.2. THE DOUBLE EXPONENTIAL DISTRIBUTION

5.2.0. Introduction. In the following, we prove that the first asymptote holds for the largest value of the exponential type and start from the initial probability of the extreme mth value. Then we study the asymptotic probability of the extreme value and compare its distribution to the normal and logarithmic normal distributions. The asymptotic distribution of the mth extreme will be considered in the section 5.3.

5.2.1. Derivations. To obtain the asymptotic distribution of the largest value for the exponential type, we expand the initial probability function for large values of the variate about the characteristic mth largest value u_m defined in 3.1.4(6), where m is now counted from the top.

(1) $$F(x) = F(u_m) + \frac{x-u}{1!}f_m + \frac{(x-u)^2}{2!}f'_m + \cdots$$
$$+ \frac{(x-u_m)^v}{v!}f_m^{(v-1)} + \cdots,$$

where
(2) $\qquad F(u_m) = 1 - m/n\ ; \qquad f_m = f(u_m)\ ; \qquad u = u_m.$

5.2.1 FIRST ASYMPTOTIC DISTRIBUTION

Multiplication and division of (1) by m/n leads to

(3) $\quad F(x) = 1 - \dfrac{m}{n} + \dfrac{m}{n}(x - u_m)\left(\dfrac{n}{m}f_m\right)$

$\qquad + \dfrac{m}{n}\dfrac{(x-u_m)^2}{2!}\left(\dfrac{n}{m}f'_m\right) + \dfrac{m}{n}\dfrac{(x-u_m)^v}{v!}\dfrac{n}{m}f_m^{(v-1)} + \cdots,$

or

(4) $\quad F(x) = 1 - \dfrac{m}{n}\left[1 - \sum_1^\infty \dfrac{(x-u_m)^v}{v!}\dfrac{n}{m}f_m^{(v-1)}\right].$

To obtain the successive derivatives $f_m^{(v-1)}$, we use L'Hôpital's Rule 4.1.3(2) which becomes, from (2),

(5) $\quad \dfrac{-f'_m}{f_m} = \dfrac{n}{m}f_m.$

We write

(6) $\quad \dfrac{n}{m}f_m = \alpha_m,$

which is a function of m and n. This designation is, of course, a generalization of the intensity function α_n defined in 3.1.5(1). It may also be written in the form

(6') $\quad \alpha_m = \dfrac{f(u_m)}{1 - F(u_m)}.$

Equation (5) may be written, from (6), $f'_m = -\alpha_m f_m$. From (6), we obtain the first derivative in the expansion (4),

(7) $\quad \dfrac{n}{m}f'_m = -\alpha_m^2.$

In the same way we obtain the higher derivatives. From (5) we see that

$$f'_m = -\dfrac{f_m^2}{1 - F_m}.$$

With the same approximation as previously used further differentiation gives

$$\dfrac{f''_m}{f_m} = \dfrac{2f_m f'_m}{-(1-F_m)f_m} - \dfrac{f_m^2}{(1-F_m)^2} = 2\alpha_m^2 - \alpha_m^2,$$

whence

(8) $\quad \dfrac{n}{m}f''_m = \alpha_m^3.$

168 FIRST ASYMPTOTIC DISTRIBUTION 5.2.1

It will now be shown that, if

$$\frac{n}{m} f_m^{(v)} = (-1)^v \alpha_m^{v+1}$$

holds for a certain v, it also holds for $v + 1$. For the proof, we start from

$$f_m^{(v)} = (-1)^v \alpha_m^v \frac{m}{n} \alpha_m = (-1)^v \alpha_m^v f_m = (-1)^v \frac{f_m^{(v+1)}}{(1-F_m)^v}.$$

For the next derivative we have

$$\frac{f_m^{(v+1)}}{f_m} = \frac{(v+1)(-1)^v f_m^v f'_m}{(1-F_m)^v f_m} + \frac{(-1)^v f_m^{v+1} v f_m}{(1-F_m)^{v+1} f_m}$$

$$= (v+1)(-1)^{v+1} \alpha_m^{v+1} - (-1)^{v+1} v \alpha_m^{v+1}$$

$$= (-1)^{v+1} \alpha_m^{v+1},$$

whence

(9) $$\frac{n}{m} f_m^{(v+1)} = (-1)^{v+1} \alpha_m^{v+2},$$

which proves the statement. If we write $v - 1$ instead of $v + 1$ and introduce

(9') $$\frac{n}{m} f_m^{(v-1)} = (-1)^{v-1} \alpha_m^v$$

into (4), the expansion of the initial probability $F(x)$ in the neighbourhood of the characteristic mth largest value is

(4') $$F(x) = 1 - \frac{m}{n} \left[1 + \sum_{1}^{\infty} \frac{(x - u_m)^v}{v!} (-1)^v \alpha_m^v \right].$$

We introduce the *reduced* mth *largest value* y_m, defined by

(10) $$y_m = \alpha_m (x - u_m),$$

as the difference of the mth largest value from the characteristic mth largest value multiplied by a factor which has the dimension x^{-1}. The transformation (10) reduces the largest value in size and changes the scale so that y_m has no dimension. Then the probability function (4') becomes.

(11) $$F(x) = 1 - \frac{m}{n} e^{-y_m}.$$

This approximation holds for any initial probability function of the exponential type and for large x. The distribution $f(x)$ becomes, under the same conditions,

(11') $$f(x) = \frac{m}{n} \alpha_m e^{-y_m}.$$

5.2.1　FIRST ASYMPTOTIC DISTRIBUTION

The initial probability $F(x)$ and the initial distribution $f(x)$ in the neighborhood of the characteristic largest value tend for $m = 1$ to

(12)　　$F(x) = 1 - e^{-y}/n; \quad f(x) = \alpha_n e^{-y}/n; \quad y = \alpha_n(x - u_n).$

This asymptotic expression stated in Table 5.1.3(1) justifies the term "exponential type" in 4.1.4. It also shows that these distributions possess all moments. Formula (12) leads immediately to the first asymptotic probability,

$$\lim_{n=\infty} \Phi_n(x) = \lim_{n\to\infty} (1 - e^{-y}/n)^n,$$

which is written, as in Table 5.1.3(2),

(I) (13)　$\Phi(x) = \exp(-e^{-y}); \quad \varphi(x) = \alpha_n \exp(-y - e^{-y});$
$$y = \alpha_n(x - u_n)$$

The probability and the distribution of the reduced extremes are, of course, parameter free. Gnedenko has shown that the condition (12) is not only sufficient, but necessary, for the existence of the asymptotic distribution (13).

The probability $\Phi(x)$ and the distribution $\varphi(x)$ as function of y, the inverse function $y = -\lg(-\lg \Phi)$, and the distribution

(14)　　　　　　　　　$\varphi = -\Phi \lg \Phi$

as function of $\Phi(x)$ have been tabulated by the National Bureau of Standards. These tables are indispensable for all uses of extreme values.

The asymptotic distribution of the smallest value is obtained from the symmetry principle 3.1.1(6). Assume first that the initial distribution is symmetrical and unlimited in both directions. Then the asymptotic distribution, $_1\varphi(x)$, of the smallest value x_1, is obtained from (I) by writing $-x$ and $-u$ instead of x and u respectively. Hence,

(15)　$_1\Phi(x) - 1 - \exp(-e^y); \quad _1\varphi(x) = \alpha_n \exp(y - e^y);$
$$y = \alpha_n(x - u_n).$$

If the initial distribution is asymmetrical, we have to write u_1 and α_1 instead of $-u_n$ and α_n, and hence obtain the same probability and distribution of the reduced value.

The largest (smallest) reduced value is not confined to the positive (negative) domain. For the sake of simplicity, formulae (I) and (15) are called, respectively, the *first* and the *second double exponential distributions*. If we write in (13)

(16)　　　　　　　　　$e^{\alpha_n u_n} = \lambda_n,$

the asymptotic probability and distribution are

(17) $\quad \Phi(x) = \exp(-\lambda_n e^{-\alpha_n x})\ ; \qquad \varphi(x) = \alpha_n \lambda_n \exp(-\alpha_n x - \lambda_n e^{-\alpha_n x})\ .$

From 4.1.7(2) the parameter λ_n increases with n. Consequently the probability for the largest value to be negative,

(18) $$\Phi(0) = e^{\lambda_n}\ ,$$

decreases with increasing n. This is also the probability that the smallest value is positive. For example, for $\alpha_n u_n = 3$ the probability $\Phi(0)$ is of the order $2\ 10^{-9}$. We do not require here that x be counted from the median as in 3.1.1.

Exercises: 1) Construct the asymptotic distribution of the smallest value by the corresponding Taylor expansion (Gumbel, 1935). 2) Verify the stability postulate for the distribution of the smallest value. 3) Calculate the distribution of the range for the double exponential distribution. For $n = 2$ the right side of the logistic distribution is obtained. 4) Prove 4.1.7(2) from (12).

Reference: Gumbel (1935).

5.2.2. The Methods of Cramér and Von Mises. A procedure used by Cramér leads to a very simple derivation of the asymptotic probability 5.2.1(13) if the initial probability, at least for large values of x, is of the form

(1) $$F(x) = 1 - e^{-h(x)}\ ,$$

where $h(x)$ increases with x monotonically and without limit. If we introduce the characteristic largest value u_n being the solution of

$$e^{h(u_n)} = n\ ,$$

the initial probability may be written

$$F(x) = 1 - \exp\left[-(h(x) - h(u_n))\right]/n\ .$$

The asymptotic probability $\Phi(x)$ of the largest value becomes for large n

$$\Phi(x) = \exp\left\{-e^{-[h(x)-h(u_n)]}\right\}\ .$$

If x remains in the neighborhood of u_n and if for $x = u_n$

(2) $$h''(x) \ll h(x)\ ,$$

the Taylor expansion leads to 5.2.2(13) with

(3) $$y = h'(u_n)(x - u_n)\ ,$$

provided that (2) also holds for the higher derivatives.

5.2.2 FIRST ASYMPTOTIC DISTRIBUTION

From (1) the derivative $h'(x)$ at the value $x = u_n$ is the α_n used in 5.2.2 (13). The condition (2) may give an indication for the sample size n necessary for the validity of this approximation. Of course, this procedure is a generalization of the method used in 5.1.1.

In 5.2.1 L'Hôpital's rule (4.1.3) was used in a form which may be written

(4) $$\lim_{x=\infty} \left[\frac{f(x)}{1 - F(x)} + \frac{f'(x)}{f(x)} \right] = 0.$$

Von Mises (1936) has given another derivation of the first asymptote, which is linked to a condition on the intensity function $\mu(x)$, namely,

(5) $$\lim_{x=\infty} \frac{d}{dx} \left[\frac{1}{\mu(x)} \right] = 0.$$

Both conditions require the existence of the first derivative $f'(x)$. Condition (5) may be written

(6) $$\lim_{x=\infty} \frac{d}{dx} \left[\frac{1}{\mu(x)} \right] = -1 - \frac{f'(x)}{f(x)} \cdot \frac{1}{\mu(x)}$$

The expression on the right side converges to zero if the critical quotient $Q(x)$ defined in 4.1.3(3) converges to unity. Therefore condition (5) is identical to the definition 4.1.4 of the exponential type and it is sufficient to check the validity of the Mises condition (5) for given initial distributions.

On the basis of (5) and the additional condition

(7) $$\lim_{n=\infty} \left[n \frac{d}{dx} \frac{1}{\mu(x)} \right]_{x = u_n} = \infty,$$

Uzgoeren gave a development of $F^n(x)$ in the neighborhood of $x = u_n$, similar to 5.2.1(4), which leads for initial distributions unlimited to the right to

(8) $$-\lg[-\lg \Phi(x)] = y + \sum_{v=2}^{\infty} \frac{y^v}{v!} \frac{\mu^{(v-1)}(u_n)}{\mu^v(u_n)},$$

where

(9) $y = \alpha_n(x - u_n);$ $\dfrac{\mu^{(v-1)}(u_n)}{\mu^v(u_n)} = \left[\left(\dfrac{d^{(v-1)}\mu(x)}{dx} \right) \Big/ \mu^v(x) \right]_{(x = u_n)}.$

The first member in (8) gives of course again the asymptotic probability function 5.2.1(13) which the author (wrongly) attributes to Fréchet.

Exercises: 1) Study the asymptotic distribution of the largest value for $h(x) = x^k$ and show that y is linear in x. (A statement to the contrary was made by Longuet-Higgins for $k = 2$.) 2) Use the Uzgören method for the same probability function. 3) Prove that condition (5) holds for the exponential, normal, log-normal, and logistic distributions.

Problem: The transition from (1) to the first asymptote must involve further conditions on $h(x)$ since the second asymptote holds for $h(x) = k \lg x$.

5.2.3. Mode and Median. The double exponential probability function contains two parameters, u_n and $1/\alpha_n$. They depend on the initial distribution *and* on the sample size n. The first parameter, u_n, is the characteristic largest value, and increases as shown in 4.1.6 as a function of the logarithm of n. The other parameter, α_n, being an intensity function, need not increase with n. It may be independent of n; it may even decrease with n, and consequently with u_n. (See, however, the conditions in 1.2.1 and 4.1.7.) The sizes of α_n and u_n depend on the initial distribution. For example, for the exponential and logistic distributions we obtained in 4.1.1 and 4.2.1 respectively,

(1) $\tilde{x}_n = u_n = \lg n$; $\alpha_n = 1$; $\alpha_n u_n = \lg n$; $\lambda_n = n$

and

(2) $\tilde{x}_n = \lg n$; $u_n = \lg(n-1)$; $\alpha_n = 1 - 1/n$; $\alpha_n u_n \sim \lg n$; $\lambda_n \sim n$.

From (1) it follows that *the modes of the reduced extreme values are zero*. In 4.1.4(3) the modal extremes converge to the characteristic extremes. In the asymptotic distribution the equality is reached:

(3) $\tilde{x}_n = u_n$; $\tilde{x}_1 = u_1$.

For small values of x, i.e., in the neighborhood of the mode, the expansion of the exponential function in (I) about u_n and u_1 leads to the normal approximations:

(4) $\varphi(x) \approx \alpha_n \exp[-1 - \alpha_n^2(x - u_n)^2/2]$;

(4′) $_1\varphi(x) \approx \alpha_1 \exp[-1 - \alpha_1^2(x - u_1^2)/2]$.

The points of inflection of the distributions $\varphi(x)$ and $_1\varphi(x)$ are the solutions of

$$e^{-y_{1,2}} = (3 \pm \sqrt{5})/2$$

The two solutions are equal in size and differ in sign. The numerical values of y_1 and y_2, the corresponding probabilities $\Phi(y_1)$, $\Phi(y_2)$ and densities $\varphi(y_1)$, $\varphi(y_2)$ of the reduced variate y are

$$y_1 = -0.96242 \qquad \varphi(y_1) = 0.19098 \qquad \Phi(y_1) = 0.07295$$
$$y_2 = 0.96242 \qquad \varphi(y_2) = 0.26070 \qquad \Phi(y_2) = 0.68252$$
$$\Delta\Phi = 0.60957 .$$

The medians $\breve{y}_n$ and $\breve{y}_1$ of the reduced extremes, and the medians $\breve{x}_n$ and $\breve{x}_1$ of the extremes themselves, are

(5) $$\breve{y}_n = -\breve{y}_1 = -\lg(\lg 2) = 0.36651$$
$$\breve{x}_n = u_n + 0.36651/\alpha_n ; \qquad \breve{x}_1 = u_1 - 0.36651/\alpha_1 .$$

5.2.4. Generating Functions. The generating function $G_n(t)$ of the reduced largest value obtained from (I) is

(1) $$G_n(t) = \int_{-\infty}^{+\infty} e^{yt - e^{-y}} e^{-y} \, dy .$$

By the transformation $e^{-y} = z$ this becomes

(2) $$G_n(t) = \Gamma(1 - t) ,$$

a formula due to Fisher and Tippett. The generating function $G_1(t)$ of the reduced smallest values is, from 1.1.6(2) and the mutual symmetry, 1.1.3(1),

(3) $$G_1(t) = \Gamma(1 + t) .$$

The generating functions $G_{x_n}(t)$ and $G_{x_1}(t)$ of the extreme value x_n and x_1, obtained from the procedure 1.1.6, are

(4) $$G_{x_n}(t) = e^{u_n t} \Gamma(1 - t/\alpha_n) ; \qquad G_{x_1}(t) = e^{u_1 t} \Gamma(1 + t/\alpha_1) .$$

The means of reduced extremes and of the extremes themselves are

(5) $$\bar{y}_n = -\bar{y}_1 = \gamma ; \qquad \bar{x}_n = u_n + \gamma/\alpha_n ; \qquad \bar{x}_1 = u_1 - \gamma/\alpha_1 .$$

Hence the quotient,

(6) $$\frac{\bar{x} - \tilde{x}}{\breve{x} - \tilde{x}} = \frac{\gamma}{.36651} = 1.57490 ,$$

is independent of the standard deviation. The central moments of the two extremes are, from 1.1.8(1),

(7) $$\mu_{2,n}\alpha_n^2 = S_2 = \mu_{2,1}\alpha_1^2 ; \qquad \mu_{3,n}\alpha_n^3 = 2S_3 = -\mu_{3,1}\alpha_1^3 ;$$
$$\mu_{4,n}\alpha_n^4 = 6S_4 + 3S_2^2 = \mu_{4,1}\alpha_1^4 .$$

The seminvariant generating function of the largest value,

$$L_n(t) = \sum_2^\infty \frac{S_v t^v}{v}, \tag{8}$$

leads to the seminvariants

$$\lambda_v = (v-1)! S_v. \tag{9}$$

The first values obtained from 1.1.8(5) are given in Table 5.2.4.

Table 5.2.4. *Seminvariants of the First Asymptote*

$\lambda_2 = \pi^2/6 = 1.64493407 \qquad \lambda_2^2 = \pi^4/36 = 2.70580809$
$\lambda_3 = 2.40411380 \qquad \lambda_2^3 = \pi^6/216 = 4.45087592$
$\lambda_4 = \pi^4/15 = 6.49339939 \qquad \lambda_2^4 = \pi^8/1296 = 7.32139739$
$\lambda_5 = 24.8862662 \qquad \lambda_3^2 = 5.77976316$
$\lambda_6 = 8\pi^6/63 = 122.081167$

The moments are related to the seminvariants by the equations 1.1.6(7) which lead to

$$\sigma = 1.2825499, \qquad \mu_2 = \lambda_2;$$
$$\mu_3 = \lambda_3; \qquad \mu_4 = 3\pi^4/20 = 14.611364 \qquad \mu_5 = 64.432353; \tag{10}$$
$$\mu_6 = 406.873470.$$

The numerical values of the Betas, defined in 1.1.5(5) are

$$\sqrt{\beta_1} = 1.1396; \qquad \beta_1 = 1.2985676; \qquad \beta_2 = 5.4 \tag{11}$$
$$\beta_3 = 21.1575319; \qquad \beta_4 = 91.4142468.$$

Exercises: 1) Derive the distribution from the generating function. (Schelling: Start from $\Gamma(2-t) = (1-t)\Gamma(1-t); \ t < 1$). 2) Prove that the distribution of the sample means of the extreme values converges toward normality.

Reference: Daniels (1941).

5.2.5. Standard and Mean Deviations. The standard deviations of the largest and smallest values are, from 5.2.4(7),

$$\sigma_n = \pi/\sqrt{6}\alpha_n; \qquad \sigma_1 = \pi/\sqrt{6}\alpha_1. \tag{1}$$

If the extremal intensity α_n increases (remains invariant or decreases) with n, the standard deviation of the extremes decreases (remains invariant or increases), the distribution of extreme values contracts (is invariant or expands), the density of probability at the modes increases (remains invariant or decreases), and the distances from the mode to the median and to the mean diminish (are invariant or increase) with the sample size n from which the extremes are taken. This supplements the knowledge of the three classes obtained in 4.1.5 and 4.1.6. The skewness and the excess remain unchanged, as shown in 5.2.4(11). For a given set of N extreme observations, we omit the index n. Then the parameter $1/\alpha$ may be estimated from the standard deviation as

(1') $$\frac{1}{\hat{\alpha}} = \frac{\sqrt{6}}{\pi} s.$$

The asymptotic variance $\sigma^2(1/\hat{\alpha}, s)$ of this estimate is from (1)

$$N\sigma^2(1/\hat{\alpha}, s) = N\sigma^2(\sqrt{6}s/\pi).$$

From 1.1.5(4) and 5.2.4(11) the variance is

(2) $$\alpha^2 N\sigma^2(1/\hat{\alpha}, s) = 1.1.$$

Combination of (1) and 5.2.4(5) leads to

(3) $$\bar{x}_n = u_n + 0.45005\sigma_n ; \qquad \bar{x}_1 = u_1 - 0.45005\sigma_1.$$

The parameters u_n, u_1, $1/\alpha_n$, $1/\alpha_1$ may thus be estimated from the sample means and standard deviations. Formulae 5.2.4(5) and 5.2.4(7) lead to the relation

$$V = \pi/\sqrt{6}(\alpha u + \gamma)$$

between αu and the coefficient of variation V. Consequently the product of the parameters,

(4) $$\alpha u = \frac{\pi}{\sqrt{6}V} - \gamma,$$

is a linear function of $1/V$. From 4.1.7(2) the coefficient of variation diminishes with n. Equation (4) may serve for an estimate of αu from the sample coefficient of variation.

From the median 5.2.3(5) and the mean 5.2.4(5) we calculate the mean deviations by the method shown in 1.1.4(6). The mean deviation θ' about the median becomes

(5) $$\alpha\theta' = 0.57722 - 0.36651 + 2\int_{-\infty}^{\tilde{y}} \exp(-e^{-y}) \, dy.$$

The substitution $z = \exp(-y)$ leads to

$$\alpha\theta' = 0.21071 - 2Ei(-\lg 2),$$

where Ei stands for the exponential integral. Consequently,

(6) $$\theta' = 0.96805/\alpha ; \qquad 1/\alpha = 1.03295\theta'$$

may be used to estimate $1/\alpha$ from the sample mean deviation taken about the central value.

The mean deviation taken about the mean, obtained by the same procedure, is

$$\alpha\theta = \int_{-\infty}^{\gamma} \exp(-e^{-y})\, dy.$$

Therefore,

(7) $$1/\alpha = 1.01731\theta$$

is nearly equal to the mean deviation of the largest value taken about the mean. Since the distribution of the smallest value is symmetrical to the distribution of the largest value, the same relations hold for the parameter $1/\alpha_1$. The asymptotic variance $\sigma^2(1/\hat{\alpha},t')$ in the estimate of the parameter from (6),

$$N\sigma^2(1/\hat{\alpha},t') = N\sigma^2(1.03285 t'),$$

becomes, from 1.1.5(4) and (6),

(8) $$\alpha^2 N\sigma^2(1/\hat{\alpha},t') = 0.75512.$$

It is smaller than the asymptotic variance of the estimate from the standard derivation. Alternative methods for estimating the parameters will be studied in the next chapter.

Problems: 1) What are the exact and asymptotic distributions of the sample standard deviation; of $x - \bar{x}_0$, of t and the quotient of two sample standard deviations? 2) What is the error of estimation involved in the use of (3)? 3) Reproduce the distribution of s^2 by the Gamma distribution after equalizing the first moments.

Reference: Lieblein (1953), Tiago de Oliveira (1957).

5.2.6. Probability Paper and Return Period. A probability paper for extreme values first proposed by Powell is shown in Graph 5.2.6. If we have observed N largest values x_m ($m = 1, 2, \ldots, N$) each taken from n independent observations where n is large, and if the initial distribution is of the exponential type, the N values plotted at the frequencies $m/(N+1)$ should be scattered about the straight line

(1) $$x = u + y/\alpha.$$

The same paper may be used for smallest values to be plotted in decreasing magnitude. In this case, the largest observation is plotted at $1/(N+1)$ and the smallest one at $N/(N+1)$. The curves traced on the Graph 5.2.6 are the exponential (4.1.1), the normal, the logistic (4.2.1), and logarithmic normal (1.1.9) probabilities. For large values of the reduced variate, the exponential probability becomes a straight line, since the extremal probability converges for large values of the variate toward the exponential function. The last curve in Graph 5.2.6 will be explained in 7.1.2.

5.2.6 FIRST ASYMPTOTIC DISTRIBUTION

The Geological Survey modified the probability paper slightly. The return period $T(x)$, defined in 1.2.2(4), is plotted on the horizontal scale. The extremal variate x, the annual flood, is plotted on the vertical scale. The scales for the reduced variate y and the probability $\Phi(y)$ are omitted.

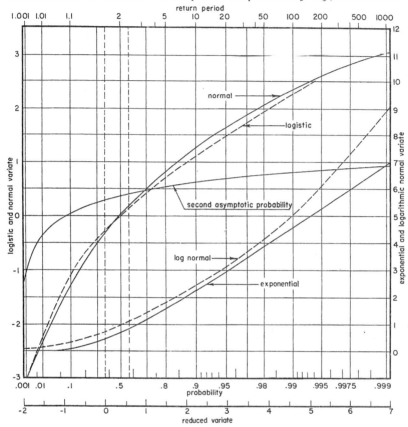

Graph 5.2.6. **Extremal Probability Paper**

The return period for $x = u$, the most probable annual flood, is $T(u) = e/(e-1) = 1.582$, while the return period of the mean annual flood is 2.328. Of course, the return period of the median annual flood is 2. The mth flood (m counted from the top) is plotted at the return period $(N+1)/m$, which corresponds to the plotting position advocated in 1.2.7. For large values of T, the return period scale gives the most probable value x_T within the period T, (not the most probable value at the time T). The first value is of interest to engineers; the second value is of no particular interest.

On the basis of the probability paper, Jenkinson derived a graphical test for the preference to be given to one of the three asymptotes for the analysis of largest observations. If we write $\exp[-e^{-y}]$ for the three probabilities of largest values, Table 5.1.3(2) leads to three relations between x and y, given in Table 5.2.6.

Table 5.2.6. *The Three Curvatures of Largest Values*

Probability	Relation of y to x	Sign of d^2x/dy^2
$\Phi^{(1)}_{(x)} = \exp[-e^{-\alpha(x-u)}]$	$y = \alpha(x - u)$	$= 0$
$\Phi^{(2)}_{(x)} = \exp\left[-\left(\dfrac{v-\varepsilon}{x-\varepsilon}\right)^k\right]$	$y = k \lg(x - \varepsilon) + \text{const}$	> 0
$\Phi^{(3)}_{(x)} = \exp\left[-\left(\dfrac{\omega-x}{\omega-v}\right)^k\right]$	$y = \text{const} - k \lg(\omega - x)$	< 0

Since $y = -\lg(-\lg \Phi)$ is the abscissa on the extremal paper, Jenkinson uses the curvature of a smooth curve drawn by hand through the observations as an empirical indication for the choice of the asymptote. Of course his method will not always lead to a clear distinction, and further reasons are needed to justify the existence of the lower and upper limits ε and ω in the second and third asymptote.

The reduced variate y in 5.2.1(1) and the return period T are related by

(2) $$y = -\lg \lg [T/(T-1)] < \lg T$$

For increasing values of T, the reduced variate converges to

(3) $$y = \lg T - 1/(2T),$$

an approximation which holds with an error of less than 0.7% from $T = 7$ onward. Inversely

(3') $$T(y) = e^y + \tfrac{1}{2}.$$

The first approximation,

(4) $$\lg T(x) \sim \alpha(x - u),$$

derived by Coutagne (1937), leads to the fundamental formula for flood control,

(5) $$x \sim u + (\lg T)/\alpha.$$

Floods increase asymptotically as the logarithm of the return periods, and $1/\alpha$ is the rate of increase. Fuller, one of the first engineers to make a systematic study of floods in this country, claimed, on a purely empirical basis from incomplete data, that floods should increase as the logarithm of the return period (which he called simply "time"). Thus formula (5) is Fuller's statement put on a probability basis. At the same time, it is an implementation of the logarithmic trend stated in 4.1.6.

5.2.7 FIRST ASYMPTOTIC DISTRIBUTION

Chow (1951) has drawn an interesting conclusion from the general theorem 3.1.2(4) as applied to floods. Their return period T and the return period T_b of a large discharge above a certain basic stage are linked by

$$T = 1/[1 - \exp(-e^{-y})] = 1/[1 - \exp(-1/T_b)],$$

whence

(6) $$\lg T_b = y = \alpha(x - u).$$

The large discharges plotted on semi-logarithmic paper are linear functions of their return periods. This technique, used on a purely empirical basis by Cross and Rantz, led to a good fit which gives an indirect justification to the use of the theory of extreme values for floods.

The largest value corresponding to the double of a given return period T is asymptotically:

$$x_{2T} \sim x_T + (\lg 2)/\alpha,$$

or, from 5.2.5(1),

(7) $$x_{2T} \sim x_T + 0.54045\sigma.$$

If we double the return period, we add about 54% of the standard deviation to the previous largest value. The return period, $_1T(x)$, for values of x below the median is

$$_1T(x) = \exp(e^{-y}),$$

whence, for small values of x,

(8) $$x \sim u + (\lg \lg {_1T})/\alpha.$$

The difference between (5) and (8) is due to the asymmetry of the distribution.

Exerise: Rewrite Table 5.2.6 for the smallest values.

Problem: Construct a numerical test for the distinction between the three asymptotes.

References: Coutagne (1930, 1952), Lieblein (1954a), Velz (1950b), Weiss (1955).

5.2.7. Comparison with Other Distributions. To get a clear insight into the nature of the two double exponential distributions, we compare them in Table 5.2.7 with the normal one.

The two double exponential distributions with a common mean γ and the normal values with standard deviation $\pi/\sqrt{6}$ are traced in Graph 5.2.7 (1). The density of probability at the mode is larger for the double exponential distributions than for the normal one. For large (small) values of the variate, the first double exponential distribution has larger (smaller) densities than the normal one. The opposite is true for the second double exponential distribution.

Table 5.2.7. Selected Probabilities for Normal and Largest Values

Value	Reduced Variate		Probabilities		Return Periods	
	Largest	Normal	Largest	Normal	Largest	Normal
$\bar{x} - \sigma$	−.70533	−1	.13206	.15866	7.57	6.30
$\bar{x} + \sigma$	1.85977	1	.85581	.84134	6.93	6.30
$\bar{x} \pm \sigma$	—	—	.72375	.68268	—	—
$\bar{x} - 2\sigma$	−1.98788	−2	.00068	.02275	1480.	43.96
$\bar{x} + 2\sigma$	3.14232	2	.95773	.97725	23.7	43.96
$\bar{x} \pm 2\sigma$	—	—	.95705	.95450	—	—
$\bar{x} - 3\sigma$	−3.27043	−3	$3.7 \cdot 10^{-12}$	.00135	$.27 \cdot 10^{11}$	741
$\bar{x} + 3\sigma$	4.42486	3	.98810	.99865	84.01	741
$\bar{x} \pm 3\sigma$	—	—	.98810	.99730	—	—

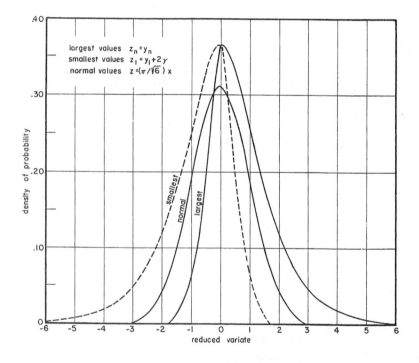

Graph 5.2.7(1). **Extreme and Normal Distributions**

In Graph 5.2.7(2) the probabilities of the largest and the smallest values and the normal probabilities for the same mean and standard deviation

5.2.7 FIRST ASYMPTOTIC DISTRIBUTION

are traced on extremal probability paper. The standardized extremes and normal values are traced as functions of the return periods for $T \geq 10$ in logarithmic scale in Graph 5.2.7(3). The standardized largest values increase more quickly and the standardized smallest values decrease more slowly with the sample size than do the normal values. The curves

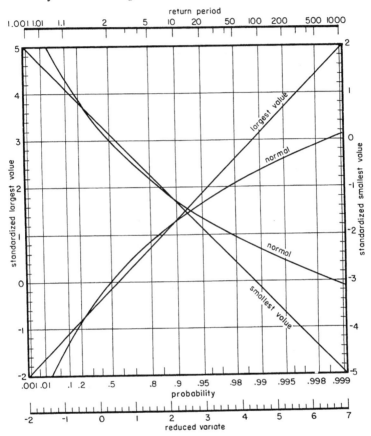

Graph 5.2.7(2). Probabilities for Extreme and Normal Standardized Values

showing the range and midrange will be explained in 8.1.6 and 8.1.4. Graph 5.2.7(4) traces the distribution as function of the probability for the largest value, the normal variate, and the range. The latter relation will be derived in 8.1.6.

In the representation of skewed distributions, the logarithmic normal one often gives satisfactory results. It has been applied also to extreme values, and especially to floods. This leads to the question of why some extreme values plotted on logarithmic normal probability paper lead to approximately straight lines.

For the logarithmic normal distribution, see 1.1.9. The skewness $\sqrt{\beta_1}$ is related to the coefficient of variation V by

(1) $$\sqrt{\beta_1} = 3V + V^3.$$

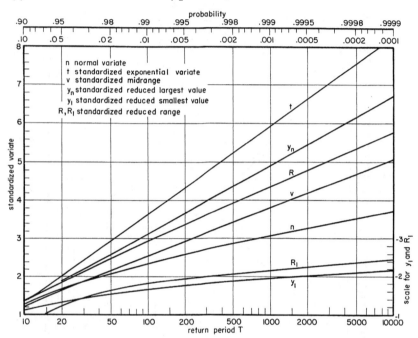

Graph 5.2.7(3). Probabilities of Standardized Extremes, Range, and Midrange

Therefore, extremal observations where $V = 0.36381$, corresponding to $\sqrt{\beta_1} = 1.1396$, will lie approximately on a straight line on logarithmic normal paper. The corresponding value of αu obtained from 5.2.5(4) is $\alpha u = 2.94811$. Graph 5.2.7(5) shows the corresponding probability function $\Phi(x)$ traced on logarithmic normal probability paper. Within the domain $2.5 < \alpha u < 3.5$, and in particular for small samples, no graphical distinction between the two theories is possible. Conversely, logarithmic normal values with $V = 0.36381$ will plot as straight lines on extremal paper.

For the Pearson type III (Gamma distribution) the corresponding relation is

(2) $$\sqrt{\beta_1} = 2V.$$

5.2.7 FIRST ASYMPTOTIC DISTRIBUTION 183

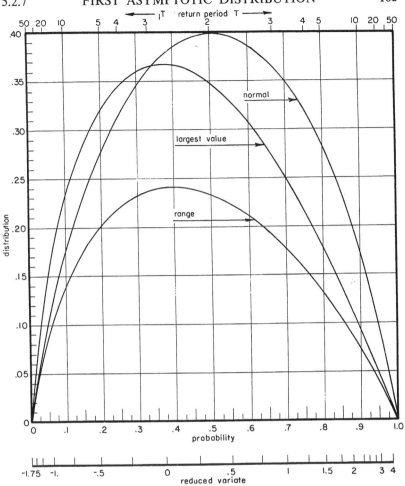

Graph 5.2.7(4). Normal and Extremal Distribution as Function of the Probability

Therefore, extremal values with $V = 0.5698$ will be well reproduced by this distribution. In this case, the parameter $v - 1$ in the Gamma distribution is 2.080, while for the extremal distribution $\alpha u = 1.67366$.

The relations (1) and (2) explain why some extremal observations have been successfully analyzed by the logarithmic normal and the Pearson type III distributions.

Exercises: 1) Trace Graph 5.2.7(3) on normal probability paper. 2) Show that the relation 1.1.9(13′) between the sample values of $\bar{x}_0/\check{x}_0$ and the coefficient of variation does not lead to a clear distinction between the logarithmic normal and the first double exponential distribution.

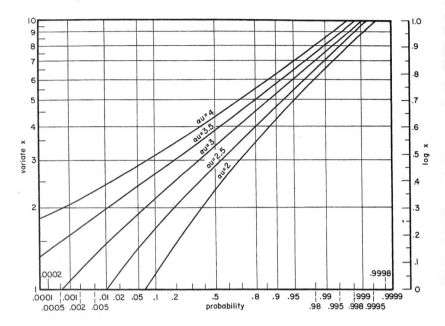

Graph 5.2.7(5). First Asymptotic Probability Traced on Logarithmic Normal Paper

References: Benham, Foster (1952), Hazen, A. I. Johnson, Moshman (1953).

5.2.8. Barricelli's Generalization. Thus far it has been assumed that the extreme values in every large sample were taken from the same population. Obviously this is not always true. In meteorological extremes, there may be a cyclical variation which influences the population mean or standard deviation. To account for changes in the mean, Barricelli assumes that the observed distribution is a convolution of the normal and the extremal distributions. The first holds for the means around which the extremes oscillate.

If the initial mean has a normal distribution, while the initial standard deviation remains constant, the characteristic largest value u becomes a normal variate. The new distribution $\varphi_B(x)$ of the largest value is obtained from the convolution of a normal distribution of u with mean ξ and standard deviation $\sigma = 1/(h\sqrt{2})$, and a double exponential distribution of x with standard deviation $\sigma = \pi/(\sqrt{6}\alpha)$. Barricelli's distribution is therefore

(1) $$\varphi_B(x) = \frac{\alpha h}{\sqrt{\pi}} \int_{-\infty}^{\infty} \exp\left[-h^2(u-\xi)^2 - \alpha(x-u) - e^{-\alpha(x-u)}\right] du,$$

5.2.8 FIRST ASYMPTOTIC DISTRIBUTION

and the probability is

(2) $$\Phi_B(x) = \frac{h}{\sqrt{\pi}} \int_{-\infty}^{\infty} \exp[-h^2(u-\xi)^2 - e^{-\alpha(x-u)}] \, du .$$

This function contains three parameters, ξ, h, and $1/\alpha$, all having the same dimension as x. The transformation

$$\alpha(x-u) = -\lg(-\lg z) ; \qquad u = x + \frac{\lg(-\lg z)}{\alpha}$$

yields

(1') $$\varphi_B(x) = \frac{h}{\sqrt{\pi}} \int_0^1 \exp[-h^2(x-\xi+\lg(-\lg z)/\alpha)^2] \, dz .$$

If a *critical quotient*, different from the quotient 4.1.3(3),

(3) $$\alpha^2/h^2 = q^2 ,$$

is introduced, the distribution becomes

$$\varphi_B(x) = \frac{\alpha}{q\sqrt{\pi}} \int_0^1 \exp[-q^{-2}(\alpha(x-\xi)+\lg(-\lg z))^2] \, dz .$$

The distribution $\varphi_B^*(y)$ of the reduced variate,

(4) $$y = h(x-\xi) ,$$

is

$$\varphi_B^*(y) = \varphi_B(x) \, q/\alpha ,$$

whence

(5) $$\varphi_B^*(y) = \frac{1}{\sqrt{\pi}} \int_0^1 \exp[-y+[\lg(-\lg z)]/q]^2 \, dz ,$$

which depends only upon y and q. Barricelli calculated this expression by graphical integration for large y and $q^2 = 1, 2, 3, 4$. If q and α are known, $\varphi_B(x)$ is obtained as

(5') $$\varphi_B(x) = \varphi_B^*(y)\alpha/q .$$

Since the distribution $\varphi_B(x)$ is obtained from a convolution, its moments are known, and the parameters may be estimated from them. The first two moments are

(6) $$\bar{x} = \xi + \gamma/\alpha$$

and

(7) $$\mu_2 = \frac{1}{2h^2} + \frac{\pi^2}{6\alpha^2} .$$

The variance consists of two parts. The first one is the variance of the mode, while the second one is the variance of the largest values proper.

The shape of the distribution depends upon the relative importance of the two constituents, i.e., the *critical quotient* (3), which is proportional to the ratio of the normal variance and the variance of the largest values. If this quotient is large, the variance of the largest values is negligible compared to the normal variance, and the combined variate is approximately normal. If the quotient approaches zero, the variance of the mean becomes negligible compared to the variance of the largest value, and the combined variate is distributed approximately as a largest value. The asymmetry is large (or small) if the variance of the normal constituent is small (or large) compared to the variance of the double exponential constituent. This scheme explains why a distribution of extreme values may be less asymmetrical than foreseen by the previous theory.

To estimate the critical quotient, Barricelli calculated the variance of the larger of two values. This requires complicated approximations and assumptions about convergence. It is simpler to consider the third central moment of the convoluted distribution (5'),

(8) $$\mu_3 = 2S_3/\alpha^3,$$

whence

(9) $$\left(\frac{2S_3}{\mu_2}\right)^{2/3} \frac{1}{\alpha^2} = 1.$$

Now, equation (7) may be written, from (4),

(10) $$2\alpha^2 \mu_2 = q^2 + \pi^2/3.$$

The product of equations (9) and (10) leads to

$$(8S_3^2/\beta_1)^{1/3} - \pi^2/3 = q^2$$

or

(11) $$q^2 = \frac{3.58922}{\beta_1^{1/3}} - 3.28987.$$

The quotient q depends only on β_1. For $\beta_1 = 0$, the critical quotient is infinite, and the resulting distribution is normal. For

$$\beta_1 = 6^3 S_3^2/\pi^6 = 1.2985676,$$

the critical quotient is zero, and the resulting distribution is extremal.

Equation (11) may be used for estimating the critical quotient from the sample value b_1. Equation (10) leads to an estimate of α. Finally, the parameter ξ is estimated from (6). Unfortunately, these estimates are subject to large errors. It would be desirable to have tables of the distribution (5). Since $\beta_1^{1/3}$ varies in the interval,

$$0 \leq \beta_1^{1/3} \leq 1.091,$$

it would be sufficient to calculate tables for

$$\beta_1^{1/3} = .88, .66, .44, .22$$

which correspond roughly to the quotients $q^2 = 0.8, 2, 5, 13$.

Problems: 1) Calculate the mode and median of the reduced variate y as a function of q. 2) Improve the estimation of the parameters. 3) Apply the Barricelli method to the second and third asymptotes. 4) Calculate the convoluted distribution if u is an exponential variable.

References: Brooks and Carruthers (p. 131), Jenkinson.

5.3. EXTREME ORDER STATISTICS

5.3.0. Problems. In 2.1.4 it was shown that the distribution of the mth central value with $m/n = 0(\frac{1}{2})$ converges to normality for n increasing. If m remains finite with n increasing, as in 5.2.1, we speak of mth extremes. Then m/n, where m is counted from the top (or bottom) for the largest (or smallest) order, is very small. Finally the rank m may increase with n so that m/n converges to unity but the difference $n - m$ remains finite.

The distribution of the mth extreme among n observations taken from an initial distribution of the exponential type must be distinguished from the distribution of the mth member of a sample consisting of N largest values, which will be considered in 6.1. It will be shown that the asymptotic distribution of the mth extreme converges with increasing rank m to normality and that the distributions of the mth extreme distances converge to an exponential distribution. Finally, the distribution of the absolute largest value will be reduced to a distribution of the largest value.

5.3.1. Distribution of the mth Extreme. The asymptotic expression for the probability $F(x_m)$ and the density of probability $f(x_m)$ of the mth value from the top, for initial distribution of the exponential type, are basic for the construction of the asymptotic distribution $\varphi_m(x_m)$ of the mth largest values. The general distribution 2.1.1(1) of the mth largest value becomes from 5.2.1(11) asymptotically for n large and m small

(1) $\quad \varphi_m(x_m) = \dfrac{n^m}{(m-1)!} \left(1 - \dfrac{m}{n} e^{-y_m}\right)^{n-m} \left(\dfrac{m}{n}\right)^{m-1} e^{-(m-1)y_m} \alpha_m \dfrac{m}{n} e^{-y_m}$

or,

(2) $\quad \varphi_m(x_m) = \dfrac{m^m \alpha_m}{(m-1)!} \exp(-my_m - me^{-y_m}) ; \qquad y_m = \alpha_m(x_m - u_m).$

The distribution of the largest value is obtained by putting $m = 1$. From the symmetry principle, the asymptotic distribution $_m\varphi(_mx)$ of the mth extreme value from the bottom is

(3) $\quad _m\varphi(_mx) = \dfrac{m^m\, _m\alpha}{(m-1)!} \exp(m\,_my - me^{_my}) ; \qquad _my = {_m\alpha}(_mx - {_mu})$,

where $_mu$, the expected mth smallest value, and $_m\alpha$ are defined by

(4) $\qquad\qquad F(_mu) = m/n ; \qquad _m\alpha = nf(_mu)/m$.

If the initial distribution is symmetrical,

(5) $\qquad\qquad _m\alpha = \alpha_m ; \qquad _mu = -u_m$.

As previously, in 5.2.3, the modes of the reduced mth extremes vanish and the modes of the mth extremes are equal to the characteristic extremes. The asymptotic distributions of the reduced second to fifth extreme values are traced about a common origin in Graph 5.3.1, which shows that the distributions of the reduced extremes contract with increasing values of m. This will be proven analytically in 5.3.3.

At the modes of the mth reduced extremes the densities of probability,

(6) $\qquad\qquad \varphi_m(\tilde{y}_m) = \dfrac{m^m e^{-m}}{(m-1)!} = {_m\varphi(_m\tilde{y})}$,

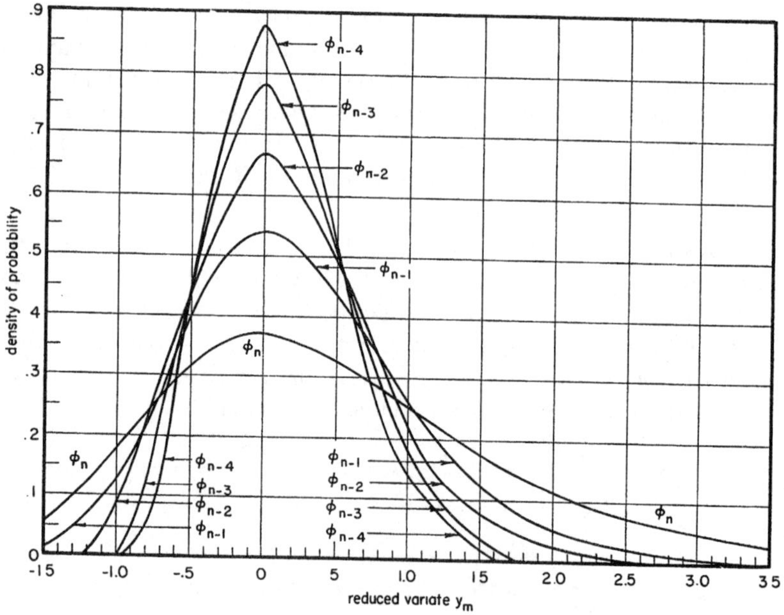

Graph 5.3.1. **Asymptotic Distributions of mth Reduced Extremes**

are equal to the probabilities of the mean in the Poisson distribution and diminish with increasing m.

If we have N series, each composed of n independent observations, and if we choose, from each series, the mth extreme observation, the N observations ought to be distributed according to (2) and (3), provided that the initial distribution is of the exponential type and n sufficiently large.

Exercise: Derive the distribution of the mth extreme by the method 5.2.2.

References: Kawata, Molina, Smirnov (1952).

5.3.2. Probabilities of the mth Extreme. The asymptotic probability $\Phi_m(y_m)$ of the mth reduced largest value is, from 5.3.1(1),

(1) $$\Phi_m(y_m) = \int_{-\infty}^{y_m} (me^{-y})^{m-1} \exp(-me^{-y}) me^{-y} \, dy / \Gamma(m) ,$$

If we introduce

$$me^{-y_m} = z ,$$

the probability leads to the incomplete Gamma function

(2) $$\Phi_m(y_m) = \int_{me^{-y_m}}^{\infty} z^{m-1} e^{-z} \, dz / \Gamma(m) .$$

From the symmetry the probability of the mth reduced smallest value is

(2') $$_m\Phi(_m y) = 1 - \Phi_m(y_m) .$$

The successive probabilities for the mth extremes can be expressed by the probability of the largest value. Integration of (2) by parts leads, after reversing the order of summation, to

(3) $$\Phi_m(y_m) = \Phi^m(y) \sum_{v=0}^{m-1} \frac{m^v e^{-v y_m}}{v!} .$$

For $m = 1, 2, 3$, we obtain for the largest, penultimate, and third largest values, after dropping the index of y,

$$\Phi(y) = e^{-e^{-y}} ; \quad \Phi_2(y) = e^{-2e^{-y}}(1 + 2e^{-y})$$

$$\Phi_3(y) = e^{-3e^{-y}}(1 + 3e^{-y} + 9/2 \, e^{-2y}) .$$

Table 5.3.2 gives the asymptotic probabilities for the first, second (Gumbel and Greenwood), and third largest reduced values. Probability points for the mth extremes, obtained from (2), are given in the probability tables of the National Bureau of Standards.

Table 5.3.2. *Asymptotic Probabilities of the Three Largest Values*

Variate y_m	Probability $\Phi(y)$	Probability $\Phi_2(y)$	Probability $\Phi_3(y)$
−1.5	.01131	.00128	.0002
−1.0	.06599	.02803	.0122
−0.5	.19230	.15891	.1293
0	.36788	.40601	.4232
0.5	.54524	.65791	.7254
1.0	.69220	.83167	.8997
1.5	.80001	.92563	.9695
2.0	.87342	.96935	.9917
2.5	.92119	.98791	.9979
3.0	.95143	.99536	.9995
3.5	.97025	.99825	.9999
4.0	.98185	.99935	1.000
4.5	.98890	.99976	1.000
5.0	.99328	.99991	1.000
5.5	.99590	.99997	1.000
6.0	.99752	.99999	1.000

The probability that a certain reduced value y is the penultimate one is not always larger than the probability that the same y is the largest one. This inequality holds only for the extreme values x themselves. Graph 5.3.2(1),

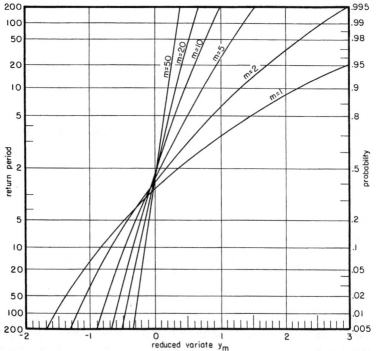

Graph 5.3.2(1). **Asymptotic Probabilities of mth Extremes Traced on Normal Paper**

5.3.2 FIRST ASYMPTOTIC DISTRIBUTION 191

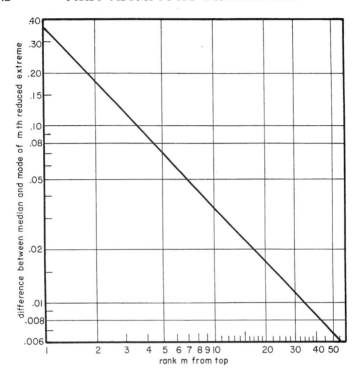

Graph 5.3.2(2). The Convergence of the Median to the Mode for mth Extremes

plotted on normal paper, shows how the successive probabilities become more and more concentrated, and indicates that the probabilities of extreme mth values tend to normality as m increases. The distance from the median of the reduced mth extreme to the mode shown in Graph 5.3.2(2) diminishes to zero with increasing m. The probabilities at the modal mth values obtained from (3) by putting $y_m = 0$ are

(4) $$\Phi_m(\tilde{x}_m) = e^{-m} \sum_{v=0}^{m-1} \frac{m^v}{v!},$$

and increase very slowly with m. The curve $\tilde{y}_m$ in graph 5.3.2(3) shows this probability obtained from Molina's tables. The curve $\breve{y}_m$ has to be read from the scale parallel to the ordinate.

The probabilities at the most probable mth value $\Phi_m(\tilde{x}_m)$ must be distinguished from

(5) $$F(u_m) = 1 - m/n.$$

In the first case, we have N series, each composed of n observations. In the

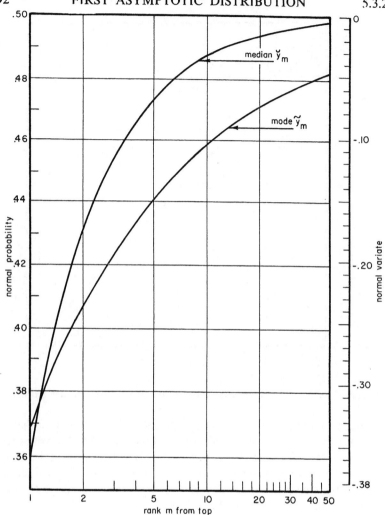

Graph 5.3.2(3). Asymptotic Mode and Median of mth Extremes

second case, we have one series of n observations. Since the two populations differ, the probabilities at the modes differ, too, although the modes are equal to the characteristic largest values.

5.3.3. Generating Functions. The moment generating function $G_m(t)$ of the mth largest reduced values obtained from 5.3.1(2),

$$G_m(t) = \frac{m^{m-1}}{(m-1)!} \int_{-\infty}^{+\infty} \exp\left[-(m-1-t)y_m - me^{-y_m}\right] me^{-y_m}\, dy_m,$$

5.3.3 FIRST ASYMPTOTIC DISTRIBUTION

becomes, by the transformation
$$me^{-y_m} = z\,; \quad me^{-y_m}\,dy_m = -dz,$$

(1) $$G_m(t) = \frac{m^t\,\Gamma(m-t)}{(m-1)!}.$$

From the mutual symmetry of the two extremes the generating function $_mG(t)$ of the mth smallest reduced value is

(2) $$_mG(t) = \frac{m^{-t}\,\Gamma(m+t)}{(m-1)!}.$$

The generating function of the mth largest reduced value may for $m \geq 2$ be written

(3) $$\lg G_m(t) = t \lg m + \sum_{v=1}^{m-1} \lg(1 - t/v) + \lg \Gamma(1-t).$$

The mean mth largest reduced values

(4) $$\bar{y}_m = \lg m - \sum_{v=1}^{m-1} 1/v + \gamma,$$

as functions of m are given in Table 5.3.3, column 2. For m increasing $\bar{y}_m$ diminishes to zero according to 1.1.8(3'). The means of the mth largest and smallest values are

(4') $$\bar{x}_m = u_m + \bar{y}_m/\alpha_m\,; \quad _m\bar{x} = {}_mu - {}_m\bar{y}/{}_m\alpha.$$

From (3) and (4) the seminvariant generating function becomes, upon expansion of the Gamma function and the logarithm,

$$L_m(t) = \sum_{k=2}^{\infty} \frac{S_k t^k}{k} - \frac{t^2}{2}\sum_{k=1}^{m-1}\frac{1}{k^2} - \frac{t^3}{3}\sum_{1}^{m-1}\frac{1}{k^3} - \cdots$$

If we write for $m \geq 2$,

(5) $$\sum_{k=m}^{\infty} 1/k^v = S_{v,m},$$

the seminvariant generating function becomes

(6) $$L_m(t) = \sum_{2}^{\infty} \frac{S_{v,m}\,t^v}{v},$$

in analogy to 5.2.4(8) valid for the largest value. However, the S_v which occur in the previous formula all decrease toward unity with increasing v. For constant m and increasing v the $S_{v,m}$ decrease rapidly to zero. From (6) it follows that the seminvariants $\lambda_{v,m}$ for the mth extreme are

(7) $$\lambda_{v,m} = (v-1)!\,S_{v,m}.$$

In particular the variance is

(8) $$\sigma_m^2 = S_{2,m}.$$

Numerical values for the different characteristics obtained from the sums of the inverse powers of the natural numbers given by Karl Pearson (1932) are given in Table 5.3.3.

Table 5.3.3. Characteristics of mth Extremes

m	Mean $\bar{y}_m$	Standard Deviation σ_m	Asymmetry $\delta = \dfrac{\bar{x}_m - u_m}{\sigma_m}$	Beta $\sqrt{\beta_{1,m}}$	$\beta_{2,m}$
1	.57722	1.28255	0.450	1.1396	2.4000
2	.27036	0.80308	0.337	0.7802	1.1875
3	.17583	0.62844	0.280	0.6210	0.7626
4	.13018	0.53275	0.244	0.5293	0.5569
5	.10332	0.47045	0.220	0.4686	0.4374
6	.08564	0.42582	0.201	0.4247	0.3597
7	.07313	0.39185	0.187	0.3911	0.3053
8	.06380	0.36488	0.175	0.3643	0.2651
9	.05658	0.34280	0.165	0.3424	0.2342
10	.05083	0.32429	0.157	0.3240	0.2098

Now the seminvariant $\lambda_{v,m}$ is of the order m^{-v+1}. Thus, if we introduce the reduced variate y_m/σ_m, the first term, $v = 2$ in the development (6) is of the order one, whereas the second term,

$$\lambda_{3,m} = 2\, S_{3,m}/(S_{2,m})^{3/2},$$

is of the order $m^{-1/2}$. Each member has the order of the preceding one multiplied by $m^{-1/2}$. Consequently, the seminvariant generating function of the doubly reduced mth extreme tends with increasing m to the normal seminvariant generating function $L(t) = t^2/2$, and the distribution of $(y_m - \bar{y}_m)/\sigma_m$ tends with increasing m to a normal distribution with zero mean and unit standard deviation.

This behavior has its analogy in the binomial distribution, which tends to a normal one as $n \to \infty$ and p remain fixed as do the order statistics in the neighborhood of the median. The binomial converges to Poisson's distribution as $n \to \infty$ and $p \to 0$ if np remains finite. This corresponds to the extreme values. If m increases without limit, the Poisson distribution converges to normality and this corresponds to the mth extreme.

References: Homma (1951), Kawata.

5.3.4. Cramér's Distribution of mth Extremes. Cramér (p. 373) has given another asymptotic distribution of the mth largest value for certain transformed variates. He introduces the transformation

(1) $$\xi_m = n[1 - F(x_m)],$$

used in 3.2.7 into the general distribution 2.1.1(1) of the mth largest among n observations. The distribution $h_n(\xi_m)$ thus obtained converges for increasing n and fixed values of m to the Gamma distribution,

(2) $$h(\xi_m) = \xi_m^{m-1} e^{-\xi_m}/\Gamma(m) .$$

Consider now the first Laplacean distribution 1.2.5(3) with mean λ. Then the variate x expressed by ξ_m is, for positive values of x,

(3) $$x = \lambda \lg (n/2) - \lambda \lg \xi_m .$$

The introduction of the further transformation,

(4) $$\lg \xi_m = -z_m ,$$

leads to the asymptotic distribution $\varphi_m^*(z_m)$ for this transformed variate z_m, where

(5) $$\varphi_m^*(z_m) = [\exp(-mz_m - e^{-z_m})]/\Gamma(m) .$$

For $m = 1$ the distribution I of the largest value is obtained. The mode

(6) $$\tilde{z}_m = -\lg m$$

is negative and decreases with m, while the density of probability at the mode increases. The distribution of the variate z_m measured from the mode, i.e., $z_m + \lg m$, is the distribution 5.3.1(2) for y_m. Therefore, numerical values for the probability $\Phi_m^*(z_m)$ are obtained from the probabilities $\Phi_m(y_m)$ by the transformation

(7) $$z_m = y_m - \lg m .$$

The probabilities $\Phi_m^*(z_m)$ calculated from the Bureau of Standards Tables are traced for $m = 1(1)6, 8, 10, 15, 20$, in Graph 5.3.4. The probability $\Phi_m^*(0)$ converges quickly toward unity, which means that the variates z_m are essentially negative. The probability at the mode is given by 5.3.2(4). The mean and the median are easily obtained from (7) and the corresponding values for y_m. The generating function is, from (7), for $z = z_m$,

(8) $$G_z(t) = \Gamma(m - t)/\Gamma(m) .$$

Hence the variance of z_m, equal to the variance of y_m is

(9) $$\sigma_m^2 = S_{2,m} .$$

For m increasing, the successive distribution curves move to the left and become more concentrated.

In the previous treatment m is fixed and n increases. Instead Kawata

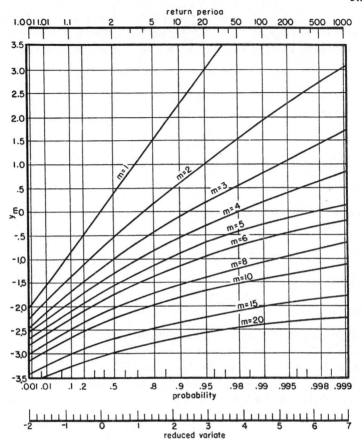

Graph 5.3.4. Cramér's Probabilities of Transformed mth Extremes

considers the case where n and m increase without limits but their difference tends to a finite positive integer, a, so that

(10) $\qquad n \to \infty \;;\qquad m \to \infty \;;\qquad n/m \to 1\;;\qquad n - m \to a$.

The transformation (1) leads again to (2) where m is replaced by $a + 1$. Now, for large x

$$\lg F(x) = \lg [1 - (1 - F(x))] \sim 1 - F(x).$$

Therefore, ξ_m may be replaced by $-n \lg F(x_m)$ and since $n/m \to 1$,

(11) $\qquad\qquad\qquad \xi_m \sim \lg F^n(x)$.

For the three types of initial distributions where an asymptotic probability

$F^n(x)$ exists, the mth extremes defined by (6) are linked to ξ_m by the three transformations

(12) $\quad \xi_m = \exp\left[-\alpha(x_m - u)\right]; \quad \xi_m = -\left(\dfrac{v}{x_m}\right)^k; \quad \xi_m = \left(\dfrac{\omega - x_m}{\omega - v}\right)^k$

taken from Table 5.1.3(2) respectively, and ξ_m has the Gamma distribution with $a + 1$ degrees of freedom.

Exercises: 1) Apply Cramér's procedure to the Cauchy and normal distributions. 2) Prove (5) for the exponential distribution.

5.3.5. Extreme Distances. The exact distributions of the $(m-1)$th distance i_{m-1} $(2 \leq m \leq n)$, (m counted from the top) were given in 2.1.8 for the exponential distribution. Now we are interested in the asymptotic distribution for the exponential type. If n is large and m is small, the probability $F(x)$ and the distribution $f(x)$ of the mth value from the top $x_m = x$ are given in 5.2.1(11) and (11'). We assume that we may write for the $(m-1)$th value from the top

(1) $\quad F(x + i) = 1 - \dfrac{m}{n} e^{-\alpha_m(x+i-u_m)}; \quad f(x+i) = \alpha_m \dfrac{m}{n} e^{-\alpha_m(x+i-u_m)}.$

This implies that $u_m - u_{m-1}$ is small compared to u_m, and $\alpha_m - \alpha_{m-1}$ compared to α_m. Then the joint distribution $w_n(x, x+i)$ of the mth value from the top and the $(m-1)$th distance $i = i_{m-1}$ is, from 2.1.8(2).

(2) $\quad w_n(x, x+i) = \dfrac{n!}{(n-m)!(m-2)!}\left(1 - \dfrac{m}{n}e^{-y_m}\right)^{n-m}$

$\qquad \times \alpha_m \dfrac{m}{n} e^{-y_m} \alpha_m \dfrac{m}{n} e^{-y_m - \alpha_m i} \left(\dfrac{m}{n} e^{-y_m - \alpha_m i}\right)^{m-2},$

where
$$y_m = \alpha_m(x - u_m).$$

Consequently, the asymptotic distribution $\varphi(i)$ of the $(m-1)$th distance is

(3) $\quad \varphi(i) = (m-1) e^{-\alpha_m i(m-1)} \displaystyle\int_{-\infty}^{+\infty} \left[\dfrac{m^m \alpha_m}{(m-1)!} \exp\left(-my_m - me^{-y_m}\right)\right] dy_m.$

Since the integral over the distribution of the mth extreme value is unity, the distribution of the $(m-1)$th distance converges, under condition (1), toward the exponential distribution

(4) $\quad \varphi(i_{m-1}) = (m-1)\alpha_m e^{-(m-1)i_{m-1}\alpha_m},$

with mean $\overline{i_{m-1}}$ and standard deviation $\sigma(i_{m-1})$ given by

(5) $\quad \overline{i_{m-1}} = \dfrac{1}{(m-1)\alpha_m} = \sigma(i_{m-1}).$

This is a generalization of the result for the exponential distribution 2.1.8(7), where $\alpha_m = 1$ and (4) is exact.

Formula (4), valid for large n and small m, was first published by J. H. Darwin (1957). He used it to calculate the 1% and 5% significance points for $m = 1(1)5$ and $n = 3(1)10$ for the normal distribution. Another approximation obtained by the saddle point method gave smaller values. This is not to be wondered at since the condition $m \ll n$ is not fulfilled by Darwin's values.

Exercises: 1) Verify condition (1) for the logistic, the double exponential, and the normal distributions. 2) Trace Irwin (1925a) probability tables for the first two normal distances on semi-logarithmic paper and compare the result with the probability function obtained from (4).

5.3.6. The Largest Absolute Value and the Two Sample Problem. If the initial distribution is symmetrical and such that for large values of x

$$F(x) = 1 - g(x, u_n, \alpha_n)/n \,,$$

then the asymptotic probabilities $\Phi(x)$ [and $\Phi^*(z)$] of the largest value [deviation] are related by

(1) $$\Phi^*(z) = \Phi^2(x) \,.$$

For the three types of initial symmetrical distributions where an asymptotic distribution of the largest value exists, the probability of the absolute largest value is equal to the probability of the larger of two independent values taken from the distribution of the largest value.
If in particular

$$g(x, u_n, \alpha_n) = \exp\left[-\alpha_n(x - u_n)\right],$$

i.e., for the exponential type the probability of the largest absolute value is equal to the probability of the largest value with a mode shifted to the right by the amount $(\lg 2)/\alpha_n$.

Instead of the larger of two values taken from the same double exponential distribution, consider now the larger of two independent values taken from two double exponential distributions. Such conditions arise if we study streams with two regimes. One series of floods is due to the melting of snow in spring, the other to autumnal rainfalls. More generally: From two samples of large size n we take simultaneously the two largest values and obtain by repetition N couples of largest values. In each couple we take the larger one, and ask for its distribution. The two initial probabilities are

(2) $$F_1(x) = \exp\left[-e^{-\alpha_1(x - u_1)}\right]: \quad F_2(x) = \exp\left[-e^{-\alpha_2(x - u_2)}\right].$$

5.3.6 FIRST ASYMPTOTIC DISTRIBUTION

The probability $\Phi(x)$ for the larger of the two values is their product, a complicated expression, since the sum of two exponential functions is not an exponential function. However, a double exponential probability is obtained, 1) if the two distributions have the same parameters, and 2) if they differ only with respect to the modes. If $\alpha_1 = \alpha_2 = \alpha$ and $u_1 \neq u_2$, the probability for the larger of the two values is

(3) $$\Phi(x) = \exp\left[-e^{-\alpha(x-u)}\right],$$

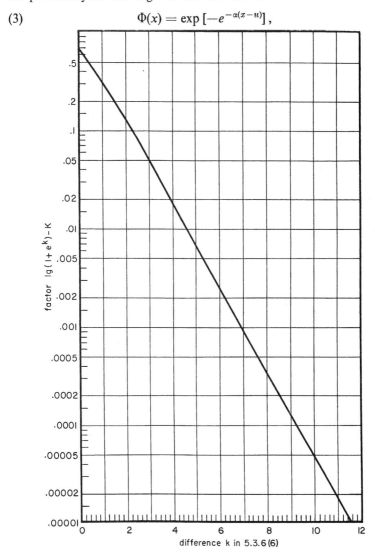

Graph 5.3.6. Correction for Obtaining the Composite Mode for Two Samples

where the new mode u is defined by

(4) $$e^{\alpha u} = e^{\alpha u_1} + e^{\alpha u_2}.$$

Since α, u_1 and u_2 are known, the mode u is easily obtained. If the two modes coincide, the mode u of the larger of the two values is

(5) $$\alpha u = \alpha u_1 + \lg 2.$$

Let

(6) $$\alpha u_2 = \alpha u_1 + k,$$

then the mode of the larger of the two values is obtained from (4) as

(7) $$\alpha u = \alpha u_1 + \lg(1 + e^k).$$

If k is large, the correction on the right side converges to k. The factor $\lg(1 + e^k)$ may be obtained from Graph 5.3.6.

The two probabilities, $F_1(x)$ and $F_2(x)$, of the largest values traced on extremal probability paper are two parallel lines passing through the modes u_1 and u_2. The probabilities, $\Phi(x)$, of the larger of the two values are represented by a third parallel passing through a mode which is larger than both u_1 and u_2. Thus, the largest value taken from two (or generally n) double exponential distributions which differ only with respect to the modes has again a double exponential distribution.

Exercises: 1) Calculate the averages of the largest deviation for the logistic distribution. 2) Apply the procedure to the smaller of two values taken from the second double exponential distribution. 3) Assume that u_1 and u_2 differ only by a small amount and calculate an approximation for u from (4).

Problems: 1) Calculate the higher moments of the largest deviation. 2) Study the distribution of the larger of two values taken from two double exponential distributions under the conditions $u_1 = u_2$, $\alpha_1 \neq \alpha_2$, and in the general case $u_1 \neq u_2$, $\alpha_1 \neq \alpha_2$. 3) Generalize these problems for n distributions.

Chapter Six: USES OF THE FIRST ASYMPTOTE

"*Il est impossible que l'improbable n'arrive jamais.*"
(*The improbable is bound to happen one day.*)

6.1. ORDER STATISTICS FROM THE DOUBLE EXPONENTIAL DISTRIBUTION

6.1.0. Problems. Starting with the distribution of the largest (smallest) values, we consider the preceding (following) mth values. The largest (smallest) of the largest reduced values is henceforth called the maximum (minimum). The properties of the maximum (minimum) of the largest values also hold for the minimum (maximum) of the smallest values. The study of the variances leads to control intervals for the extremes. The distribution of the mth extreme taken from the double exponential distribution is a special case of the general distribution, 5.3.1(2), of the mth extreme for the exponential type. This leads to control intervals for the second and third largest values.

6.1.1. Maxima of Largest Values. We study the extremal properties of the initial distribution,

(1) $\qquad f(y) = \exp[-y - e^{-y}]; \qquad F(y) = \exp[-e^{-y}]$.

From the stability postulate 5.1.2, it follows that the distribution of the maximum is the initial distribution, shifted to the right by $\lg N$. Thus, the modal, median and mean maxima are, from 5.2.3(3), 5.2.3(5) and 5.2.4(5),

(2) $\quad \tilde{y}_N = \lg N < \breve{y}_N = .36651 + \lg N < \bar{y}_N = .57722 + \lg N$,

and all central moments remain as in the initial distribution. B. F. Kimball (1942) proposed to use the probability of the mean mth value as plotting position (See 1.2.6.). Now, the return period $T(\bar{y}_N)$ of the mean maximum among N largest values is

$$T(\bar{y}_N) = 1/(1 - \exp[-e^{-\gamma}/N]).$$

If we neglect $0(N^{-2})$, the return period becomes

(3) $\qquad\qquad T(\bar{y}_N) = Ne^{\gamma} + \tfrac{1}{2}$.

The proposed plotting position attributes to the maximum among N largest values a return period which is 78% larger than the number of observations.

Numerical values of the occurrence intervals are given in column 5 of Table 6.1.1. From (2) the occurrence interval converges to

(3') $$T(\tilde{y}_n) = N + \tfrac{1}{2}.$$

The extremal characteristics u_N and α_N are defined by

(4) $$\exp[-e^{-u_N}] = 1 - 1/N; \quad \alpha_N = Nf(u_N).$$

For given values of N the characteristic maximum u_N is obtained from the Bureau of Standards tables. Since

(4') $$u_N = -\lg[-\lg(1 - 1/N)] \sim \lg N - 1/2N - 5/24N^2,$$

a formula which is quite accurate from $N = 100$ onward, we have

(5) $$u_N < \lg N = \tilde{y}_N.$$

The modes, $\lg N$, and the characteristic largest values, u_N, are given in Table 6.1.1., columns 2 and 3, as function of N (column 1). For increasing N, the characteristic maxima increase of course with N and tend to the modal maxima.

The extremal intensity function α_N obtained from (4),

(6) $$\alpha_N = N(1 - 1/N)[-\lg(1 - 1/N)],$$

given in column 4 of Table 6.1.1, increases with N towards unity.

Table 6.1.1. Extreme Values for the Double Exponential Distribution

1	2	3	4	5	6
Number of Extreme Observations	Modal Largest Value	Characteristic Largest Value	Extremal Intensity	Occurrence Interval	Probability
N	$\tilde{y}_N$	u_N	$\mu_N = \alpha(u_N)$	$T(\tilde{y}_N) - N$	F
2	0.69315	0.36651	.69315	.54149	.5000
4	1.38629	1.24590	.86305	.52081	.7500
5	1.60944	1.49994	.89257	.51667	.8000
8	2.07944	2.01342	.93472	.51041	.8750
10	2.30259	2.25037	.94824	.50833	.9000
20	2.99573	2.97020	.97457	.50417	.9500
25	3.21888	3.19853	.97973	.50333	.9600
40	3.68888	3.67625	.98739	.50208	.9750
50	3.91202	3.90194	.98993	.50167	.9800
80	4.38203	4.37654	.99382	.50104	.9875
100	4.60517	4.60015	.99498	.50083	.9900

6.1.2 USES OF THE FIRST ASYMPTOTE

The intensity function and the probability F are related by

$$(7) \qquad \mu = \frac{-\lg F}{1/F - 1}.$$

The limit is obtained from

$$\lim_{y=\infty} \frac{\mu(y)}{F(y)} = \lim_{y\to\infty} \frac{e^{-y}}{1 - \exp[-e^{-y}]} = 1.$$

Consequently, the intensity tends to the probability function and converges toward unity with increasing y. Finally, the critical quotient Q, defined in 4.1.3(3) becomes, from (6), for $y = u_N$,

$$\frac{\alpha_N}{-f'(u_N)/f(u_N)} = \frac{1 - 1/(2N) - 1/(6N^2) - \cdots}{1 + \lg(1 - 1/N)}.$$

It follows that

$$Q = 1 + 1/(2N) + 5/(6N^2) + \cdots$$

converges to unity as N increases.

Thus, the double exponential distribution belongs to the first class of distributions of the exponential type. Consequently, the distribution of the maximum again converges to the double exponential distribution. This seems to contradict the stability for any number N of samples. The contradiction is only apparent, because the asymptotic distribution of the largest value contains parameters α_N and u_N, which differ from but converge to α and $\lg N$.

6.1.2. Minima of Largest Values. The first (second) double exponential distribution is stable with respect to the largest (smallest) value. The converse is true in the following sense: The distribution of the largest (smallest) value taken from the second (first) double exponential distribution converges toward the first asymptotic distribution. For the proof, consider the critical quotient Q, for the second double exponential distribution.

(1) $f(y) = \exp[y - e^y]$: $\quad F(y) = 1 - \exp[-e^y]; \quad \mu(y) = e^y;$
$$-f'(y)/f(y) = e^y - 1.$$

It follows that

$$(2) \qquad Q = e^y/(e^y - 1).$$

By virtue of 4.1.3(4) the distribution of the largest (smallest) value taken from the second (first) double exponential distribution belongs to the first class of distributions of the exponential type.

Since the double exponential distribution is asymmetrical, the distribution of the minimum, taken from the first double exponential distribution, behaves differently from the distribution of the maximum. The characteristic minimum $\hat{y}_1$ is

(3) $$\hat{y}_1 = -\lg (\lg N),$$

an expression which decreases very slowly with N as shown in Table 6.1.2(1).

Table 6.1.2(1). Characteristic Minima

v	$N = e^v = {}_1T(\hat{y}_1)$	$-\hat{y}_1$	$\Phi(\hat{y}_1)$
1	2.718	0.0000	.36788
2	7.389	0.6932	.13534
3	20.09	1.0986	.04979
4	54.60	1.3863	.01832
5	148.4	1.6094	.00674
6	403.4	1.7918	.00248
7	1097	1.9459	.00091

From the density of probability at the characteristic minimum, it follows that

(4) $$\alpha_1 = N\varphi(\hat{y}_1) = \lg N$$

increases with N. The distribution of the minimum contracts with increasing sample size. The modal minimum $\tilde{y}_1$ is, from 3.1.6(2), the solution of

$$N - 1 = \varphi_1'(1 - \Phi_1)\varphi_1^{-2},$$

where $\quad \Phi_1 = \Phi(\tilde{y}_1); \quad \varphi_1 = \varphi(\tilde{y}_1).$

From 5.2.1(I) the right side may be written

$$\frac{(1 - \Phi_1)(e^{-\tilde{y}_1} - 1)}{e^{-\tilde{y}_1}\Phi_1} = \frac{1}{\Phi_1} - \frac{e^{\tilde{y}_1}}{\Phi_1} + e^{\tilde{y}_1} - 1.$$

Consequently,

(5) $$N = \frac{1}{\Phi_1} + \frac{1}{\Phi_1(-\lg \Phi_1)} + \frac{1}{(-\lg \Phi_1)}$$

or

$$\frac{1}{N} = \frac{\Phi_1(-\lg \Phi_1)}{-\lg \Phi_1 - 1 + \Phi_1}.$$

We choose certain values of Φ_1, and obtain N. Finally $\tilde{y}_1$ is obtained as the solution of

(6) $$-\tilde{y}_1 = \lg(-\lg \Phi_1)$$

from the Bureau of Standards tables. The values of $\Phi(\tilde{y}_N)$, $\Phi(\tilde{y}_1)$, $\tilde{y}_1$, and $T(\tilde{y}_1)$ are given in Table 6.1.2(2).

6.1.2 USES OF THE FIRST ASYMPTOTE

Table 6.1.2(2). Probabilities of Maxima and Minima of Largest Values

1	2	3	4	5
	Probability of Most Probable		Smallest Value	
Number of Largest Values N	Maximum $\Phi(\tilde{y}_N)$	Minimum $\Phi(\tilde{y}_1)$	Occurrence Interval $T(\tilde{y}_1)$	Mode $-\tilde{y}_1$
2	.60653	.2356	4.24	.369
4	.77880	.1405	7.12	.676
6	.84648	.1013	9.87	.828
8	.88250	.0795	12.6	.929
10	.90484	.06565	15.2	1.002
15	.93551	.04600	21.7	1.125
20	.95123	.03554	28.2	1.205
25	.96079	.02902	34.5	1.264
30	.96720	.02456	40.7	1.310
35	.97182	.02130	46.9	1.348
40	.97531	.01882	53.2	1.379
45	.97802	.01687	59.5	1.408
50	.98020	.01529	65.4	1.430
60	.98346	.01289	77.6	1.470
70	.98581	.01131	88.4	1.500
80	.98758	.00982	102	1.531
90	.98895	.00878	114	1.555
100	.99005	.00795	126	1.576
200	.99501	.00409	244	1.705
300	.99667	.00277	361	1.773
400	.99750	.00209	478	1.820
500	.99800	.00168	595	1.855
750	.99867	.00144	877	1.913
1000	.99900	.00086	1163	1.954

The second formula in 5.2.4(4), with u_1 and α_1 given by (3) and (4), leads to the asymptotic generating function $G_1(t)$ of the minimum y_1,

(7) $\qquad G_1(t) = (\lg N)^{-t} \Gamma(1 + t/\lg N).$

The asymptotic mean $\bar{y}_1$, and the variance σ_1^2 of the minimum become

(8) $\qquad \bar{y}_1 = -\lg(\lg N) - \gamma/\lg N; \qquad \sigma_1^2 = \pi^2/(6 \lg^2 N).$

The absolute value of the mean increases and the variance decreases very slowly with N. Table 6.1.2(3) is useful for the calculation of the averages for the minimum and for very large N.

Table 6.1.2(3). The Iterated Natural Logarithm

$\lg(\lg N)$	2.4	2.5	2.6	2.7	2.8	2.9	3.0	3.1
N	$5.99 \cdot 10^4$	$1.95 \cdot 10^5$	$7.03 \cdot 10^5$	$2.90 \cdot 10^6$	$1.39 \cdot 10^7$	$7.81 \cdot 10^7$	$5.28 \cdot 10^8$	$4.37 \cdot 10^9$

The exact mean minima (maxima) for the distribution of the largest (smallest) value were calculated by the National Bureau of Standards for $N = 2(1)10(5)20$. The approach to the asymptotic values is rapid, since for $N = 20$ the difference is only 5%. The averages of the maxima increase as $\lg N$; the averages of the minima decrease as $\lg(\lg N)$. The occurrence interval of the maxima differs only slightly from N; the occurrence interval of the minima exceeds N and converges to it very slowly. The standard deviation of the maximum is equal to the initial standard deviation, while the standard deviation (8) of the minimum is much smaller. The distribution of the maxima is stable. The distribution of the minima contracts with increasing sample sizes.

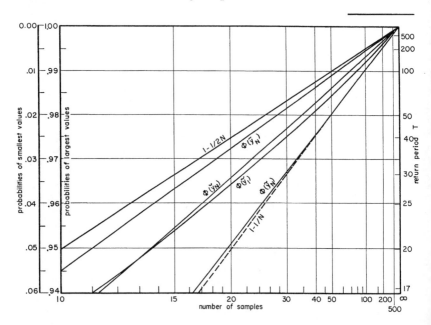

Graph 6.1.2. Probabilities of Average Extreme Values

The probabilities at the modal, median, and mean maxima, and the probabilities at the modal minima obtained from 6.1.1(2) and Table 6.1.2(2), are traced against the number of extremes N in Graph 6.1.2. It shows that Hazen's plotting position, 1.2.6(3), the probabilities $\Phi(\bar{x}_N)$ and $\Phi(\check{x}_N)$, and the probabilities at the most probable minima lead to return periods which are far larger than the sample size.

Exercise: Calculate the generating function from the asymptotic distribution of the minimum.

References: Barricelli, Kimball (1946a, b, 1949).

6.1.3. Consecutive Modes. The modes $\tilde{y}_m(N)$ of the mth largest and $\tilde{y}_{m'}(N)$ of the mth smallest values among N observations obtained from the recurrent procedure 3.1.6(6) are traced in Graph 6.1.3.

6.1.3　USES OF THE FIRST ASYMPTOTE

We now change the designation and write $y_{N-m+1}(N)$ instead of $y_m(N)$. The recurrent procedure 3.1.6(6) may be used to obtain approximate expressions for the modes $\tilde{y}_{N-m+1}(N)$ of the mth largest values for N observations and their occurrence intervals. We ask for a number N_m so that

$$\tilde{y}_{N-m+1}(N) = \lg(N) \tag{1}$$

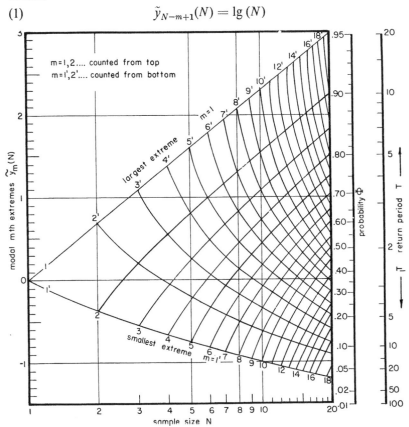

Graph 6.1.3.　Modal mth Extremes $\tilde{y}_m(N)$

satisfies the equation 2.1.2(3) for the mode of the mth largest value for N_m observations,

$$\frac{N_m - m}{\Phi}\varphi - \frac{m-1}{1-\Phi}\varphi + \frac{\varphi'}{\varphi} = 0. \tag{2}$$

Since for the mode of the maximum,

$$\Phi = e^{-1/N}; \qquad \varphi = \frac{1}{N}e^{-1/N}; \qquad \varphi'/\varphi = -1 + \frac{1}{N},$$

equation (2) becomes after expansion

(2') $$N_m \sim Nm + (m-1)/2.$$

If $m \ll N$, the modal mth largest value in Nm observations is equal to the modal largest value in N observations. To obtain $\tilde{y}_{N-m+1}(N)$, we write $N_m = N'$, and express N by N'. From (2) it follows that

$$N = (2N' - m + 1)/2m,$$

whence

(3) $$\tilde{y}_{N'-m+1}(N) = \lg \frac{2N' - m + 1}{2m}.$$

Finally, when the prime for $m = 1, 2, 3, 4$ has been dropped, the modal mth largest values become asymptotically

(4) $$\tilde{y}_N(N) = \lg N; \qquad \tilde{y}_{N-1}(N) = \lg \left(\frac{N}{2} - \frac{1}{4}\right);$$

$$\tilde{y}_{N-2}(N) = \lg \left(\frac{N}{3} - \frac{1}{3}\right); \qquad \tilde{y}_{N-3}(N) = \lg \left(\frac{N}{4} - \frac{1}{8}\right).$$

Formula (3) leads to an approximation for the occurrence interval. Dropping the prime in (3),

$$1/T_{N-m+1} = 1 - \exp\left[(-2m)/(2N - m + 1)\right],$$

whence

(5) $$T_{N-m+1} \sim \frac{N+1}{m} - \frac{1}{2m}.$$

The occurrence intervals approach the return periods $(N+1)/m$, which correspond to the plotting positions $m/(N+1)$. The first occurrence intervals for large N and small m are

$$T_N = N + \tfrac{1}{2}; \qquad T_{N-1} = \frac{N}{2} + \frac{1}{4}; \qquad T_{N-2} = \frac{N}{3} + \frac{1}{6};$$

$$T_{N-3} = \frac{N}{4} + \frac{1}{8}; \ldots.$$

6.1.4. Consecutive Means and Variances. The distribution $\phi_{N-m+1}(y)$ of the mth largest ($m \geq 2$) among N values $y_{N-m+1} = y$, taken from the first double exponential distributions, becomes from 2.1.1(1) after expansion of the binomial,

$$\varphi_{N-m+1}(y) = \binom{N}{m} m \sum_{v=0}^{m-1} \binom{m-1}{v} (-1)^v \exp\left[-y - (N - m + v + 1)e^{-y}\right].$$

6.1.4 USES OF THE FIRST ASYMPTOTE

This is the sum of initial distributions of linearly transformed largest variates shifted along the horizontal axis. The generating function $G_{N-m+1}(t) = G(t)$ is

(1) $\quad G(t) = \binom{N}{m} m \sum_{v=0}^{m-1} \binom{m-1}{v} (-1)^v \int_{-\infty}^{\infty} \exp[-y + yt$
$$- (N - m + v + 1)e^{-y}]\, dy\,.$$

If we write
$$(N - m + v + 1)e^{-y} = e^{-z},$$
the integral becomes
$$\frac{(N - m + v + 1)^t}{N - m + v + 1} \int_{-\infty}^{\infty} \exp[-z + zt - e^{-z}]\, dz$$
$$= \frac{(N - m + v + 1)^t}{N - m + v + 1} \Gamma(1 - t)\,.$$

Thus the generating function of the mth value from above can be expressed by the generating function of the variate itself as

(2) $\quad G(t) = \Gamma(1 - t) \sum_{0}^{m-1} \frac{N(N - 1) \ldots (N - m + 1)}{v!(m - v - 1)!} (-1)^v$
$$\frac{(N - m + v + 1)^t}{N - m + v + 1}\,.$$

For $m = N$, i.e., for *the smallest value*,
$$G_1(t) = \Gamma(1 - t) \sum_{0}^{N-1} \frac{N!}{(N - v - 1)!\,(v + 1)!} (-1)^v (v + 1)^t\,.$$

If we put $v + 1 = \lambda$, this expression becomes

(3) $\quad G_1(t) = -\Gamma(1 - t) \sum_{1}^{N} \binom{N}{\lambda} (-1)^\lambda \lambda^t\,,$

whence

(4) $\quad \bar{y}_1 = \gamma - \sum_{1}^{N} \binom{N}{\lambda} (-1)^\lambda \lg \lambda\,,$

a formula due to Barricelli. The generating functions for the penultimate ($m = 2$) and the preceding largest values are

(5) $\quad G_{N-1}(t) = \Gamma(1 - t)(N(N - 1)^t - N^t(N - 1))$

and

(6) $\quad G_{N-2}(t) = \dfrac{\Gamma(1 - t)}{2!}\,(N(N - 1)(N - 2)^t - 2N(N - 1)^t(N - 2)$
$$+ N^t(N - 1)(N - 2))\,.$$

With enough patience, the moments can be calculated. For the penultimate value, the first derivative of the generating function is

$$G'_{N-1}(t) = \Gamma'(1-t)(N(N-1)^t - N^t(N-1)) + \Gamma(1-t)(N(N-1)^t \\ \times \lg(N-1) - N^t(N-1)\lg N).$$

For $t = 0$, the mean,

(7) $\qquad \bar{y}_{N-1} = \gamma + N\lg(N-1) - (N-1)\lg N,$

becomes, after expansion of the second member for large N,

(8) $\qquad \bar{y}_{N-1} = \lg N + \gamma - 1 - 1/2N + 0(N^{-2}).$

The second moment becomes, after expansion of the logarithm and for large N,

$$\overline{y^2_{N-1}} \sim \gamma^2 + \pi^2/6 + 2\gamma(\lg N - 1 - 1/(2N)) \\ + \lg^2 N - 2\lg N - (\lg N)/N + 1/N + 0(N^{-2}).$$

We subtract $\overline{y_{N-1}}^2$ and obtain the variance σ^2_{N-1} of the penultimate value.

(9) $\qquad \sigma^2_{N-1} = \dfrac{\pi^2}{6} - 1 + 0(N^{-2}).$

If we pass from the last to the penultimate reduced value, the variance decreases by unity for large N, and the standard deviation decreases from 1.28255 to .80308, i.e., by about 30%. The mean of the third largest value obtained from (6) and (7) is

(10) $\qquad \bar{y}_{N-2} = \gamma + \lg N - 3/2 - 1/(3N) - 0(N^{-2}).$

The distance from the mean to the mode diminishes with m. The respective distances are, for $m = 1, 2, 3,$

(11) $\qquad \bar{y}_N - \tilde{y}_N = \gamma = 0.57722;$

$\qquad\qquad \bar{y}_{N-1} - \tilde{y}_{N-1} = \gamma - 0.30685 = 0.27037;$

$\qquad\qquad \bar{y}_{N-2} - \tilde{y}_{N-2} = \gamma - 0.40139 = 0.17583.$

This result indicates that the asymptotic distributions contract with increasing rank m.

B. F. Kimball (1949) has put the expression (2) into a form which can be handled easily by automatic computing machines. The generating function (5) of the penultimate value can be written

$$G_{N-1}(t) = \Gamma(1-t)\binom{N}{2} 2\Delta^{(1)}[(N-1)^{t-1}],$$

where $\Delta^{(v)}$ stands for the difference of order v. In the same way,

$$G_{N-2}(t) = \Gamma(1-t)\binom{N}{3} 3\Delta^{(2)}[(N-2)^{t-1}].$$

6.1.4 USES OF THE FIRST ASYMPTOTE

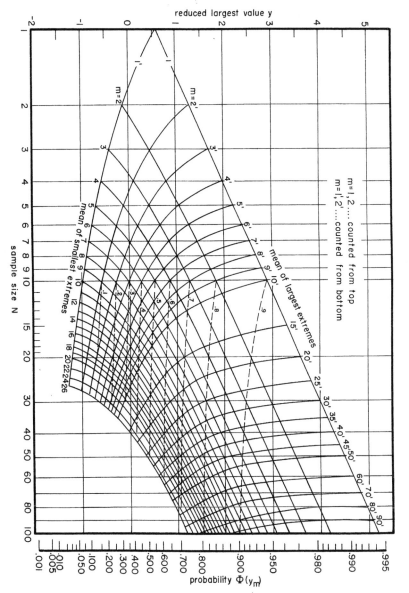

Graph 6.1.4. Mean mth Extremes $y_m(N)$

From induction it follows that the generating function of the mth value from the top is

(12) $\quad G_{N-m+1}(t) = \Gamma(1-t) \binom{N}{m} m \Delta^{(m-1)} [(N-m+1)^{t-1}]$.

The mean mth values,

(13) $\quad \bar{y}_{N-m+1} = \gamma + \binom{N}{m} m \Delta^{(m-1)} \left[\dfrac{\lg(N-m+1)}{N-m+1} \right]$,

calculated by the National Bureau of Standards for $N = 2(1)10(5)50(10)100$, $m = 1(1)26$ are traced in Graph 6.1.4. The first dotted curve joins the mean largest value among 10 to the mean second largest among 20 observations and so on, up to the 10th among 100 observations. The last dotted curve joins the mean 9th among 10 observations to the 18th among 20 observations, etc. The mean smallest value is so close to the asymptotic formula 6.1.2(8) that it cannot be shown on the graph.

J. Lieblein (1953) derived closed formulae for the covariance of the order statistics. The formulae, simpler than those for the normal distribution, involve the moments of the mth values and the so-called Spence function, for which tables by Newman exist. In the subsequent brochure (1954a) he gave the numerical values of $\text{Cov}_n(y_i, y_j)$ for $n = 2(1)6$ and $i, j = 1(1)6$.

Exercises: 1) Calculate the variances of the third and fourth largest values. 2) Express the variance of the mth largest value by the mth differences of the logarithms.

Problem: Prove analytically that the mean kmth largest among kN observations converges toward the reduced largest value y corresponding to the probability $\Phi(y) = m/N$.

Reference: J. Lieblein and H. E. Salzer (1957). Table of the first moment of ranked extremes, J. Res. Natl. Bur. Stand., 59: 203.

6.1.5. Standard Errors. The reduced standard error σ_m for the mth central value obtained from 2.1.4(6) and 5.2.1(14) is given by

(1) $\quad \alpha \sqrt{N} \sigma_m = \sqrt{1/\Phi - 1}/(-\lg \Phi)$.

Numerical values are given in Table 6.1.5. The most probable reduced value, $y = 0$, has the control interval $\pm \sqrt{(e-1)/N}$. The intervals about $y = 0$ corresponding to $P = 0.68269$ and $P' = 0.95450$, traced in Graph 6.1.5. as functions of $1/\sqrt{N}$, are straight lines. Two additional scales show the probabilities $\Phi(y)$ and the return periods T. The graph may be used as a *homogeneity test for floods* in the following way: Different rivers in a certain region with catchment areas of similar sizes show different most probable annual floods. They are estimated by plotting the floods on extremal probability paper and their mean for the N stations considered is

6.1.5 USES OF THE FIRST ASYMPTOTE

taken. Then the return periods of these means differ for the different stations. However, the region may be considered as homogeneous if, say, 95% of the observed return periods lie within the control interval. Those which depart markedly are considered inhomogeneous. This procedure, a modification of a method devised by Langbein and used by the U.S.

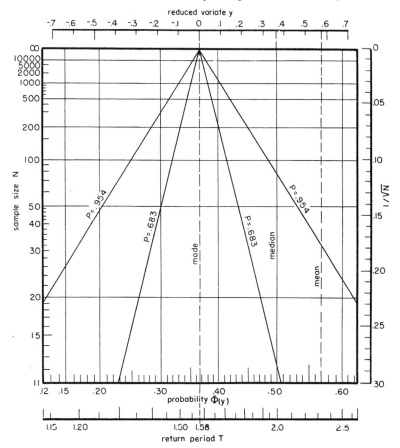

Graph 6.1.5. Homogeneity Test for the Most Probable Annual Flood

Geological Survey (Carter, Robinson and Bodhaine), may also be used for other extremes. No estimation of the parameters is required.

The order statistic with minimum variance 2.1.5(1′) is obtained from (1), after differentiating σ_m^2 with respect to y, as the solution y of

$$2(\exp e^{-y} - 1) = \exp(-y + e^{-y}),$$

whence

(2) $\quad \hat{y} = -0.56601; \quad \Phi(\hat{y}) = .20319; \quad \sqrt{N}\sigma_m = 1.24263.$

The standard errors $\sigma(x_m)$ of the mth values x_m are obtained from the reduced values $\sigma_m\sqrt{N}$ as

(3) $$\sigma(x_m) = (\alpha\sigma_m \sqrt{N})/(\alpha\sqrt{N}).$$

This leads to the control band for the mth central values, described in 2.1.6, valid as long as the order statistics are normally distributed, say, from $\Phi = .15$ to $\Phi = .85$.

Table 6.1.5 Standard Errors of Reduced Central Values*

1	2	3	4	5
	Reduced		Reduced	
Probability	Standard Error	Reduced Variate	Standard Error	Standard Error
Φ	$\sqrt{N}\alpha\sigma_m$	y	$\sqrt{N}\alpha\sigma_y$	Mississippi River
.01	(2.1607)	−1.52718	1.854	
.02	(1.7894)	−1.36405	1.718	
.05	(1.4550)	−1.09719	1.511	
.10	(1.3028)	−0.83403	1.332	
.15	1.2548	−0.64034	1.223	45.9
.20	1.2427	−0.47588	1.151	45.4
.25	1.2494	−0.32663	1.106	45.6
.30	1.2687	−0.18563	1.082	46.3
.35	1.2981	−0.04862	1.078	47.4
.40	1.3366	0.08742	1.092	48.8
.45	1.3845	0.22501	1.125	50.6
.50	1.4427	0.36651	1.177	52.7
.55	1.5130	0.51444	1.248	55.3
.60	1.5984	0.67173	1.339	58.4
.65	1.7034	0.84215	1.452	62.2
.70	1.8355	1.03093	1.592	67.0
.75	2.0069	1.24590	1.765	73.3
.80	2.2408	1.49994	1.982	81.8
.85	2.5849	1.81696	2.269	94.4
.90	(3.1639)	2.25037	2.678	
.95	(4.4721)	2.97020	3.382	
.98	(7.0710)	3.90194	4.320	
.99	(10.000)	4.60015	5.033	

* Column 2 is obtained from (1). The figures in brackets apply to values outside of $.15 < \Phi < .85$ interval. Columns 4 and 5 will be explained in 6.2.3(6) and 6.3.1 respectively.

In addition, we construct intervals so that the extreme observations have a probability, $P = .68269$ (or $P' = .95450$) to be contained therein. As for the central values, we simplify the task by using the asymptotic instead of the exact distribution of the extreme. From the infinity of intervals defined by P, we determine one by requiring that the modal extreme values be situated in the middle. The reduced length Δ of this interval is obtained from the stability as the solution of

(4) $$\Phi(\Delta) - \Phi(-\Delta) = P.$$

6.1.5 USES OF THE FIRST ASYMPTOTE

For $P = .68269$ (or $P' = .05450$) the Bureau of Standards Tables lead to the reduced values,

(5) $\qquad \Delta = 1.14078\ ;\qquad \Delta' = 3.0669\ .$

From 6.1.1(6) and 6.1.2(4) the asymptotic intervals about the largest and smallest values are

(6) $\qquad \Delta_N = 1.14078/\alpha \qquad \Delta_1 = 1.14078/(\alpha \lg N)$
$\qquad\qquad \Delta_N' = 3.0669/\alpha \qquad \Delta_1' = 3.0669/(\alpha \lg N)\ .$

If the values $\Delta_1(\Delta_N)$ are added to and subtracted from the values $x_1(x_N)$ situated on the theoretical straight line at the abscissa $1/(N+1)$; $[1 - 1/(N+1)]$, control intervals $x_1 \pm \Delta_1 [x_N \pm \Delta_N]$ are obtained. The ends of these intervals are joined to the central control band, which now extends from the smallest to the largest value. The control band to be used for the extrapolation starting with the largest value consists of two straight lines parallel to the theoretical line at the vertical distances $\pm 1.14078/\alpha$, i.e., from 5.2.5(1), at about 89% of the initial standard deviation.

Since the y scale on the probability paper corresponds, for large values, to a logarithmic scale for T, the parallel straight lines expand considerably if measured in return periods. The return period corresponding to a reduced largest value $y + \Delta$ is, from 5.2.6(3'), for $\Delta < y - 2$ and $y > 2$,

(7) $\qquad\qquad T(y + \Delta) = T(y)e^{\Delta} + (1 - e^{\Delta})/2\ ,$

whence, from (5),

(8) $T(y + \Delta) = 3.12918 T(y) - 1.06459$:
$\quad T(y - \Delta) = 0.31957 T(y) + 0.34021\ .$

These control intervals may be used for the largest flood of T years and for the extrapolation and *do not require the estimation of the parameters.* From $T \geq 25$ the T year flood has a probability .68268 of occurring in the interval $0.32T$ to $3.13T$. Of course, we may also choose other probabilities P. However, for $P = .95$, the limit becomes so broad as to be practically meaningless. In contrast to the usual control intervals, an increase in the sample size N does not lead to a reduction of the control interval.

It is interesting to investigate the extent of the error in using the formula (3) for the maximum $m = N$, although it holds only for the central mth values. Since then

$$\Phi = 1 - 1/N,$$

formula (3) becomes

(9) $\qquad\qquad \sigma(x_N) = \dfrac{1}{\alpha N \sqrt{1 - 1/N}\,[-\lg(1 - 1/N)]} \sim 1/\alpha$

if $O(N^{-2})$ is neglected. This result differs from the value (6) by 14%.

In the same way, the application of (3) to the smallest value leads to

$$(9') \qquad \sigma(x_1) = \frac{N\sqrt{1-1/N}}{\alpha N[-\lg(1/N)]} \sim \frac{\sqrt{1-1/N}}{\alpha \lg N},$$

which differs from (6) by the same percentage

Exercises: Calculate the relations corresponding to (8) for the 1) exponential, 2) logistic, 3) normal, 4) logarithmic normal distribution, and for large values of the variate.

References: Dalrymple (1950), Dick and Darwin, Levert (1954), Lieblein (1954a).

6.1.6. Extension of the Control Band.

In the following, it will be shown how the exact distribution of the mth largest value taken from the first double exponential distribution converges for $m \ll N$ to the asymptotic distribution 5.3.1(2) of the mth largest value. The result will be used to estimate the factors Δ_{N-m+1} necessary for the construction of the control intervals for the mth extreme values. In the exact distribution $\varphi_N(x_m)$, 2.1.1(1) of the mth largest value x_m among N observations taken from the first double exponential distribution, we consider the expansion

$$(1) \qquad 1 - \Phi(x) = e^{-y} - e^{-2y}/2! + e^{-3y}/3! - \ldots$$

for $y = y_m$ in the neighborhood of the mode $\tilde{y}_m$. The order of magnitude of the mode obtained from the plotting position of the mth largest value is

$$\tilde{y}_m = -\lg[-\lg(1 - m/(N+1))].$$

Hence, for large N,

$$O(\tilde{y}_m) = \lg(N/m).$$

It follows that the higher orders of e^{-y} in the development (1) may be neglected, and that

$$(2) \qquad 1 - \Phi(x) \sim \exp[-y - \tfrac{1}{2}e^{-y}],$$

a formula which introduces only an error of the magnitude $e^{-2y}/24$. From this formula, the distribution $\varphi_N(x_m)$ becomes approximately

$$(3) \qquad \varphi_N(x_m) \sim \binom{N}{m} m\alpha \, \exp\left[-(N-m)e^{-y} \right.$$
$$\left. - (m-1)y - \frac{m-1}{2}e^{-y} - y - e^{-y} \right].$$

The exponent may be written

$$-my - \left(N - \frac{m-1}{2}\right)e^{-y}$$
$$= -my - m\exp\left[-y - \lg\left\{\frac{N}{m} - \frac{m-1}{2m}\right\}\right].$$

6.1.6　USES OF THE FIRST ASYMPTOTE

Repetition of the same procedure leads to the approximation

$$(4) \quad \varphi_N(x_m) \sim \binom{N}{m} m\alpha \left(\frac{m}{N - (m-1)/2}\right)^m$$
$$\exp\left[-m\left[y - \lg\left\{\frac{N}{m} - \frac{m-1}{2m}\right\}\right] - me^{-[y - \lg\{\frac{N}{m} - \frac{m-1}{2m}\}]}\right].$$

The constant factor is for large N, i.e., if we neglect $O(N^{-2})$,

$$\frac{\alpha N^m m^m (1 - 1/N)(1 - 2/N) \cdots (1 - (m-1)/N)}{(m-1)! N^m (1 - (m-1)/2N)^m} \sim \frac{\alpha m^m}{(m-1)!}.$$

Thus, the distribution $\varphi_N(x_m)$ becomes asymptotically for $m = O(1) \ll N$,

$$(5) \quad \varphi_N(x_m) = \frac{\alpha m^m}{(m-1)!} \exp\left[-m\left\{y - \lg\left(\frac{N}{m} - \frac{m-1}{2m}\right)\right\}\right.$$
$$\left. - me^{-\{y - \lg(\frac{N}{m} - \frac{m-1}{2m})\}}\right],$$

which is the asymptotic distribution 5.3.1(2) with

$$y_m = y - \lg\left(\frac{N}{m} - \frac{m-1}{2m}\right).$$

If we now count m from the bottom,

$$(6) \quad \alpha_{N-m+1} = \alpha; \quad u_{N-m+1} = u + \frac{1}{\alpha} \lg\left(\frac{N}{m} - \frac{m-1}{2m}\right).$$

Thus, the two parameters in the asymptotic distribution of the mth largest values for the first double exponential distribution considered as the initial one are obtained from α, the observed number of largest values N, and the chosen rank m of the extreme. For $m = 1, 2, 3$, the modes of the largest and the two preceding values are

$$u_N = u + 1/\alpha \lg N; \quad u_{N-1} = u + 1/\alpha \lg\left(\frac{N}{2} - \frac{1}{4}\right);$$
$$(7) \quad u_{N-2} = u + 1/\alpha \lg \frac{N-1}{3}.$$

The first formula is simply the stability postulate. The difference between these values and the values

$$(7') \quad u_{N-m+1} = u - 1/\alpha \lg\left(-\lg\left(1 - m/(N+1)\right)\right),$$

resulting from the plotting positions of the observed mth extremes, are negligible for $N = 100$, and $m = 1, 2, 3$; for $N = 50$, and $m = 1, 2$, and even for $N = 25$ and $m = 1$. If N is greater than, say, 100, formula (6) may be used for the 2nd, 3rd, etc. largest values. Thus the control curves valid in the neighborhood of the median can be extended to the mth largest values. Table 5.3.2 leads the control band given in Table 6.1.6.

Table 6.1.6 The Extreme Control Band

Interval	Probability		Mississippi River at Vicksburg (6.3.1)
	$P = .68269$	$P = .95450$	$P = .68269$
Δ_N	$1.14078/\alpha$	$3.0669/\alpha$	322.7
Δ_{N-1}	$0.75409/\alpha$	$1.7820/\alpha$	216.9
Δ_{N-2}	$0.589/\alpha$	$1.35/\alpha$	171.9
Δ_{N-3}	$0.538/\alpha$	$1.17/\alpha$	—
Δ_1	$1.14078/(\alpha \lg N)$	$3.0669/(\alpha \lg N)$	78.8

On probability paper, the lengths Δ_m are added to and subtracted from the theoretical value x_m, situated on the straight line. If the ends of these intervals are joined, this curve, together with the curve traced for the central values 6.1.5(3) and the extrapolation 6.1.5(6), gives a control band covering the whole observed range. If we pass from $P = .683$ to $P = .955$, the interval for the central mth value is doubled; for the extreme mth values the increase is stronger.

6.1.7. The Control Curve of Dick and Darwin. Another system of control curves corresponding to a fixed probability is due to I. D. Dick and J. H. Darwin. They state that the distribution of

$$t = (u + y_T/\alpha - \bar{x}_0)/s$$

is independent of u and α. Then they use the normal distribution as an approximation for the distribution of $\bar{x}_0 + ts$. This leads to two values t_1 and t_2, so that the probability for the largest value x_T to be contained in the interval

$$\bar{x}_0 + t_1 s \leq x_T \leq \bar{x}_0 + t_2 s$$

is 2/3. Here t_1 and t_2 are obtained as the solutions of

$$[(y_T - \gamma)/\sigma - t]^2 = (1 + 1.1396t + 1.1t^2)/N.$$

The authors give a table of t_1 and t_2 for $N = 20$ and 50 and $T = 1.05$, 1.5, 2.5, $.5 \, 10^v$, $1 \, 10^v$, $2 \, 10^v$ ($v = 1, 2, 3$). The control curves thus obtained agree fairly well with those given in Table 6.1.5. In 240 samples of size 40 taken at random for 5 values of T, more values fell inside of the control curves than was to be expected. Other systems of control curves will be shown in 6.2.2 and 6.2.3.

The expressions for the control interval hold only for the distributions

6.2.1 USES OF THE FIRST ASYMPTOTE

of the exponential type. For distributions of the Pareto type, the standard errors of the extremes need not exist and must be replaced by the probable errors or similar quantiles.

Problems: Prove the statements of Dick and Darwin.

6.2. ESTIMATION OF PARAMETERS

"A net is a series of holes connected by a string."

6.2.0. Problems. In the first paragraph, we compare computed distributions of extremes taken from specified initial distributions to their asymptotic distributions. These methods are developed to permit comparing observed extremes, taken from unknown initial distributions of the exponential type, with the asymptotic distribution.

The estimation of parameters has two aspects: If the initial distribution, its parameter, and the large sample size n from which each largest observation is taken are known, the parameters u_n and $1/\alpha_n$ are obtained from their definition 3.1.4(1) and 3.1.5(1), as shown for the exponential, logistic, normal, and double exponential distributions in 4.1.1, 4.2.1, 4.2.2, and 6.1.1. In practical applications, the initial distribution and its parameter are usually unknown, and even the number n may be unknown. If, however, we have reasons to believe that the initial distribution is of the exponential type, we need not worry about its analytic expression and we may estimate the parameters u and $1/\alpha$ from the N observed largest values alone. The first estimate of parameters is based on order statistics. Then the general methods developed in 1.2.8 are adapted to the probability paper. Since Mr. Kimball has done considerable work on the estimation of parameters by the method of maximum likelihood, the author has asked him to present his own methods in paragraphs 6.2.4, 6.2.5, and 6.2.7.

6.2.1. Exponential and Normal Extremes. The asymptotic and the exact probabilities of exponential, logistic, and normal extremes for given sample sizes n are traced in Graphs 6.2.1(1), (2), and (3). Even for $n = 10$, the differences are barely visible for the exponential distribution. The same quick convergence holds, of course, for the extreme value taken from a Laplacean distribution. The approach of the exact probabilities of extreme value to the asymptotic probabilities is also very quick for the logistic distribution.

For the normal distribution, however, the approach is very slow. The curves for $n = 100$, 200, 500, and 1000 taken from Tippett (1925) depart sensibly from a straight line, if we go outside the interval .05 to .95. Thus, for a small number N of extremes, say, $N \leq 20$, the asymptotic theory may

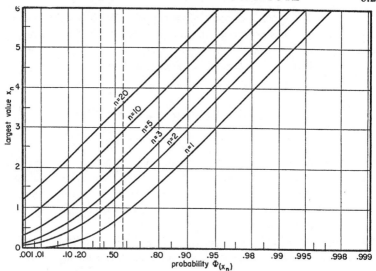

Graph 6.2.1(1). Probabilities of Extreme Exponential Values

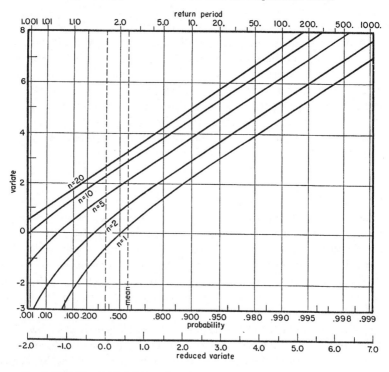

Graph 6.2.1(2). Probabilities of Extreme Values for the Logistic Variate

6.2.1 USES OF THE FIRST ASYMPTOTE

be used. But for large samples the calculated extremes of the extremes depart from the asymptotic values, especially if we are interested in the return periods.

The calculated distribution $\Delta\Phi(x_n)$ of extreme normal values for $n = 10$, 100, 1000, obtained as differences of the probabilities given by Tippett

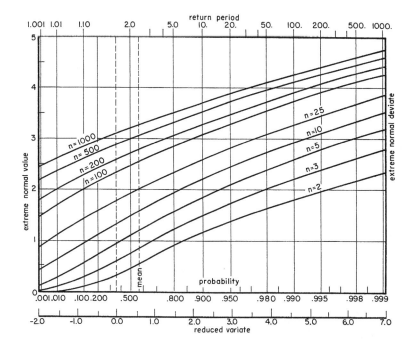

Graph 6.2.1(3). **Probabilities of Normal Extremes and Extreme Deviates**

(1925), are traced in Graph 6.2.1(4). To obtain the asymptotic distributions we take the values u_n and $1/\alpha_n$ for $n = 10, 100, 1000$, from Table 4.2.2(1), and calculate the extremes $x_n = u_n + y/\alpha_n$. The difference of probabilities $2(\Phi(y + 1/2\alpha) - \Phi(y))$ divided by the corresponding lengths Δx_n are theoretical counterparts to the calculated values $\Delta\Phi(x_n)$ and are also traced in Graph 6.2.1(4). The differences between the calculated and the theoretical distribution are barely visible from $n = 100$ onward.

Fisher and Tippett stated that the distribution of the largest normal values converges only very slowly towards the first type. Their argument, based on the numerical values of the two Betas which are strongly influenced by the tail of the distribution, does not contradict the shape of the Graphs 6.2.1(4).

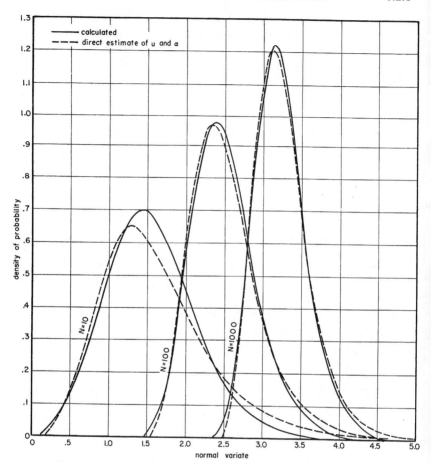

Graph 6.2.1(4). Calculated and Asymptotic Distributions of Normal Extremes for 10, 100, 1,000 Observations

The method of Uzgoeren (5.2.2) leads from the values given in 4.2.3 to the probability function $\Phi(x)$ for normal extremes,

$$-\lg(-\lg \Phi(x)) = y + \frac{y^2}{2u^4} + \frac{y^3}{3u^6} + \cdots$$

If, instead of the extreme value, we use the extreme deviate $x_n - \overline{x_0}$ calculated by Nair and Grubbs, the approach of the exact to the asymptotic probabilities shown in Graph 6.2.1(3) becomes much quicker. For $n = 25$ the differences between the calculated curve and the straight line are of the same order as the differences for the extreme value for $n = 100$, It would be desirable to extend the tables of Grubbs to $n > 25$. Since the asymptotic theory may be used for relatively small samples for the extreme deviate, the deviate is preferable to the extreme itself.

Problems: 1) What is the analytic reason for the quick approach of the probabilities of the normal extreme deviate to the asymptotic probabilities? 2) Is the same true for other distributions of the exponential type? 3) Compare Uzgoeren's approximations to the numerical values for $n - 10$, 100, 1000.

6.2.2. Use of Order Statistics. An extremal quantile defined by a given probability Φ may be estimated from the order statistics x_m by the use of the most probable rank $\tilde{m}$, 2.1.2(6). For N extreme observations the relation 5.2.1(14) between the reduced distribution and the probability leads to

(1) $$\tilde{m} = N\Phi + (1 - \Phi)/(-\lg \Phi).$$

The mode u considered as a quantile is defined by $\varphi' = 0$, whence $y = 0$, $\Phi = 1/e$, and

(2) $$\tilde{m}_u = 0.36788N + 0.63212.$$

This equation can be used to estimate the parameter u. For the median where $\Phi = 1/2$ and $\breve{y} = .36651$, and for the quantile with minimum variance defined in 6.1.5(2) where $\Phi = .20319$ and $y = -.46601$, the corresponding ranks $\tilde{m}_c$ and $\tilde{m}_q$ are

(3) $$\tilde{m}_c = N/2 + 0.72135 \; ; \qquad \tilde{m}_q = 0.20319N + 1.26985.$$

The sum $u + 1/\alpha$, for which $y = 1$ may be estimated as the order statistic corresponding to the probability $\Phi = .69220$. If u has been estimated from (2) the other parameter $1/\alpha$ may be estimated as the difference of the two order statistics with the help of

(3') $$\tilde{m}_{(y=1)} - 0.69220N + 0.11323.$$

For the estimation of u from (2), linear interpolation for x is sufficient, since $\Phi(y)$ is linear in the neighborhood of the mode u. For the interpolation between x_m and x_{m+1}, we calculate y_m and y_{m+1} from the plotting positions and a table of $\Phi(y)$ and interpolate x_m linearly from y_m.

The standard errors of estimates are asymptotically, from 2.1.6(3) and 5.2.1(14),

(4) $$\alpha\sigma_m\sqrt{N} = \sqrt{1/\Phi - 1}/(-\lg \Phi),$$

as long as the order statistics are normally distributed, i.e., in the interval

$-.64 < y < 1.82$. The standard errors for the sample median $\hat{c}$, mode $\hat{u}$, and quartile $\hat{q}$ of minimum variance are

(5) $\quad \alpha\sigma_{\hat{c}}\sqrt{N} = 1.443 > \alpha\sigma_{\hat{u}}\sqrt{N} = 1.331 > \alpha\sigma_{\hat{q}}\sqrt{N} = 1.243$.

The standard error of the arithmetic mean,

(5') $\quad \alpha\sigma(\bar{x}_o)\sqrt{N} = \pi/\sqrt{6} = 1.28255$,

is larger than the standard error of the quantile having least variance. Therefore, the single observation $x(\tilde{m}_q)$ contains more information than the arithmetic mean based on all observations. The order statistics $x(\tilde{m}_u)$, $x(\tilde{m}_c)$, and $x(\tilde{m}_q)$ and the sample mean estimate u, $u + .36651/\alpha$, $u - .46601/\alpha$, and $u + .57722/\alpha$ respectively. Therefore, the order statistics and the mean can be used for the estimation of u only if $1/\alpha$ is known, a condition which is hardly fulfilled, or if α is estimated by another procedure.

The parameter $1/\alpha$ may be estimated independently of u as a difference of order statistics. Let x_l and x_m be the order statistics corresponding to the probabilities Φ_l and Φ_m; then the linear relation between x and y leads to the estimate

(6) $$\frac{1}{\hat{\alpha}} = \frac{x_l - x_m}{y_l - y_m}.$$

For $x = u \pm 1/2\alpha$, the reduced variates are $y = \pm 1/2$ and the probabilities are $\Phi_m = .19230$; $\Phi_l = .54523$. Thus, $1/\alpha$ can be estimated independently of u as the difference of the corresponding order statistics or as one-half of the difference of the order statistics corresponding to $y \pm 1$; $\Phi_m = 0.06599$; $\Phi_l = 0.69220$.

The asymptotic variance of these estimates $1/\hat{\alpha}$ becomes, from 2.1.7(3),

(7) $\quad \alpha N\sigma^2(x_l - x_m)$
$= (1/\Phi_m - 1)e^{2y_m} + (1/\Phi_l - 1)e^{2y_l} - 2(1/\Phi_l - 1)e^{y_m + y_l}$.

Lieblein (1954a) devised estimation procedures especially adapted to very small samples of extremes in order to obtain the greatest amount of information from costly data. Instead of studying the parameters u and $1/\alpha$ separately, he considers the probability points

(8) $$\xi_\Phi = u + y_\Phi/\alpha.$$

Thus, the estimation of the parameters and the forecast are embodied in the same quantity. He proposes the use of a linear function of the order statistics x_v,

(9) $$L = \sum_{v=1}^{N} (a_v + b_v y_\Phi) x_v,$$

where the weights,

$$w_v = a_v + b_v y_\Phi,$$

6.2.2 USES OF THE FIRST ASYMPTOTE

should be determined in such a way that the expectation of L is ξ_Φ and that the variance $\sigma^2(L)$ is minimum, i.e., that L is unbiased and as efficient as possible. Comparison of (8) and (9) leads to the single estimates:

(10) $$\hat{u} = \sum_1^N a_v x_v \; ; \qquad 1/\hat{\alpha} = \sum_1^N b_v x_v \; .$$

The variances $\alpha^2 \sigma^2(L)$ become quadratic expressions in y_Φ. The standard deviations $\sigma(L)$ are proposed as control band. The constants a_v and b_v are to be obtained as the solutions of a system of $N + 2$ linear equations whose coefficients require the knowledge of the covariances of all order statistics. (See 6.1.4.)

Unfortunately the analytical and numerical difficulties involved in this aim are such that the numerical values for a_v, b_v, and the coefficients in $\sigma^2(L)$ are given only for sample sizes $N = 2$ to $N = 6$. The values for $N = 4, 5, 6$ are shown in Table 6.2.2.

Table 6.2.2 The Weights in Lieblein's Estimator for $N = 4, 5, 6$

N		x_1	x_2	x_3	x_4	x_5	x_6
4	a_v	.51100	.26394	.15368	.07138	—	—
	b_v	−.55862	.08590	.22392	.24880	—	—
5	a_v	.41893	.24628	.16761	.10882	.05835	—
	b_v	−.50313	.00653	.13045	.18166	.18448	—
6	a_v	.35545	.22549	.16562	.12105	.08352	.04887
	b_v	−.45928	−.03599	.07319	.12673	.14953	.14581

The variances are, respectively,

(11) $$\alpha^2 \sigma^2(L) = \begin{cases} 0.22528 y_\Phi^2 + 0.06938 y_\Phi + 0.29346 \; ; & N = 4 \\ 0.16665 y_\Phi^2 + 0.06798 y_\Phi + 0.23140 \; ; & N = 5 \\ 0.13196 y_\Phi^2 + 0.06275 y_\Phi + 0.19117 : & N = 6. \end{cases}$$

Of course most sample sizes are much larger than 6. For these sample sizes Lieblein advocates a procedure taken from quality control and used by Grubbs and Weaver for the estimation of the standard deviation of a normal distribution. If the sample size N is a multiple $N = km$ of $m = 5$ or 6, and if the data are truly random, sub-groups of these sizes are formed in the order in which the data are observed, then sub-estimators $L_i = \sum_{j=1}^m w_j x_j$ are formed and the sample total estimate is

(12) $$\bar{L} = \sum_{i=1}^k L_i / k \; .$$

If the sample size is not a multiple of 5 or 6, estimators for the remaining 1, 2, 3, 4 observations are used to form a total sample estimator.

Lieblein's system leads to a single control curve, while in the methods shown in 6.1.5 and 6.1.6 the control curve is obtained from the composition of three parts. The quadratic expressions for the variance lead to control curves spreading for large values of y, while the procedure advocated in 6.1.5 leads to parallels which expand only if measured in return periods. Lieblein (1956) generalized Table 6.2.2 for the case of missing data.

If the observations are not available in the original historical order, or if they are obtained in ordered form, Lieblein's methods require a previous randomization, a procedure now used in many types of surveys. In consequence, different investigators may obtain different results, especially different extrapolations from the same observations. This will not encourage engineers to base their decisions on statistical procedures.

Exercises: 1) Calculate the quantile of minimum variance for the asymptotic distribution of smallest values. 2) Calculate the mth values which have the same variance as the arithmetic mean. 3) Calculate the most probable mth values corresponding to $y = \pm 1/2$ and $y = \pm 1$ and the standard errors of these estimations of $1/\alpha$. 4) Estimate the parameters from the median and the mean deviation about median. 5) Estimate the parameters from the order statistic with minimum variance and the mean deviation. 6) Use $m = .20319$ and calculate l in such a way that the variance of the estimation of $1/\alpha$ obtained from (7) becomes a minimum. 7) Estimate $1/\alpha$ from the observed largest value x_N, the sample mean, and the standard deviation, assuming that x_N is the most probable largest value for N observations. 8) Compare Lieblein's control curves for $N = 6$ with the control curves given in 6.1.5, 6.1.6, and 6.1.7.

Problems: 1) Use Mosteller's (1946) technique to estimate ξ_ϕ for large sample sizes from the order statistic 6.1.5(2) with minimum variance and two order statistics x_m and x_l so that this difference has minimum variance. Calculate m and l to five decimals. (A solution to two decimals was given by Lieblein, 1954a.) 2) Estimate $1/\alpha$ from range. 3) What is the distribution of $(x_N - \bar{x}_0)/s$?

References: H. L. Jones (1953), Kase (1955), Tiago de Oliveira (1957).

6.2.3. Estimates for Probability Paper.

The general method, 1.2.8, of fitting straight lines on probability paper applied to extreme values leads to the estimates

(1) $$1/\hat{\alpha} = s/\sigma_N \; ; \qquad \hat{u} = \bar{x}_o - \bar{y}_N/\hat{\alpha} \, .$$

The population values σ_N and $\bar{y}_N$ as functions of N calculated by the Watson Computing Laboratories, Columbia University, to five decimals are given in Table 6.2.3. For increasing values of N, the mean $\bar{y}_N$ and the standard deviation σ_N converge, of course, to the population mean γ and standard deviations $\pi/\sqrt{6}$.

6.2.3 USES OF THE FIRST ASYMPTOTE

The quotients $\bar{y}_N/\gamma$ and $\sigma_N\sqrt{6}/\pi$ are traced in Graph 6.2.3 on extremal probability paper as a function of lg N, which is the most probable largest value for N observations. For reasons unknown, the relation is approximately linear. Hence, linear interpolation on the probability scale can be used to obtain other values of $\bar{y}_N/\gamma$ and $\sigma_N\sqrt{6}/\pi$. This leads to the other figures in Table 6.2.3, wherein linear interpolation is permitted for $N > 60$. The values given may be in error in the last decimal.

The theoretical straight line obtained from (1) is

$$(2) \qquad x = \bar{x}_0 + s(y - \bar{y}_N)/\sigma_N .$$

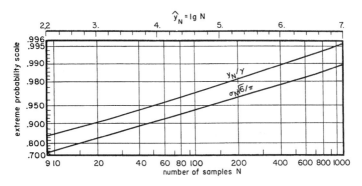

Graph 6.2.3. Population Mean $\bar{y}_N$ and Standard Deviation σ_N as Function of N

L. L. Weiss calculated a table of $Z(n,T) = (y_T - \bar{y}_N)/\sigma_N$ as a function of the return period T ($= 2, 5, 10, 25, 50, 100$) and the sample size $N = 10, 15, 20, 30, 40, 60, 100$) and used it for the construction of a nomogram which gives, for a sample of size N, the value of x_T corresponding to a desired return period T in the form $x_T - \bar{x}_0 = sZ(n,T)$ with an error of about one percent of the values computed in the usual way.

The method of moments leads to the estimates,

$$(3) \qquad 1/\hat{\alpha}' = s\sqrt{6}/\pi ; \qquad \hat{u}' = \bar{x}_0 - \gamma/\hat{\alpha}',$$

whence,

$$(4) \qquad x = \bar{x}_0 + s(y - \gamma)\sqrt{6}/\pi .$$

Since $\sigma_N < \pi/\sqrt{6}$, and since, as seen from Table 6.2.3, $\bar{y}_N/\sigma_N > 0.45005$, the increase of the largest values with y, i.e., with lg T, is stronger in the first method. It leads to a larger forecast than is the case with the second one.

Formula (4) may be used to construct control curves which cover the whole domain of variation. The variance of the sum of the sample mean and standard deviation given in 1.1.5 (4) becomes from the values of the Betas, 5.2.4(11), as in 6.1.7.

(5) $\quad N\sigma^2(x)/\sigma^2 = 1 + 1.1396(y-\gamma)\sqrt{6}/\pi + 1.1(y-\gamma)^2 6/\pi^2$.

The variances of the reduced values become therefore for large N

(6) $\quad \alpha^2 N\sigma^2(y) = \pi^2/6 + 1.1396(y-\gamma)\pi/\sqrt{6} + 1.1(y-\gamma)^2$.

They are again quadratic functions in y. The corresponding standard deviations may be used as control curves. Numerical values were given in

Table 6.2.3 Means and Standard Deviations of Reduced Extremes

N	$\bar{y}_N$	σ_N	N	$\bar{y}_N$	σ_N
8	.4843	.9043	49	.5481	1.1590
9	.4902	.9288	50	.54854	1.16066
10	.4952	.9497	51	.5489	1.1623
11	.4996	.9676	52	.5493	1.1638
12	.5035	.9833	53	.5497	1.1653
13	.5070	.9972	54	.5501	1.1667
14	.5100	1.0095	55	.5504	1.1681
15	.5128	1.02057	56	.5508	1.1696
16	.5157	1.0316	57	.5511	1.1708
17	.5181	1.0411	58	.5515	1.1721
18	.5202	1.0493	59	.5518	1.1734
19	.5220	1.0566	60	.55208	1.17467
20	.52355	1.06283	62	.5527	1.1770
21	.5252	1.0696	64	.5533	1.1793
22	.5268	1.0754	66	.5538	1.1814
23	.5283	1.0811	68	.5543	1.1834
24	.5296	1.0864	70	.55477	1.18536
25	.53086	1.09145	72	.5552	1.1873
26	.5320	1.0961	74	.5557	1.1890
27	.5332	1.1004	76	.5561	1.1906
28	.5343	1.1047	78	.5565	1.1923
29	.5353	1.1086	80	.55688	1.19382
30	.53622	1.11238	82	.5572	1.1953
31	.5371	1.1159	84	.5576	1.1967
32	.5380	1.1193	86	.5580	1.1980
33	.5388	1.1226	88	.5583	1.1994
34	.5396	1.1255	90	.55860	1.20073
35	.54034	1.12847	92	.5589	1.2020
36	.5410	1.1313	94	.5592	1.2032
37	.5418	1.1339	96	.5595	1.2044
38	.5424	1.1363	98	.5598	1.2055
39	.5430	1.1388	100	.56002	1.20649
40	.54362	1.14132	150	.56461	1.22534
41	.5442	1.1436	200	.56715	1.23598
42	.5448	1.1458	250	.56878	1.24292
43	.5453	1.1480	300	.56993	1.24786
44	.5458	1.1499	400	.57144	1.25450
45	.54630	1.15185	500	.57240	1.25880
46	.5468	1.1538	750	.57377	1.26506
47	.5473	1.1557	1000	.57450	1.26851
48	.5477	1.1574	∞	.57722	1.28255

Table 6.1.5, column 4. For increasing probabilities they spread considerably. The difference between (6) and 6.2.2(11) is due to the fact that Lieblein introduces weights to minimize the error of estimation of the parameters. In both methods a unique control curve is obtained. However, the probabilities of the control intervals are not constant, while this is the case for the composition of the control curves proposed in 6.1.5 and 6.1.6.

Problems: 1) Extend Table 6.2.3 to $100 < N < 1000$. 2) Why do the quotients $\bar{y}_N/\gamma$ and $\sigma_N\sqrt{6}/\pi$ traced on probability scale approximate a linear function of the most probable largest value?
Reference: Botts.

6.2.4. Sufficient Estimation Functions, by B. F. Kimball.

The problem of the best determination of the parameters α and u, has been discussed at some length in a paper published in 1946. A primary requirement that a set of statistical estimation functions such as $\bar{x}$ and s^2, be *sufficient* for estimating two parameters, such as α and u, from a sample is as follows: Consider the probability density function of the complete sample,

(1) $$P(O_N) = \alpha^N \cdot e^{-\alpha\Sigma(x_i-u)} \exp\left(-\Sigma e^{-\alpha(x_i-u)}\right)$$

where the summation signs refer to all indices from 1 to N. It should be possible to express this as a function of x, s^2, α, and u only. In other words, the sample values x_i should be completely absorbed by the estimation functions $\bar{x}$ and s^2. (For a case where less than the complete set of parameters is estimated, see Kimball, 1946.) Since it is impossible to express $P(O_N)$ in terms of $\bar{x}$, s^2, α, and u, only, these estimates are not "sufficient." They do not use the information furnished by the sample in such a way as to *fix* the probability density function relative to the sample values observed. (In this and the following paragraph $\bar{x}$ stands for the sample mean and E for expectations. E. J. G.)

If we introduce

(2) $$z_0 = e^{-\alpha u},$$
$$z_i = e^{-\alpha x_i}, \quad \bar{z} = (\Sigma e^{-\alpha x_i})/N,$$

we can write the probability density function of the sample as

(3) $$P(O_N) = \alpha^N \cdot e^{-N\bar{z}/z_0} \cdot e^{-N\alpha(\bar{x}-u)}.$$

This points to the use of estimates based on $\bar{x}$ and $\bar{z}$. Unfortunately, $\bar{z}$ is not a "statistic," since it involves one of the unknown parameters. Hence, the concept of "a set of sufficient statistical estimation functions" is required (see Kimball, 1946). If these estimation functions are chosen as

(4) $X = \sqrt{N}[\alpha(\bar{x} - u) - \gamma]$, $\gamma =$ Euler's Constant $= E(y)$,

$Y = \sqrt{N}[\bar{z}/z_0 - 1]$,

we can write

(5) $$P(O_N) = \alpha^N \cdot e^{-N(1+\gamma)} \cdot e^{-\sqrt{N}(X+Y)}.$$

Hence, X and Y in relation to α and u will completely determine the probability distribution of the sample. The reason for the specific relations of X and Y to $\bar{x}$, u, $\bar{z}$, and z_0, used in (4), is that a study of the bivariate sampling distribution of (X, Y) (see Kimball, 1946, pp. 304–5) shows that

(6) $$E(X) = E(Y) = 0,$$

and that the variance-covariance matrix is

(7) $$\begin{Vmatrix} \pi^2/6 & -1 \\ -1 & 1 \end{Vmatrix}$$

for all values of N. This means that, for all values of N,

(8) $E(X^2) = \pi^2/6$, $E(Y^2) = 1$, $E(XY) = -1$.

Furthermore, it is proved that the asymptotic distribution is the normal bivariate distribution possessing the same variance-covariance matrix.

Thus a "sufficient" estimate of parameters α and u is obtained by setting $X = Y = 0$, resulting in the relations

(9) $u + \gamma/\alpha = \bar{x}$,

$e^{\alpha u}(\Sigma e^{-\alpha x_i})/N = 1$.

Substituting the first of these equations in the second, they become

(10) $\Sigma e^{-\alpha(x_i - \bar{x})} = Ne^{\gamma}$,

$u = \bar{x} - \gamma/\alpha$.

It should be noted that the first of these equations is independent of the parameter u.

It will be further noted that while the second equation of (10) is the same as the second equation obtained by the earlier method 6.2.3(3), the first equation of (10) acts to give statistical weights to the deviations $(x_i - \bar{x})$, which are *exponentially* heavier for deviations at the lower end of the data series than for those at the upper end, while the earlier method involves only the squares of the deviations.

6.2.5. Maximum Likelihood Estimations, by B. F. Kimball. It is perfectly possible to have more than one "sufficient" estimate of a set of parameters (see Kimball, 1946, §3). In general, the best asymptotically normal estimate (which is also sufficient) is that obtained by the method of maximum likelihood.

The maximum likelihood estimate is found by setting the partial derivatives of the logarithm of $P(O_N)$ with respect to the parameters sought, equal to zero. Denoting this logarithm by L, we can write, from 6.2.4(3),

(1) $$L = N[\lg \alpha - \bar{z}/z_0 - \alpha(\bar{x} - u)]$$

and the partial derivatives with respect to u and α are†

(2) $$L_u = -N\alpha[\bar{z}/z_0 - 1]$$

$$L_\alpha = N[1/\alpha - (\bar{x} - u) - \partial(\bar{z}/z_0)/\partial\alpha].$$

Now

(3) $$\partial(\bar{z}/z_0)/\delta\alpha = (\bar{z}/z_0)(u + \bar{z}_\alpha/\bar{z})$$

and

$$\bar{z}_\alpha/\bar{z} = -\Sigma x_i e^{-\alpha x_i}/\Sigma e^{-\alpha x_i}.$$

Hence, setting $L_u = 0$, we note that

$$\bar{z}/z_0 = 1.$$

Making this substitution in $L_\alpha = 0$, the conditions $L_u = L_\alpha = 0$ reduce to

(4) $$e^{\alpha u}\Sigma e^{-\alpha x_i} = N,$$

$$(\Sigma x_i e^{-\alpha x_i})/(\Sigma e^{-\alpha x_i}) + 1/\alpha = \bar{x}.$$

Furthermore,

$$(\Sigma x_i e^{-\alpha x_i})/(\Sigma e^{-\alpha x_i}) = (\Sigma x_i e^{-\alpha(x_i-u)})/(\Sigma e^{-\alpha(x_i-u)}),$$

and

$$e^{-\alpha(x_i-u)} = -\lg \Phi(x_i).$$

Hence, (4) can be written

(5) $$-\Sigma \lg \Phi(x_i) = N,$$

$$(\Sigma x_i \lg \Phi(x_i))/(\Sigma \lg \Phi(x_i)) = \bar{x} - 1/\alpha.$$

The set of equations (4) constitutes the working equations for obtaining the critical values of α and u sought. It will be noted that the second of these

† The equations recorded at the bottom of page 305 of Kimball (1946) involve two corrections in sign. In the second equation of (5.2), $+\partial(\bar{z}/z_0)/\partial\alpha$ should read $-\partial(\bar{z}/z_0)/\partial\alpha$, and in the equation for Y in (5.4), $e^{-\alpha u}$ should read $e^{\alpha u}$. These typographical errors were not involved in the numerical computations leading up to results quoted on pp. 307–309.

equations is independent of u. If α_1 denotes the first trial value of α used in a trial and error solution process, it will be found that a fairly rapid convergence to the true value of α can proceed as follows:

Using the second equation of (4), compute the first term with $\alpha = \alpha_1$. Solve for $1/\alpha$ and let $\alpha = \alpha_2$ denote the second value of α thus obtained. The true value (giving the solution) will lie between α_1 and α_2. In a few trial studies that the present writer has made, convergence is facilitated by taking the next value of α, α_3 from the relation

(6) $$\alpha_3 = \alpha_2 + (\alpha_1 - \alpha_2)/3 .$$

In making the first choice of α for trial, it might be noted that, from 2.1.3(1),

(7) $$E[-\lg \Phi(x_m)] = 1/m + 1/(m+1) + \ldots + 1/n , \quad m \leq N .$$

Thus, if the sum of the reciprocals noted here be substituted for $-\lg \Phi(x_1)$ in (5) a first approximation to α may be obtained.

References: Hotelling, Kimball (1956), Neyman (1949), Wilks (1943).

6.2.6. Approximate Solutions. Another approximate solution of the maximum likelihood equations 6.2.5(5) is as follows: The second equation 6.2.5(5) is written

(1) $$1/\alpha = \bar{x}_0 + \Sigma(x_i \lg \Phi_i)/N .$$

This equation contains the sample values x_i, the sample mean $\bar{x}_0$, and the (unknown) probabilities Φ_i. To eliminate them, we may be tempted to replace them by the plotting positions

(2) $$-\lg \Phi_i = -\lg (i/(N+1)) .$$

However, this procedure would contradict the first equation 6.2.5(5), since the sum of the second member divided by N,

$$-\frac{\lg N!}{N} + \lg (N+1) = 1/A_N ,$$

is smaller than unity. Therefore, the probabilities Φ_i in (2) are replaced by

(3) $$\lg \Phi_i = A_N \lg i - A_N \lg (N+1) .$$

Consequently, equation (1) becomes now

(4) $$1/\alpha = A_N \Sigma x_i \lg i/N - \bar{x}_0(A_N \lg (N+1) - 1) ,$$

and is compatible with the first equation 6.2.5(5). The numerical values of

(5) $$B_N = A_N - 1$$

6.2.6 USES OF THE FIRST ASYMPTOTE

are traced in Graph 6.2.6. For large values of N, Stirling's formula leads to

(6) $$B_N = [\lg(2\pi N) - 2]/2N,$$

which may be used for values of N which exceed those shown in the Graph 6.2.6. The first approximation $\hat{\alpha}_1$, obtained from (4) with the help of the

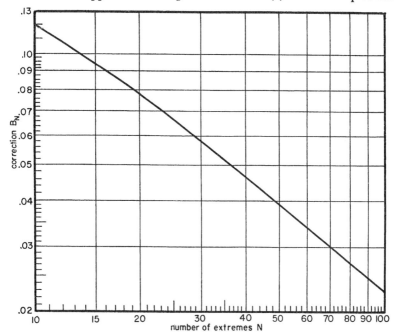

Graph 6.2.6. Correction B_N for Maximum Likelihood Method

graph or (6), is introduced into the second equation 6.2.5(4) and leads to approximations $\hat{\alpha}_2$ and $\hat{\alpha}_3$ by the procedure outlined above. Finally, u may be estimated from

(7) $$e^{-\alpha_3 \hat{u}} = \sum_{1}^{N} e^{-\alpha_3 x_i}/N.$$

In a later article (1956) Kimball showed that the estimation of u from the maximum likelihood method is biased even under the assumption that the scale parameter is known. For the estimation of $1/\alpha$ Kimball modified the methods given in 6.2.5 and reached an approximation,

(8) $$1/\hat{\alpha}_0 = \bar{x}_0 + \sum_{m=1}^{N} x_m \overline{\lg \Phi_m}/N.$$

Since the expected value $\overline{\lg \Phi_m}$ is given by 6.2.5(7) the estimate is simply

(9) $\quad 1/\hat{\alpha}_0 = \sum_{m=1}^{N} c_m x_m \, ; \quad c_m = [1 - \sum_{i=m}^{N} 1/i]/N \, ; \quad \sum_{1}^{N} c_m = 0 \, .$

The bias in this estimate is obtained from 6.2.5(7) and

$$\alpha \overline{(1/\hat{\alpha} - 1/\alpha)} = -(1-\gamma) + \sum_{m=1}^{N} \bar{y}_m \, \overline{\lg \Phi_m}/N \, .$$

A procedure similar to 6.1.4(13) leads to the *unbiased estimate:*

(10) $\quad\quad\quad\quad 1/\hat{\alpha} = 1/\hat{\alpha}_0 \, b_N \, .$

Kimball gives a table (4D) of the correction b_N for $N = 2(1)\,112$. For $N = 10, 20, 50, 100$, the values are respectively 1.23, 1.11, 1.05, 1.02.

The estimator (10) is a linear function of the sample values, where the values at the lower end are much more heavily weighted than those near the upper extreme. In all practical applications, however, we are interested in a good fit of the upper end of the series. Therefore, the maximum likelihood estimate may be worse than an estimate which gives less weight to the lower part of the distribution.

Problem: What is the bias in the estimation (4)?

6.2.7. Asymptotic Variance of a Forecast, by B. F. Kimball.

Using the maximum likelihood estimation functions for estimating α and u, it is possible to make an asymptotic estimate of the variance of a maximum value read from the fitted curve at a given frequency. Thus, at a frequency, say, of $\Phi = .99$, a maximum value X is read from the fitted curve. What is the variance of this estimate X due to the statistical uncertainty of fitting the sample? In other words, what is the sampling variance of X?

The present writer has obtained the following simple formula for the asymptotic value of this variance (see Kimball, 1949). If we let y denote the value of the reduced variable corresponding to a critical frequency such as .99,

$$\Phi = .99, \quad y = 4.60015 \, ,$$

which is to be held constant, then the asymptotic sampling distribution of the corresponding $x = X$ read from the fitted curve at this frequency is shown to be normal and have a variance given by

(1) $\quad\quad \sigma^2(\sqrt{N}\,X) = (1/\alpha^2)[1 + (1 - \gamma + y)^2](\pi^2/6) \, ,$

where γ is Euler's Constant.

This formula takes into account that α is not known, but is also estimated from the sample. It is obtained from what is called the "marginal

6.2.7 USES OF THE FIRST ASYMPTOTE

distribution" of one of the parameters of the asymptotic bivariate normal distribution.

Thus, for finite N we approximate the variance of X by

(2) $$\sigma_x^2 = [1/(N\alpha^2)][1 + (1 - \gamma + y)^2/(\pi^2/6)].$$

By taking σ positive and negative, using various values of y, a confidence band for the corresponding $x = X$ is easily delineated.

It should be noted that such a confidence band is based on the theoretical sampling distribution of *the complete sample*.

It represents *the sampling variance of the "X coordinate" of the best fitting distribution under the assumption that the universe from which the sample is drawn is truly a double exponential distribution of maximum values*.

It is the type of variance whose square root is what has been known as "the standard error of forecast" (Hotelling).

In order to forecast the variance of a single observation in a future sample to be taken under similar conditions, one should *add* to the above variance *the variance of a single mth maximum value at the proper frequency*.

It would be well to test whether the curve of maximum values used is a satisfactory fit to the sample data or not, before making any assertion about a forecast (see 6.1.6).

For example if we wish to forecast the maximum annual flood to be expected in the next 10 years, and to assign a variance to this estimate, we might proceed as follows: Find

(3) $$E(y_{10}) = \gamma + \lg 10 = 2.8798,$$

and the variance of this 10th maximum value, which happens to be independent of size of sample, is given by

(4) $$\sigma_x^2 = \sigma_y^2/\alpha^2 = (\pi^2/6)/\alpha^2 = (1.6449)/\alpha^2.$$

Using

$$y = E(y_{10}) = 2.8798,$$

(5) $$E(x_{10}) = u + (2.8798)/\alpha.$$

This is the maximum annual flood which is forecast.

From formula (2), taking $y = 2.8798$ and N equal to the size of *the sample used in determining α and u*, the variance of X due to uncertainty in fitting is obtained. To this variance is added the variance obtained in (4). The sum of these two variances represents the variance of the forecast, provided that the double exponential distribution of maximum values has been found to fit the data reasonably well.

6.3. NUMERICAL EXAMPLES

*"Wenn der Hahn kräht auf dem Mist,
Ändert sich's Wetter oder bleibt wie's ist."*

(*When the cock crows from the manure,
The weather will change or may endure.*)

6.3.0. Problems. The first asymptotic distribution of extreme values leads to the solution of important problems arising in engineering. As examples, we select floods, extreme meteorological phenomena, and the breaking strength of material. Other applications may be seen in the Bureau of Standards brochure. A main problem arises from the scarcity of data. In the analysis of floods, this is due to the small number of years in which continuous, reliable, and comparable records have been kept. Unfortunately, this number does not necessarily increase with time, since any new construction may destroy the comparability of the records.

In the breaking strength of material, most engineers are unwilling to make a sufficiently large number of systematic tests, although the financial consequences of a single major failure exceed the costs of such experiments.

6.3.1. Floods. The floods of rivers of different magnitudes can be brought into a common scale by plotting either the values $(x - \bar{x}_0)/s$ or the discharges divided by the catchment area. The latter procedure is usual among engineers.

The $N = 52$ flood discharges of the Mississippi River at Vicksburg, Mississippi, 1898–1949 measured in 10^3 c.f.s., are traced in Graph 6.3.1. The mean and standard deviations are $\bar{x}_0 = 1362.68$, $s = 332.30$. The values $\bar{y}_N$ and σ_N in Table 6.2.3. and equations 6.2.3(1) lead to the straight line

(1) $\qquad x = 1206.51 + 282.89y$,

drawn in Graph 6.3.1. The observations are neatly scattered about this line. The control curves in the neighborhood of the median obtained from $1/(\hat{\alpha}\sqrt{N}) = 36.52$ and the control curves for the extremes are given in Tables 6.1.5, column 5, and 6.1.6, column 4, respectively. All observations are contained within the control band. Within the next hundred years we may expect a flood as high as 2,650 10^3 c.f.s.

The floods of hundreds of rivers observed under different climatic conditions, plotted on extremal probability paper by Shuh, the Ohio Water Resources Board, Bulletin No. 7 (Cross); the U. S. Geological Survey, Water Supply Paper 1080 (Rantz and Riggs), Circulars 100, 191, and 204

6.3.1 USES OF THE FIRST ASYMPTOTE

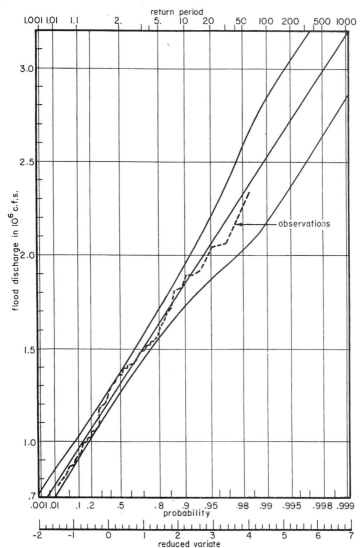

Graph 6.3.1. Floods. Mississippi River, Vicksburg, Mississippi, 1898–1949

(Carter, Robinson and Bodhaine); the Soil Conservation Service (Potter); and the Hydro-Electric Design Office, Ministry of Works, New Zealand (Benham) all lead to approximately straight lines. The good fit holds even for relatively small numbers of years, from $N = 20$ onward. The theory claims a constant value $\sqrt{\beta_1} = 1.139$ of the skewness for all rivers. This was checked (Gumbel, 1952): 108 values $\sqrt{b_1}$ calculated by Hazen for different rivers and different small lengths of records (about 25 years) plotted on normal paper are closely scattered about a straight line with mean 1.237. The difference of the mean from the expected value is not significant at the usual level.

It must be admitted that the good fit cannot be foreseen from the theory, which is based on three assumptions: 1) the distribution of the daily discharges is of the exponential type; 2) $n = 365$ is sufficiently large; 3) the daily observations are independent. Assumptions 1) and 2) cannot be checked since the analytical form of the distribution of the discharges is unknown. The third assumption does not hold and the number of independent observations is certainly less than 365. This argument is weakened by the theorem given in 5.1.4. The good fit is probably due to the fact that the distribution of the daily discharges is often so close to the distribution of floods that the relatively small number of independent discharges is sufficient to reproduce the double exponential distribution of floods.

Benson has shown how gaps in the records can be filled and, thus, the records can be lengthened. Circulars 100 and 191 combine records for different rivers on the basis of the homogeneity test 6.1.5, and show that the parameter $1/\alpha$ can be used to characterize a homogeneous region. Finally, Dalrymple has used the double exponential distribution for the design of bridge waterways.

Problem: How do u and $1/\alpha$ increase with the size of the watershed?

References: Am. Soc. Eng. *Handbook*, Am. Soc. Eng. *Review*, Baird and Potter, Barrow, Beard (1942), Bednarski, Boyer, Carter, Chamayou, Chow, (1951, 1952), Coutagne (1930, 1937, 1938, 1952c), Cragwell, Creager, Dalrymple (1950), Dick and Darwin, E. Foster, Goodrich, Grassberger (1936), Gumbel (1940, 1941, 1945a, 1949b, 1950, 1952, 1954b, 1956), Izzard, Jarvis, Kimball (1955), Kinnison and Colby, Kinnison and Conover and Bigwood, Kresge, Langbein (1953), Mitchell, Moran, National Bureau of Standards (1954), L. B. Pierce, Potter (1949b), R. Price, Rantz, Robinson, Schwob, Shuh-Chai Lee, Standish, President's Water Policy, U. S. Geol. Survey, Water Supply Comm., G. S. Watson (1954), Wemelsfelder. *Consult also:* Gumbel (1941). The return period of flood flows, Ann. Math. Stats., 12: 163.

6.3.2. The Design Flood.

"*Half a loaf is better than none.*" (*Half a bridge* . . .)

The main purpose of the calculations shown in the previous paragraph is for extrapolation with the help of the return period scale. The asymptotic

6.3.2   USES OF THE FIRST ASYMPTOTE

formula 5.2.6(4) for the floods as a function of the return period gives meaning to a cookbook rule which is widespread among engineers. "To tame the river, let us build the dam so strong that it can withstand the double of the largest flood observed until now," i.e., within the last N years. This rule may be interpreted in terms of return periods. The return period $T(2x_N)$ of the flood $2x_N$ is obtained from 5.2.6(4), as

(1) $$2x_N \sim u + \lg T(2x_N)/\alpha ,$$

whence

(2) $$T(2x_N) \sim T^2(x_N)\, e^{\alpha u} .$$

The return period of the double of the largest flood is the square of the return period of the largest flood multiplied by the factor $e^{\alpha u}$, which is larger than unity. For the Mississippi River at Vicksburg, $N = 52$, $x_N = 2.334$ million c.f.s., while, from 6.3.1(1), $\alpha u = 4.26$. Therefore, the return period of a flood 4.668 million c.f.s. is of the order 210,000 years. Of course the task of providing for such epochs is completely out of the question. This way of choosing the design flood ignores the different variability of different rivers. Since the flood increases with time, this method may lead for small N to design floods which are too small. Finally, there is no possibility of evaluating the risk involved.

The method of calculated risks, 1.2.2, leads to a more reasonable solution. The unfavorable event against which we seek protection is the occurrence of a design flood x_ω. We want the probability of failure W_ω, that this event with return period T_ω will happen before N_ω years, to be small and N_ω, the design duration, to be large. Thus x_ω is, from 1.2.2(11) and 5.2.6(4), asymptotically,

(3) $$x_\omega \sim u + (\lg N_\omega - \lg W_\omega)/\alpha .$$

For practical use, the parameters u and $1/\alpha$ are estimated from the sample values $\bar{x}_0$ and s. This procedure is sufficiently accurate, since we are mainly interested in the order of magnitude, and leads from 6.2.3(4) to

(4) $$x_\omega \sim \bar{x}_0 + [0.77970 \lg (N_\omega/W_\omega) - 0.45005]s .$$

If we write for the term in brackets, i.e., for the multiple of the standard deviation.

(5) $$z_\omega = 0.77970 \lg (N_\omega/W_\omega) - 0.45005 ,$$

the design flood becomes, from (4),

(6) $$x_\omega \sim \bar{x}_0 + z_\omega s .$$

We choose $z_\omega = 1.5(0.5)10$ and obtain y_ω and T_ω from (5) and 5.2.6(4) as

(7) $$y_\omega \sim 1.28255 z_\omega + 0.57722 ; \qquad T_\omega = e^{y_\omega} .$$

Graph 6.3.2 traces z_ω as function of the design duration N_ω and this probability of failure W_ω. For example, if we want a structure to withstand 7.4 years with probability .001 of failure, or 74 years with probability .01, the design flood obtained from (5) is the mean annual flood plus 6.5 times the standard deviation of the floods. Inversely, if a design flood x_ω is already chosen, formula (6) leads from the knowledge of the mean annual flood and the standard deviation of the floods to the multiple z_ω, and

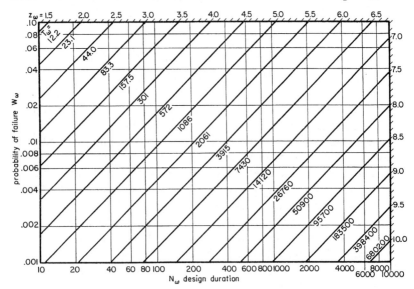

Graph 6.3.2. Design Flood for Given Risk

Graph 6.3.2 leads to an interpretation in terms of N_ω and W_ω. For example, if z_ω is chosen to be 10, there is a probability .001 that the dam will break within 600 years.

The design flood, $x_\omega = 2x_N = 400{,}000$ c.f.s., was actually used for the Colorado River. The other hypothetical examples (Gumbel, 1941) in Table 6.3.2 indicate that doubling the largest flood may lead to overbuilding.

Table 6.3.2. *The Standardized Floods z_ω for Some Rivers*

River	Location	Years	Mean Flood $\bar{x}_0$ in 1000 c.f.s.	Standard Deviation	$2x_N$	z_ω
Colorado	Black Canyon, Col.	1878–1929	100.1	45.9	400	6.5
Mississippi	Keokuk, Iowa	1878–1937	177.7	35.8	720	15.2
Mississippi	Vicksburg, Miss.	1890–1949	1362.7	332.3	4668	10.0
Cumberland	Nashville, Tenn.	1838–1931	122.4	26.7	406	11.5
Tennessee	Chattanooga, Tenn.	1874–1934	207.4	56.9	722	9.0
Columbia	The Dalles, Ore.	1858–1952	611.2	176.7	2480	10.6

6.3.3 USES OF THE FIRST ASYMPTOTE

Our methods permit the forecast of the most probable flood to be reached within a period of years, and lead to the knowledge of the return period for a chosen design flood appropriate for the construction of bridges, dams, and hydro-electric plants. The control curves give the time within which a flood exceeding a given size will return with a given probability. The calculated risk should replace the arbitrary safety factors which are now used and form a basis for actuarial calculations balancing the cost of a structure against the risk involved. These methods lead to an appraisal of the priority of water resources programs, and to criteria basing the feasibility of water resources projects on a given upper bound to the cost.

Exercise: Trace a diagram similar to 6.3.2 for the normal distribution.

References: Altman and Jebe, Dalrymple (1950), Dantzig, H. A. Foster (1952), Gumbel (1941), Moran.

Erratum: The numbers on the abscissa in Graph 6.3.2 should be divided by 10.

6.3.3. Meteorological Examples. The extreme values observed in meteorology, such as pressure, precipitations, snowfalls, rainfalls, and temperatures, constitute a further field of application.

Barricelli compared the maxima and minima of pressures and temperatures observed in Bergen, Norway, 1857–1926, for the first 10, 35, and 70 years, with the corresponding theoretical values. He found excellent agreement for the maxima of largest pressures, and good agreement for the minima of largest pressures. However, the three theoretical maxima of smallest pressures were all less than observed values, and the three theoretical maxima and minima of the largest temperatures exceeded the observations. The author believes that this test is too strong, since the extremes for 70 years may also happen within the first 10 to 35 years. It seems better to compare all observed extremes to the theoretical ones. Graph 6.3.3(1) shows the annual maxima of temperatures in Bergen, Norway, during 1857–1926, taken from Birkeland. The curve resembles the normal probability traced in Graph 5.2.6. Therefore, Barricelli concluded that there is a normal component hidden in these extreme values. Consequently, the difference of the monthly maxima of temperatures and the mean annual temperature was calculated. Graph 6.3.3(1) shows that the theory gives a better fit for the differences of temperatures than for largest temperatures proper.

The calculation of the parameters based on $\bar{x}_0 = 18.28714$; $s = 2.30049$ and Table 6.2.3 leads to the theoretical straight line,

$$x = 17.218 + 1.927y,$$

traced in Graph 6.3.3(1).

Barricelli calculated the most probable maxima and minima obtained

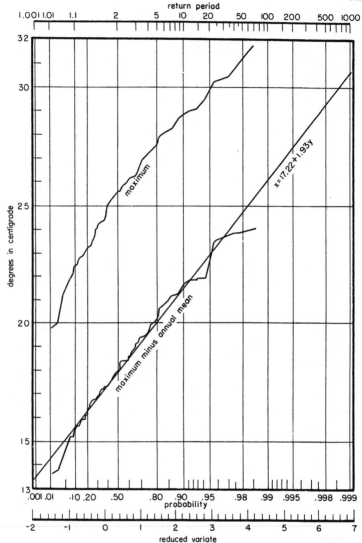

Graph 6.3.3(1). Annual Maxima of Temperatures and Their Differences from the Annual Mean. Bergen, Norway, 1857–1926

from his distribution 5.2.8(5) as a function of the number of observations N and showed that these values come nearer to the observed maxima and minima of the temperatures and pressures than the values obtained from the distribution of the extreme values alone.

The largest mean daily precipitations in inches, observed during the

6.3.3 USES OF THE FIRST ASYMPTOTE

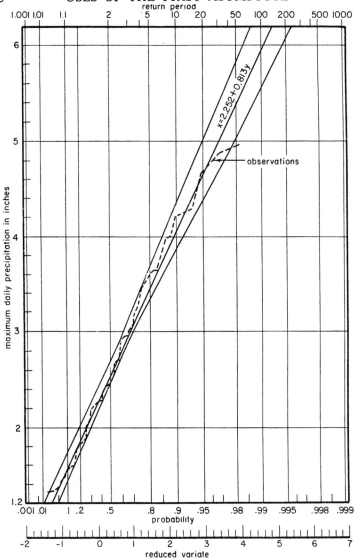

Graph 6.3.3(2). Largest One-day Precipitation. Barakar Catchment Area, India

monsoon season (July 1 to October 31) for 58 years in the Barakar River Catchment Area, India, given by Nag and Dutta, are traced in Graph 6.3.3(2). The mean and standard deviations are 2.7007 and 0.9530 respectively. Table 6.2.3 leads to

$$\bar{y}_{58} = 0.5515; \qquad \sigma_{58} = 1.1721 .$$

Hence the parameters estimated from 6.2.3(1) are

$$1/\hat{\alpha} = 0.813; \quad \hat{u} = 2.252.$$

The fit of the theoretical straight line is very good, although the minimum and the maximum lie just on the border line of the control curves obtained from Table 6.1.5, column 4. The fit is better than the one obtained by Nag and Dutta, who used the second asymptote to represent these observations. Other meteorological examples such as highest and lowest temperatures and greatest rainfalls, may be found in the author's Bureau of Standards brochure (1953) and the report with Court.

Thom (1956) applied the first asymptotic probability of the smallest value to the minimum annual temperatures observed for 30 years in 23 United States cities and obtained in all cases a very good fit on the extremal probability paper. New winter outside design temperatures of heating systems constructed on this basis have led to the discovery of striking inconsistencies in the design temperatures used up to now.

Arnold Court (1953a) analyzed the lowest winter temperature in Moran, Wyoming, for 39 years and in Border, Wyoming, for 49 years and found good agreement with the theory. In addition, he applied the theory to the highest and lowest temperatures of each year at 100 stations in the United States, and compared the expected extremes for 30 years with the observed values. The agreement found was very good. Only six of the observed maxima and nine of the observed minima lay outside the control interval calculated for probability 0.68. A further analysis shows, moreover, that some of these differences were due to questionable data. The calculations were used to extrapolate for the 100 years' extremes valid under the assumption that no systematic or cyclical change in climate occurs within this period. The results are traced on maps of the United States, showing the isotherms for the expected 100 years' highest and lowest temperatures in intervals of 5°. These isotherms turned out to be much smoother than the isotherms based on the observed extremes, since the latter depend upon the length of the observed periods. This important publication gives a convincing proof of the usefulness of the theory of extreme values in climatology.

Court (1953b) also analyzed the strongest 5 minutes' wind speed, 1912–48, at 25 places in the United States and found good agreement with the theory of extreme values. Since the strongest wind of record increases with the length of the record, the comparison of different stations should be based on a common length, say, the strongest wind expected within 25 years. The calculated risk (6.3.2) should be used for the design of structures to withstand the strongest wind.

References: Am. Soc. Civ. Eng. Review, Anonymous, Arsdel, Baur,

6.3.4   USES OF THE FIRST ASYMPTOTE

Brakensick, Brooks, Chow (1952, 1953), Court (1951), E. Foster (1948), Gillette, Gumbel (1942b), Gumbel and Lieblein, Hesselberg and Birkelund (1940, 1941, 1943), Huss, Kennedy, Kimball (1955), Kincer, Ogawara (1955), Potter (1949c, 1957), U.S. Dept. of Commerce. Weather Bureau, Yarnell.

6.3.4. Application to Aeronautics. A new and very promising field for the use of the theory of extreme values is offered by numerous problems that exist in aeronautics. Since most of the numerical data is recorded automatically, it is relatively easy to obtain large sets of observations. For the design of airplanes, the gust velocities are of particular interest, since they may cause structural damage. Previous analysis based on Pearson's Type III curves did not lead to satisfactory results, especially the probabilities for exceeding the large values were too low. The first application in aeronautics of the theory of extreme values was made by H. Press, who analyzed gust velocities and gust loads under specific test conditions and commercial transport operations.

Y. C. Fung has outlined an ambitious program of applications of the theory of extreme values to some problems of airframe design. In many dynamic stress problems the knowledge of the physical phenomena is not precise enough to justify exact predictions of individual observations. Logical statements of the forcing functions are therefore statistical, and the chief object of an analysis is to draw valid statistical inferences for the responses. This method is applied to the landing-load and gust problems.

In the landing impact, a large number of factors are beyond the control of an airplane designer, such as the softness of the ground, the tire pressure, the change of temperature, etc. When the secondary stresses due to the variations of the neglected variables are not negligible, we may either introduce a sufficiently large factor of safety (or a higher load factor), or, what is more effective, study the variations and introduce some allowances based on statistical inferences.

Considerable variations exist in the maximum impact force, bending moment, acceleration at the fuselage, and the shape of the impact-force curves. The uncertainty of predicting the forcing function leads to the problem of stating the main features of the forcing function and the response in terms of distributions and their averages.

The effective gust speed will be different for each particular type of airplane, and the measurements of the effective gust speeds cannot keep pace with the designs. A theoretically more satisfactory procedure is to measure the gusts and then design the structures accordingly. Thus, the gust-loading problem is reduced to a well-defined problem of finding the response of an elastic airplane to some specified gust profiles.

If a number of V. G. records (velocity against acceleration) are available, each representing approximately the same number of hours of flight, the distribution of the extreme values of the acceleration is obtained and the probability of exceeding the effective gust velocity in each traverse can be found. The application of the theory of extreme values to the ensemble averages leads to an envelope of the most probable largest stress in a structure. If the gust responses are considered as a stationary random process, the time average is the same as the ensemble average. When the distribution of the largest stress over space is known, measurements of the intensity of the largest stress at a few points on the structure, taken over a long period of time, will determine the time interval in which a specified maximum intensity of stress will occur *once*, in association with a given probability, Thus, the life expectancy of the aircraft can be determined for a given specification of the largest stress.

References: Gumbel and Carlson (1954), Lieblein (1954b), Peiser, Peiser and Walker, Peiser and Wilkerson, Press.

6.3.5. Oldest Ages.

> "*Omnium versatur urna serius ocius sors exitura.*"
> (*Age at death is a chance variable.*)

In the following it will be shown that the Gompertz formula for the life table is the first asymptotic probability of smallest values. This relation will be used to calculate the characteristic oldest ages at death.

If we disregard infant mortality, the distribution of the ages x at death rises from a low value at about $x = 10$, passes through a maximum $\tilde{x}$, about 76, and approaches zero for ages above 100. Gompertz assumed that the force of mortality increases for $x \geq 10$ as an exponential function of the age at death x,

(1) $$\mu(x) = \beta e^{\gamma x}.$$

The parameters β and γ are related to the modal age at death. From 1.2.1 (4) and (1),

(2) $$\gamma = \mu(\tilde{x}) = \beta e^{\gamma \tilde{x}}.$$

The parameter γ is the force of mortality at the modal age of death. From (1) and (2) it follows that

(3) $$\mu(x) = \mu(\tilde{x}) \exp\left[\mu(\tilde{x})(x - \tilde{x})\right],$$

where the parameters β and γ in (1) are replaced by the modal age at death and its force of mortality.

The probability $l(x)$ of dying after age x is related to the intensity function by 1.2.1(2), which leads to

(4) $$\frac{l(x)}{el(\tilde{x})} = \exp\left[-e^{\gamma(x-\tilde{x})}\right].$$

6.3.5 USES OF THE FIRST ASYMPTOTE

Since the intensity function of the first asymptotic distribution 5.2.1(15) of the smallest value is

(5) $$\mu(x) = \alpha e^{\alpha(x-u)},$$

the Gompertz distribution of the ages at death about the modal age is the asymptotic distribution of the smallest value. Thus, at each age death takes away the weakest members. The difference between the forms 5.2.1(15) and (4) is due to the different boundary conditions. The variable in 5.2.1(15) is unlimited in both directions, whereas in (4) we conserve the observed cumulative frequency $l(\tilde{x})$ of reaching the modal age $\tilde{x}$.

The characteristic oldest age ω_N at death for a given population and a given period, i.e., for N observations, is the characteristic largest value taken from a distribution of the smallest value,

(6) $$l(\omega_N) = 1 - 1/N.$$

The age ω_N is obtained from (4) in analogy to 6.1.2(3) as

(7) $$\omega_N = \tilde{x} + \frac{1}{\gamma} \lg \lg [Ne\, l(\tilde{x})].$$

The parameters involved are the life table values $\tilde{x}$, $l(\tilde{x})$, γ, for the modal age where the observations are reliable. The number N is the total population multiplied by the annual death rate and by the number of years for which we want to estimate the characteristic oldest age. *Under constant* hygienic conditions. i.e., if $\tilde{x}$, $l(\tilde{x})$, γ, and the annual death rate are fixed values, *the characteristic oldest age at death increases* as the iterated logarithm of the population, thus *extremely slowly*, as shown in Table 6.1.2(3). If we take the square of the number of observations, we add only about 9 months to the expected oldest age. The oldest age increases with the modal age and decreases for increasing intensities at the modal age.

The modal values $\tilde{x}$, $l(\tilde{x})$, and $\mu(\tilde{x}) = \gamma$ for the United States life table, 1939-41, are given in the lines 2, 3, and 5 of Table 6.3.5. Several regularities appear which have been checked by numerous other tables: If we pass from populations living under unfavorable conditions to populations living under favorable hygienic conditions (say, Negro to white), the modal age at death increases. The same development takes place if a population passes from unfavorable to favorable conditions. However, *the increase of the modal age at death is compensated by an increase in its intensity of mortality*.

From the observed common increase of $\tilde{x}$ and $\mu(\tilde{x})$ it follows that, for the same number of deaths, we have to expect *a larger oldest age for a population living under unfavorable than under favorable conditions*, as claimed in 3.1.1. Formula (7) is used in Table 6.3.5, which leads to the characteristic oldest ages given in line 11.

Table 6.3.5. *Modal Values and Characteristic Oldest Ages, U.S.A., 1939–41*

1 Notion	Symbol	White Male	White Female	Negro Male	Negro Female
2 Modal Age	$\tilde{x}$	76.65	78.33	58.20	59.80
3 Probability	$l(\tilde{x})$	0.316	0.342	0.469	0.493
4 Factor	$1/(el(\tilde{x}))$	1.16	1.08	0.786	0.746
5 Intensity	$\mu(\tilde{x}) = \gamma$	0.088	0.094	0.038	0.035
6 Size in millions		59.45	58.77	6.613	6.841
7 Death Rate per 1000		11.62	9.91	15.14	12.56
8 Number of Observations in 1000	$Nel(\tilde{x})$	595.6	539.5	127.4	115.2
9 Iterated Logarithm	$\lg \cdot \lg [Nel(\tilde{x})]$	2.588	2.580	2.464	2.456
10 Reciprocal of intensity	$1/\gamma$	29.41	27.45	64.85	70.16
11 Characteristic Oldest Age in Years	ω	105	106	123	130

The Gompertz formula assumes that there is no fixed limit for the age at death, and leads to quite reasonable estimates of the oldest ages at death. The oldest ages of Negroes exceed by far the oldest ages of whites. The highest age reached is not a measure of hygienic progress. If the hygienic progress continues along the same lines as in the past, the mean and the modal ages at death will rise, but the extreme age at death may even decrease by a small amount.

References: Boehm, Freudenberg, Greville, Gumbel (1937a and the bibliography given therein, 1953, 1955), Hooker and Longley-Cook, Insolera, Mather, Vincent (1950, 1951).

6.3.6. Breaking Strength. The survivorship function is also the fundamental tool for application of the theory of extreme values to numerous problems arising in the study of breaking strength, e.g. the weight under which a given material breaks in a controlled experiment. The weight per unit of surface is called stress. In the classical theory, the breaking strength is a fixed value to be calculated by purely physical considerations. However, in the fracturing of metals, textiles, and other materials under applied force, the strength varies considerably from specimen to specimen, even under constant conditions. The interpretation of the test results is made difficult by the fact that progressive damage, which finally leads to failure, is a highly structure-sensitive process. The result, therefore, is a very wide scatter, and for the rational analysis of this the character of the statistical distribution must be known.

It has often been assumed that the probability of survival may be analyzed by the logarithmic normal probability function. Then the probability of survival as function of the logarithm of the stress approaches unity in the same way as it approaches zero. This consequence is contradicted by the experimental fact that for a large stress a small increase results in a large decrease of the probability of survival (the straw that broke the camel's back), while for a small stress a considerable decrease is necessary in order to increase the probability. Instead of the logarithmic normal theory, we use in the following the theory of extreme values, which may be justified as follows: The difference between the calculated and the observed strength resides in the existence of weakening flaws. Therefore, a different amount of force will be needed to fracture the body at one or another point, and the strength of a given specimen is determined by its weakest point. A good exposition of the statistical methods applied to fracture problems has been given by Epstein (1948), who was the first to use the asymptotic theory of extreme values.

In these applications a number of assumptions must be made: The flaws of different sizes are distributed independently at random in the body; the probability density of the sizes of flaws is of the exponential type; the number of flaws is very large; finally, the breaking strength is a linear function of the size of the largest flaw alone, whatever its size may be, and diminishes with increasing size. Under these conditions, the distribution of the *largest flaws* is the first asymptotic distribution of the largest values and the distribution of the breaking strength is the asymptotic distribution of the *smallest* values. Thus the question about the initial distribution of the flaws becomes immaterial. Graph 6.3.6 shows the frequencies of survival for 200 rubber specimens observed by Mr. Shigeo Kase of the Togawa Rubber Co. in Osaka, Japan, under increasing loads. The frequencies are traced on the extremal probability scale, the loads S in kg cm^{-2} on the linear scale. The Gompertz formula for the probability of survival, the first asymptotic probability of smallest values, is

(1) $$l(S) = 1 - {}_1\Phi(S) = \exp\left[-e^{\alpha(S-u)}\right].$$

From the sample mean $\bar{S}_0 = 103.05$ and standard deviation $s = 11.45458$, the two parameters are estimated with the help of Table 6.2.3 as $1/\hat{\alpha} = 9.268$, $\hat{u} = 108.307$. The fit is so good that there is no necessity for constructing control curves.

References: Afanassiev, Bonnet, Breny, Chapin (1880), S. M. Cox, Daniels (1945), Epstein (1948a, 1955), Epstein and Brooks, J. C. Fisher and Hollomon, Frenkel and Kontorova, Freudenthal (1946), Freudenthal and Gumbel (1956), Fung, Gaede, O. Graf and Weise, Griffith (1920, 1924), Himsworth, J. W. Johnson, Kase (1953), Kontorova, Leme, Lévi, L'Hermite, Miklowitz, Pierce (1938), Plum, W. H. Price, Prot (1949), Tanenhaus, Torroja, Trautwine, Tucker (1941, 1945), Wing. *Consult also:* J. Lieblein (1954). Two early papers on the relation between extreme values and tensile strength, Biometrika, 41: 559.

6.3.7. Breakdown Voltage. Large high voltage transformers have oil as the major electrical insulation component. Because of this and other

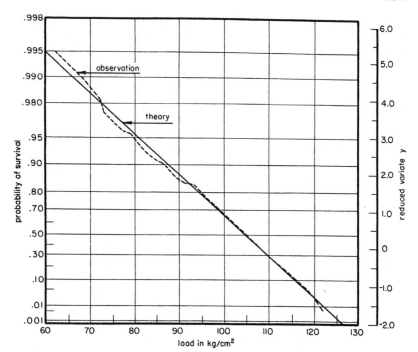

Graph 6.3.6. Survivorship Function for Rubber Specimen

uses of oil in electrical equipment, the knowledge of the dielectric strength of oil and its dependence on electrode size and operating conditions is of immediate practical importance in securing the optimum utilization of the oil.

Weber and Endicott built an automatic dielectric strength tester which measured and tabulated digitally the breakdown voltage to the nearest kilo volt. Their articles analyse 1,600 breakdown voltages for horizontal and vertical positions for each of four electrodes. Thus each of the eight sets consists of 200 observations. Data of such size, which did not exist until now, permit a systematic study of important phenomena.

In the following we indicate only the main results: the orientation of the electrode surface has no influence on the breakdown voltage. Therefore the distributions for horizontal and vertical field positions can be combined into one series. The distribution for four electrode pairs of different areas refute the normal distribution used until now and indicate clearly that the first asymptotic distribution of *smallest* values is an adequate tool of analysis. The data for four areas plotted on extremal paper are closely scattered about parallel straight lines fitted by the method described in

6.2.3. The breakdown voltage for the minima of successive groups of 10 for four sizes also yields parallel straight lines.

For a study of the areal effect the authors compare 400 individual breakdown voltages plotted on extremal probability paper with 40 minima of groups of 10 and 4 minima of groups of 100. The straight lines are parallel as foreseen by the stability of the distribution of extreme values. The breakdown voltage decreases linearly as the logarithm of the electrode area increases. For the particular oil used a ten to one increase of area results in a 16 per cent decrease in dielectric strength.

The extreme value theory yields equations for predicting the breakdown voltage for electrodes of any area for a uniform field, once the two parameters are estimated for a given area and for the oil used. The extremal nature of oil breakdown will provide the basis for the development of a logical and consistent dielectric strength theory.

6.3.8. Applications to Naval Engineering. At present the shipbuilding industry knows less than any other construction industry about the service conditions under which it must operate. Only small efforts have been made to establish the stresses and motions and to incorporate the result of such studies into design. This is due to the complexity of the problems caused by the extensive variability of the sea and the corresponding responses of the ships.

In oceanography different quantities are used for describing the state of the sea, namely, the mean height of the waves, the root mean square height, and the mean height of the highest one third of all waves (called the significant height), and finally the mean and the most probable largest wave height. Connected quantities in naval engineering are the wave-induced pitch, roll and heave motions and stresses in ships and the maxima of these quantities.

Of course, simplifying assumptions on the operating conditions must be made in the analysis. Longuet-Higgins considers a spectrum of waves consisting of a single narrow frequency band due to a large number of independent random contributions from different regions, all about the same frequency and of random phase. Under these conditions the wave heights x follow the Rayleigh distribution, which is a special case of the third asymptotic distribution of the *smallest* value, Table 5.1.3(2). Rayleigh obtained it for the amplitude of sound derived from many independent sources. The distribution is also connected with the problem of random walk. The probability $F(x)$, distribution $f(x)$, intensity $\mu(x)$, and return period $T(x)$ are

(1) $\quad F(x) = 1 - \exp[-x^2/a^2]; \quad f(x) = 2xa^{-2} \exp[-x^2/a^2]$

$\quad\quad \mu(x) = 2xa^{-2}; \quad T(x) = \exp[x^2/a^2].$

Since x^2/a^2 is equal to $\lg[1 - F(x)]$ semilogarithmic paper can be used for the plotting of observations, a procedure due to Jasper. The mean $\bar{x}$, the mode $\tilde{x}$, the variance σ^2, and the maximum likelihood estimate for the single parameter a are

(2) $\qquad \bar{x} = a\sqrt{\pi}/2; \quad \tilde{x} = a/\sqrt{2}; \quad \sigma^2 = a^2(1 - \pi/4); \quad \hat{a} = \sqrt{\overline{x^2}}.$

Let P be the probability of a value exceeding x, then the mean $\bar{x}(x_p)$ of the values surpassing a quantile x_p is

(3) $\qquad \bar{x}(x_p)/a = \sqrt{-\lg P} + \sqrt{\pi}[1 - F(\sqrt{-2 \lg P})]/2P,$

where F stands for the Gaussian integral. The expression (1) decreases with increasing P. For $P = 1$, formula (3) leads back to the mean (2). $P = 1/3$ leads to the significant wave height. Longuet-Higgins gives a table for $\bar{x}(x_p)/a$ for different values of P. Observations on wave heights at Atlantic and Pacific coast stations for different values of P showed good agreement with the theoretical values. This result encourages a study of the largest wave heights. Longuet-Higgins obtained the mean $\bar{x}_n$ of the largest among n wave heights.

(4) $\qquad \bar{x}_n/a = n(\sqrt{\pi}/2) \sum_{v=0}^{n-1} \binom{n-1}{v}(-1)^v(v+1)^{-3/2}.$

Since the distribution (1) belongs to the exponential type with respect to the largest value (see 4.1.4), this expression converges from 5.2.4(5) to

(5) $\qquad \bar{x}_n/a = u_n + \gamma/\alpha_n.$

From the definitions 3.1.4(1) and 3.1.5(1),

(6) $\qquad u_n = \sqrt{\lg n}, \quad \alpha_n = 2\sqrt{\lg n}.$

Hence the mean largest value becomes asymptotically

(7) $\qquad \bar{x}_n/a = \sqrt{\lg n} + \gamma/(2\sqrt{\lg n}),$

a formula which the author derived independently. The most probable largest among n value, $\tilde{x}_n$, is the solution $z = \tilde{x}_n/a$ of

(8) $\qquad n = e^{z^2}[1 - 1/(2z^2)] + 1/(2z^2).$

The first approximation is

(9) $\qquad z = u_n = \sqrt{\lg n}.$

The author gives a table (3D) for $\sqrt{\lg n}$, the values of (4) and (8) and the asymptotic expression (7) for $n = 1 \cdot 10^v$, $2 \cdot 10^v$, $5 \cdot 10^v$, $[v = 0(1)5]$.

6.3.8 USES OF THE FIRST ASYMPTOTE

Observations of maximum wave heights agreed fairly well with the theoretical values.

The work of Longuet-Higgins was applied to naval engineering by Jasper, who studied more than one hundred distributions of wave-induced pitch, roll, and heave motions of ships and hull stresses (longitudinal hull girder bending moments) on seven ships of different length and displacements, destroyers and aircraft carriers, at different states of the sea and ship speeds. He assumes that under steady environmental conditions the response of the ship to the wave heights, i.e. the distributions of motions and stresses are the same as the distributions of the wave height themselves.

Jasper applied this method especially to pitch angles and stresses. Each sample was large, the size varying from 136 to 308 observations. It is remarkable that his graphs show those patterns which evidence the poorest agreement between the experimental and Rayleigh's distributions. Even in these cases, the fit of the theoretical straight line to the observed data is excellent. None of the distributions of ship motions and hull stresses indicate any systematic deviations. The comparison of the 27 largest values of the pitch angles to the characteristic largest values (6) led to satisfactory results.

For long-term movements where the sea conditions, ship heading, and speed vary, Jasper assumes that the parameter a in (1) has again a Rayleigh distribution, and states that therefore the logarithmic normal distribution represents the significant wave heights and wave-induced motions and hull stresses under varying conditions. The observations plotted on logarithmic normal paper seem to justify the conclusion.

Finally, Jasper uses the first asymptotic distribution for the largest stresses encountered during 19 trips, each averaging 121 hours at sea. Tracing of the observations on extremal probability paper shows an excellent agreement with the theory. Thus the most probable amplitude of the motions of ships and the extreme values of the ships' responses can be predicted, and the capacity for which shipboard stabilization equipment should be designed and the endurance strength of a ship structure can be estimated.

Exercises: 1) Show that the distribution of $\lg x$ is the first asymptote. 2) Calculate a table of n as function of $\tilde{x}_n/a$ from (8). 3) Prove that the Gnedenko conditions 5.1.3(1) is fulfilled for a variate with the probability function $F(x) = 1 - \exp[-x^k]$; $x \geqq 0, k > 0$. 4) Calculate the generating function for the Rayleigh distribution. 5) Calculate the convolution of two Rayleigh distributions.

Problems: 1) What is the distribution of x, if the parameter a also follows Rayleigh's distribution? 2) Construct a bivariate Rayleigh distribution.

Reference: Cartwright and Longuet-Higgins. *Consult also:* N. H.

Jasper (1955). Service stresses and motions of the Esso Ashville tanker including a statistical analysis of experimental data, Navy Dept., David W. Taylor Model Basin Report 960. Washington, D. C.

6.3.9. An Application to Geology. Unusually large cobbles or boulders observed in gravel deposits raise the question whether individuals of such size may be considered as part of the pebble population or as erratics, transported by ice or other extraneous processes. Large cobbles normally present in fluvial gravels presumably are related to the action of the transporting stream. Krumbein and Lieblein analyze data on sandy gravel of fluvial origin at Gordon's Corner, Maryland, of average size of about 1.6 cm. Two sampling experiments were made on the base of the slope. In the first, 50 largest cobbles were picked up at an interval of two paces (a conventional procedure in sedimentary studies). It was estimated that each largest cobble came thus from a sample of about 25,000 pebbles. In the second experiment, another sample of 50 largest values was obtained by excavation with a shovel, which yielded about 2,000 pebbles.

The two series of largest sizes of cobbles plotted on extremal paper lead to nearly parallel straight lines with $\hat{u}_1 = 9.38$ cm, $1/\hat{\alpha}_1 = 1.09$ cm ($\hat{u}_2 = 7.43$; $1/\hat{\alpha}_2 = 0.99$ cm) for the first (second) experiment. Despite the approximate nature of the measurements and the crude estimate of the sample size used, the agreement of the data with the theory is very good. All observations are contained within the control bands. Hence the authors conclude that all large particles observed belong to the alluvial deposit, and that there is no need to assume that external forces have contributed to the existence of the large boulders. The extreme value theory may thus be used for the study of stream deposits. A study of largest pebbles concentrated in particular beds may disclose a history of specific geological events associated with the coarse layers in sediments.

Problem: Construct an analysis of variance for the extremal distributions.

Chapter Seven: THE SECOND AND THIRD ASYMPTOTES

"Les extrêmes se touchent."
(The three asymptotes are closely related.)

7.1. THE SECOND ASYMPTOTE

7.1.0. Problems. The first asymptotic probability $\Phi(x)$ of the extreme holds for the exponential type. The second and third asymptotes introduced in 5.1 hold for certain limited initial distributions, respectively. The second asymptote, which is less important for practical applications, may be obtained by various methods. 1) We may follow Fréchet, who starts from the stability postulate and looks for a class of initial distributions which satisfies certain requirements. 2) We may start from the asymptotic properties of a certain type of initial distributions and calculate the asymptotic distribution of the extremes. This procedure shows that the second asymptotic distribution holds for certain distributions, unlimited to the right, lacking all or the higher moments. 3) We may obtain the second asymptote from the first one by a logarithmic transformation. This leads to the construction of a probability paper and to the knowledge of those moments that exist.

7.1.1. Fréchet's Derivation. Fréchet considers n distributions which have a common analytic form, but differ with respect to the parameters. Each observation comes from a different distribution, and he asks for the distribution of the largest value. The initial distributions are defined by three properties: 1) The initial variates z are non-negative. 2) The probability function depends upon the relative size of the variate with respect to a certain average A, to be fixed later. 3) The distribution of the largest value is required to be stable. In the same way as in 5.1.2, Fréchet looks for the initial distribution which fulfills these conditions.

For the sake of simplicity, first consider a sample of two independent observations. Let the unknown initial probabilities be $F(z/A_1)$ and $F(z/A_2)$ respectively. Then the probability that both values are below a certain z is their product. This probability should again be of the form $F(z/A)$. Thus,

(1) $$F(z/A) = F(z/A_1) \cdot F(z/A_2),$$

where A is a function of A_1 and A_2, say,

(2) $$A = \lambda(A_1, A_2).$$

The problem is to determine the two unknown functions λ and F. Equation (1) becomes, for $z = 1$,

(3) $$\lg F(1/A) = \lg F(1/A_1) + \lg F(1/A_2).$$

We write

(4) $$-\lg F(1/A) = a > 0,$$

and introduce correspondingly $a_1 > 0$; $a_2 > 0$. If A increases from zero to infinity, F decreases from unity to zero, and a increases from zero to infinity. Let the inverse function be

(5) $$A = \Theta(a);$$

then equation (3), reads, from (4) and (5),

$$a = a_1 + a_2; \quad A = \Theta(a_1 + a_2),$$

and equation (1) may be written

(6) $$-\lg F\left[\frac{z}{\Theta(a_1+a_2)}\right] = -\lg F\left[\frac{z}{\Theta(a_1)}\right] - \lg F\left[\frac{z}{\Theta(a_2)}\right],$$

valid for any value of $z \geqq 0$. Consider this equation with respect to a_1 and a_2. If we write†

(7) $$-\lg F[z/\Theta(a)] = g(a,z)$$

equation (6) becomes

(8) $$g(a_1 + a_2, z) = g(a_1, z) + g(a_2, z);$$

hence, for any z the function g is linear and homogeneous in a and

$$g(a,z)/a = \varphi(z)$$

is independent of a, whence

(9) $$g(a,z) = a\varphi(z).$$

The probability function (7) obtained from (9) and (4),

(10) $$-\lg F[z/\Theta(a)] = -[\lg F(1/A)]\,\varphi(z)$$

is a product of two functions, one depending only on A and the other only on z. We will show now that the two functions are the same. The function $\varphi(z)$ is obtained by setting $A = 1$, whence from (5)

$$-\lg F(z) = -[\lg F(1)]\,\varphi(z).$$

† The modifications of the original publication were made by M. Fréchet in a letter to the author.

7.1.1 SECOND AND THIRD ASYMPTOTES

We introduce a positive constant h, independent of A, by writing

(11) $$-1/\lg F(1) = h.$$

Then

(12) $$\varphi(z) = -h \lg F(z),$$

and at the admissible value $z = 1/A$,

(13) $$\varphi(1/A) = -h \lg F(1/A).$$

Equation (12) may be written

(13') $$\varphi(z/A) = -h \lg F(z/A),$$

or from (10) and (13)

(14) $$\varphi(z,A) = \varphi(z)\varphi(1/A).$$

This is a functional equation for $\varphi(z)$. The two functional equations (1) and (2) will now be reduced to a single one. For the solution of (14) we write

(15) $$z = e^\zeta, \quad A = e^{-\eta}.$$

Then (14) becomes

$$\lg \varphi(e^{\zeta+\eta}) = \lg \varphi(e^\eta) + \lg \varphi(e^\zeta).$$

By writing

(16) $$\lg \varphi(e^\zeta) = \psi(\zeta),$$

the preceding equation becomes

$$\psi(\zeta + \eta) = \psi(\zeta) + \psi(\eta).$$

The solution is

$$\psi(\zeta) = k_1 \zeta,$$

whence, from (16),

$$\lg \varphi(e^\zeta) = k_1 \zeta; \quad \lg \varphi(e^\eta) = k_1 \eta$$

and, from (15),

$$\lg \varphi(z) = k_1 \lg z; \quad \varphi(z) = z^{k_1}; \quad \varphi(1/A) = A^{-k_1}$$

or from (14)

$$\varphi(z/A) = (z/A)^{k_1}.$$

Equation (13') leads finally to the probability function,

(17) $$F(z/A) = \exp\left[-\frac{1}{h}\left(\frac{z}{A}\right)^{k_1}\right].$$

Since $F(z)$ must increase with z, the factor k_1 must be negative, say,

$$-k_1 = k > 0,$$

whence

(18) $$F(z/A) = \exp\left[-\frac{1}{h}\left(\frac{A}{z}\right)^k\right].$$

This is the initial probability that permits the probability of the largest of n observations to be stable. It fulfills the two other conditions stated above. The proof given here for two initial distributions can easily be extended to n distributions.

The three parameters in (18) can be reduced to two. The parameters A_v are certain averages of the initial distributions. Assume that A_v be the medians, then, from (18),

$$\lg 2 = 1/h$$

and (18) becomes

(18′) $$F(z/A) = 2^{-(A/z)^k},$$

where A is the median of z. This is the form used by Fréchet. To obtain the form used in Table 5.1.1 for $\varepsilon = 0$, let v be the value of z so that

$$F(v/A) = e^{-1}.$$

Then, from (18),

$$1/h = (v/A)^k.$$

Thus, the probability may be written

(19) $$F(z) = \exp\left[-\left(\frac{v}{z}\right)^k\right].$$

We now establish the relationship between Fréchet's form (18′) and the functional equation (1). Let A be the median for the initial distribution; then the solution for equation (2) obtained from (18′) is

(20) $$A^k = \sum_{1}^{n} (A_v)^{k_v}.$$

If the initial distributions have the form (18′) with different medians, the distribution of the largest value is the same, and its median is given by (20). If all initial distributions have the same median $A_v = A_1$ and $k_v = k$, we have in effect taken n samples from the same distribution. Then the median A of the largest value is related to the initial median A_1 by

(21) $$A = A_1 n^{1/k}.$$

The median of the largest value increases as the kth root of the sample size. Fréchet showed that the stability also holds if n tends to infinity provided that the sum (20) converges. Finally he proved that the product $z^k [1 - F(z)]$ converges, for large z to a finite positive value. This will be our starting point in the next paragraph.

Since z is non-negative, it may also be interpreted as the absolute amount of the deviation from a certain average, say, the median, and (19) becomes the distribution of the largest absolute value in a distribution which is unlimited in both directions.

Problems: 1) Construct, in the same way, the double exponential distribution from the functional equation $F(x - A_1) \, F(x - A_2) = F(x - A)$. 2) Calculate the convolution of two distributions of Fréchet's type.

7.1.2. The Cauchy Type. We now show that the asymptotic probability 7.1.1(19) holds for initial distributions of the Pareto and Cauchy types.

The Pareto type was defined in 4.3.3 by the property

(1) $$\lim_{z=\infty} z^k[1 - F(z)] = A > 0 \,; \quad k > 0 \,; \quad z \geqq 0$$

and the Cauchy type by the additional property

(2) $$\lim_{z=-\infty} F(z)(-z)^{k_1} = A_1 > 0 \,; \quad k_1 > 0 \,; \quad z \leqq 0 \,.$$

Another form of condition (1) for the Pareto type is

(3) $$\lim_{z=\infty} z^{k+1} f(z) = Ak \,,$$

whence, from (1),

(4) $$\lim_{z=\infty} z\mu(z) = k \,.$$

This equation links the parameter k to the extremal intensity function.

To obtain the asymptotic distributions of the extremes for the Cauchy type, we use the characteristic largest among n values, v_n defined in 3.1.4(1). If n is sufficiently large, the limiting condition (1) leads asymptotically to

$$1/n \sim A v_n^{-k} \,,$$

whence, from (1),

(5) $$\lim_{z=\infty} F(z) = 1 - \frac{1}{n}\left(\frac{v_n}{z}\right)^k \,.$$

Consequently, the asymptotic probability $\Pi(z)$ and the distribution $\pi(z)$ for the largest value are

II—(6) $\quad\quad\quad \Pi(z) = \exp\left[-(v_n/z)^k\right]$

(7) $\quad\quad\quad \pi(z) = \dfrac{k}{v_n}\left(\dfrac{v_n}{z}\right)^{k+1} \exp\left[-\left(\dfrac{v_n}{z}\right)^k\right]$

as stated in 5.1.1. The variate z is non-negative, while it was unlimited in both directions for the first asymptote. Since the moments diverge from the kth onward, the second asymptote will show certain similarities to the first one if k is large. The third parameter, the lower limit ε in Tables 5.1.1 and 5.1.3, will be introduced later. It is immediately verified that the Cauchy type fulfills Gnedenko's condition, 5.1.3. From the symmetry principle 3.1.1(6), we obtain for the smallest value

$$1 - {}_1\Pi(z) = \exp\left[-(v_1/z)^{k_1}\right];$$

(8) $\quad\quad {}_1\pi(z) = \dfrac{k_1}{-v_1}\left(\dfrac{v_1}{z}\right)^{k_1+1} \exp\left[-\left(\dfrac{v_1}{z}\right)^{k_1}\right]; \quad \begin{array}{l} v_1 < 0 \\ z < 0 \end{array}; \quad k_1 > 0,$

where v_1 is the characteristic smallest value, and k_1 is used in (2). If the initial distribution is symmetrical, we have of course $-v_1 = v_n$ and $k_1 = k$. All properties of the largest value also hold for the absolute amount of the smallest value.

The first two asymptotic distributions of the largest value are linked by the transformation 5.1.1(1).

(9) $\quad\quad\quad x = \lg z, \quad u = \lg v\,; \quad \alpha = k, \text{ or } \left(\dfrac{v}{z}\right)^k = e^{-y}.$

Therefore, the logarithms of the largest values taken from a distribution of the Cauchy type are distributed as the largest values taken from the exponential type. In the designation of Table 5.1.3(2) the relation reads for the largest (smallest) reduced value:

(10) $\quad\quad\quad \Phi^{(1)}(\lg x) = \Phi^{(2)}(x)\,; \quad {}_1\Phi^{(1)}(-\lg x) = {}_1\Phi^{(2)}(x).$

To obtain the numerical values for the probability function, we write $\Pi(z) = \Phi(y)$ and calculate z from (9) as

(11) $\quad\quad\quad\quad\quad z = ue^{y/k}.$

The probability $\Pi(v) = .36788$ is independent of k. For $z > v$ ($z < v$) the probabilities $\Pi(z,k)$ increase (decrease) with increasing values of k, as shown in Graph 7.1.2(1). Graph 7.1.2(2) shows the distributions for $k = \tfrac{1}{2}$, $k = 1$, and $k = 2$, the variate being expressed in units of v.

7.1.2 SECOND AND THIRD ASYMPTOTES 261

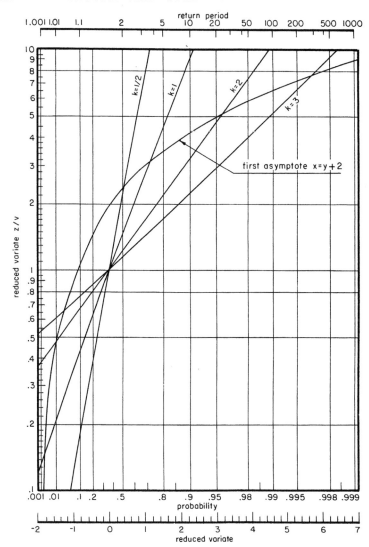

Graph 7.1.2(1). Fréchet's Probability (II) for Different Values of k

For comparison's sake, the first asymptotic probability for the variate $x = y + 2$ is also traced in Graph 7.1.2(1). The inverse procedure of tracing $\log x$ on the extremal paper designed for the first asymptote was used in Graph 5.2.6. The use of probability papers leads to a clear distinction between the first and the second asymptote if the quotient x_N/x_1 is large, say 10. However, no graphical distinction is possible if it is small, say 2, because then the plotting of the logarithms of the observations is not sufficiently different from the plotting of the observations themselves.

R. von Mises brought the first and second asymptotes into a common form which shows that the first asymptote is a sort of limit for the second one. To this end he introduces the transformation

(12) $$z/v_n = 1 + y/k \; ; \quad k > 0$$

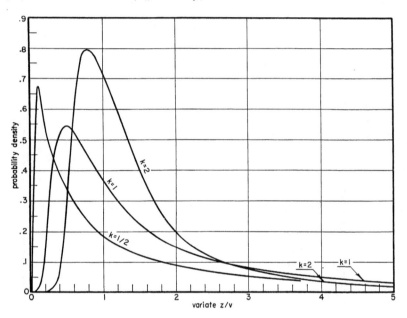

Graph 7.1.2(2). The Second Asymptotic Distribution of Largest Values

into the second asymptote (6). This leads to the probability function $V(y)$ for the variate y,

(13) $$V(y) = \exp\left[-(1 + y/k)^{-k}\right],$$

which is a generalized form of the asymptotic probability of the largest value traced in Graph 7.1.2(3). This corresponds to the interpretation of k as the order of the lowest moment which diverges and to the fact that the probability vanishes for $y = -k$. For $y = 0$ the probability becomes $1/e$ independent of k. For $k = 0$ the probability degenerates into $V(y) = 1/e$ independent of y. The moments of order $l \geq k$ do not exist. For $k = \infty$ the first asymptote is reached.

7.1.2 SECOND AND THIRD ASYMPTOTES

If we introduce a linear transformation,

(14) $\quad y = (x - a)/b \, ; \quad a \geqq 0 \, ; \quad b > 0,$

the probability for x may be written

$$\Phi^{(2)}(x) = \exp\left[-\left(\frac{bk}{bk - a + x}\right)^k\right]$$

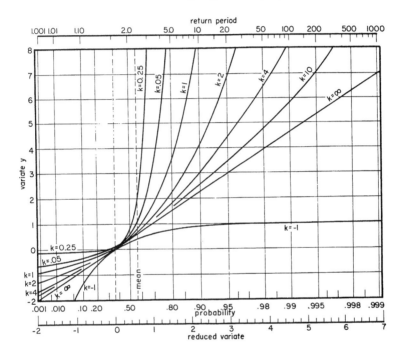

Graph 7.1.2(3). The Generalized Probability $V(y)$

or with $bk = v - \varepsilon$, $a - bk = \varepsilon$,

(15) $\quad \Phi^{(2)}(x) = \exp\left[-\left(\frac{v - \varepsilon}{x - \varepsilon}\right)^k\right],$

which is the second asymptote as used in Table 5.1.3(2).

Exercise: Show that the mode and the probability and density at the mode and the median and its density derived from (13) converge with increasing k to the corresponding values of the first asymptote.

References: Balaca (p. 51), Bernier, Jenkinson, Kawata, Leme.

7.1.3. Averages and Moments. The median $\check{z}$ and the mode $\tilde{z}$ of the distribution 7.1.2(6) are, respectively, after dropping the index n,

(1) $$\check{z}^k/v^k = 1.44629 > \tilde{z}^k/v^k = k/(k+1).$$

The position of the three averages is

(2) $$\check{z} > v > \tilde{z}.$$

The mode differs here from the characteristic largest value, while both coincide for the first asymptote. The median and the mode increase with v, the median decreases and the mode increases with increasing k. The quotient $\check{z}/v$ decreases, the quotient $\tilde{z}/v$ increases with k, and both quotients converge to unity. The mode and the median converge to the characteristic largest value with increasing k (See Graph 7.1.3.), i.e., if the initial distribution approaches the exponential type. The probability $\Pi(\tilde{z})$ at the mode,

(3) $$\Pi(\tilde{z}) = \exp\left[-(1 + 1/k)\right],$$

increases with increasing k and approaches $1/e$, the value for the first asymptote.

The densities of probability $\pi(v)$, $\pi(\check{z})$, and $\pi(\tilde{z})$ at the characteristic largest value, the median and the mode,

(4) $$\pi(v) = k/ve\,; \quad \pi(\check{z}) = \frac{1}{2}\frac{k}{v}(\lg 2)^{1+1/k}\,; \quad \pi(\tilde{z}) = \frac{k}{v}\left(\frac{1+1/k}{e}\right)^{1+1/k},$$

diminish with increasing v. The densities $\pi(v)$ and $\pi(\check{z})$ increase with increasing k. The influence of k on the density $\pi(\tilde{z})$ is obtained from

(5) $$\frac{d\lg\pi(\tilde{z})}{dk} = \frac{1}{k} - \frac{1}{k^2}\lg(1 + 1/k).$$

Therefore, the density of probability at the mode decreases (increases) with increasing k, if $k < 0.80647$ ($k > 0.80647$). The distribution of the largest value spreads with increasing v. An increase of u has no such influence on the first asymptote. The parameter k is an inverse measure of dispersion, a property it shares with the parameter α. However, α has the dimensions of x^{-1}, while k has no dimension.

To obtain those moments which exist, we use the logarithmic transformation 7.1.2(9). From the generating function $G_y(t)$ of the reduced variate y for the first asymptote 5.2.4(2), and from 1.1.6(12), the moments $\overline{z^l}$ of order l are

(6) $$\overline{z^l} = v^l \Gamma(1 - l/k).$$

7.1.3 SECOND AND THIRD ASYMPTOTES

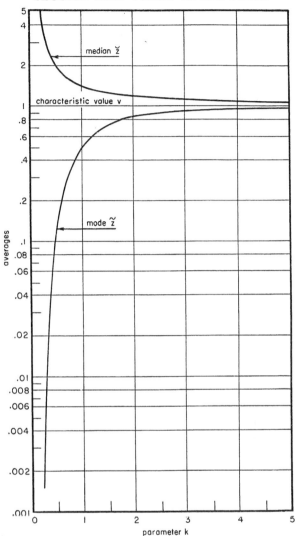

Graph 7.1.3. Averages for the Second Asymptote

The moments exist provided that $k > l$. The reciprocal moments of order $l > 0$,

$$\overline{z^{-l}} = v^{-l} \int_0^\infty k(v/z)^{k+1-l} \exp\left[-(v/z)^k\right] dz/v ,$$

become, from the transformation $(v/z)^k = x$,

(7) $$\overline{z^{-l}} = v^{-l}\Gamma(1 + l/k) .$$

In contrast to the moments all reciprocal moments exist.

Exercises: 1) Calculate the averages and moments for the distribution of smallest values. 2) Construct the distribution of the *m*th largest value taken from an initial distribution of the Cauchy type.

7.1.4. Estimation of the Parameters. If we know that an observed series of N largest values is taken from a (non-specified) distribution of the

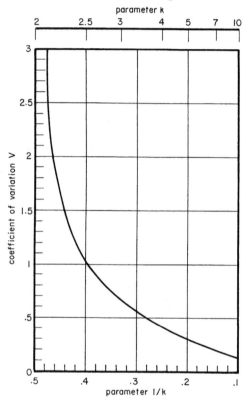

Graph 7.1.4(1). Estimate of k from Coefficient of Variation

Cauchy type, an estimate of k leads to the order of those moments that exist in the initial distribution. a) If it is known that $k > 2$, the two parameters v and k may be estimated from 7.1.3(6), i.e., from

(1) $$\bar{z} = v\Gamma(1 - 1/k) \; ; \quad 1 + V^2 = \frac{\overline{z^2}}{\bar{z}^2} = \frac{\Gamma(1 - 2/k)}{\Gamma^2(1 - 1/k)},$$

by introducing the sample values on the left side. The larger k, i.e., the

7.1.4 SECOND AND THIRD ASYMPTOTES

higher the order of moments that exist, the smaller becomes the coefficient of variation V. This relation is shown in Graph 7.1.4(1) which may be used to estimate k. The remaining parameter v is then estimated from (1). b) The logarithmic transformation 7.1.2(9) may be used for another estimate. From the first two moments 5.2.4(5) and (7) of the first asymptote, the first two geometric moments g_1 and g_2 for the second asymptote are

(2) $\quad \lg g_1 = \lg v + \gamma/k \; ; \quad \lg g_2 - \lg^2 g_1 = \pi^2/6k^2 \;.$

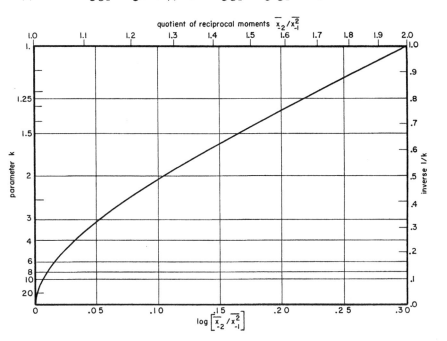

Graph 7.1.4(2). Estimate of k from the Reciprocal Moments

If the left side is replaced by the sample values, we may estimate $1/k$ and $\lg v$. c) Another estimate of the parameters is based on the reciprocal moments $\overline{z^{-l}}$. From 7.1.3(7)

(3) $\quad \overline{z^{-1}} = v^{-1}\, \Gamma(1 + 1/k) \; ; \quad \overline{z^{-2}} = v^{-2}\, \Gamma(1 + 2/k) \;,$

whence

(4) $\quad \overline{z^{-2}}/(\overline{z^{-1}})^2 = \Gamma(1 + 2/k)/\Gamma^2(1 + 1/k) \;.$

The right side of this equation is traced as a function of $1/k$ in Graph 7.1.4(2). If the left side is replaced by the sample values, we may estimate $1/k$. Finally, v is obtained from the first equation in (3).

d) If k is known, the remaining parameter v can be estimated from the ranked observations in the same way as u was estimated in 6.2.2(2). The parameter v is interpreted as an order statistic corresponding to the probability $1/e$. The equation for the most probable rank $\tilde{m}_v$ becomes, if the values for $z = v_m$, namely,

(5) $$\Pi = 1/e \; ; \quad \pi/\Pi = k/v \; ; \quad -\pi'/\pi = 1/v \, ,$$

are introduced into 2.1.2(6)

(6) $$m_v = 0.36788 N + 0.63212(1 + 1/k) \, .$$

Thus, v can be estimated as an order statistic if k is known.

e) For the maximum likelihood method, consider

(7) $$\lg \pi(z) = \lg k + k \lg v - (k+1) \lg z - (v/z)^k \, .$$

If we introduce a new parameter,

(8) $$v^k = U,$$

differentiation of (7) with respect to the two parameters k and U and summation over all observations leads to

(9) $$\frac{1}{k} + \frac{U}{N} \sum z_v^{-k} \lg z_v = \overline{\lg z} \; ; \quad \frac{1}{U} = \frac{\sum z_v^{-k}}{N} \, .$$

The first equation becomes, from the second one,

(10) $$\frac{1}{k} + \frac{\sum z_v^{-k} \lg z_v}{\sum z_v^{-k}} = \overline{\lg z} \, .$$

This equation contains only one unknown parameter k and can therefore be used for an estimate by successive approximations. To find a first approximation, the quotient is multiplied by v^k. Then (10) becomes, from (9) and (8),

(11) $$\frac{1}{k} + \frac{\sum (v/z_v)^k \lg z_v}{N} = \overline{\lg z}$$

or

$$\frac{1}{k} - \frac{1}{N} \sum \lg \Pi_v \lg z_v = \overline{\lg z} \, .$$

Now the probabilities Π_v are replaced by the frequencies, i.e., the plotting positions of the vth values. This leads, after use of the correction A_N given in 6.2.6, to

(12) $$\frac{1}{k} = \frac{A_N}{N} \sum_1^N \lg v \cdot \lg z_v - [A_N \lg (N+1) - 1] \overline{\lg z}$$

7.1.5 SECOND AND THIRD ASYMPTOTES

and hence to a first approximation for k, say $\hat{k}_1$. Then equations (9) and (8) lead to a first approximation $\hat{v}_1$, which is introduced to (11) and leads to a second approximation k_2, and so on. The numerical work involved is prohibitive for the usual equipment. However, the procedure could be carried through with an automatic computer.

Exercises: 1) Calculate a table for V^2 as function of k (>2) from (1). 2) Calculate the geometric moments without the use of the generating function. 3) Calculate the geometric moments without the use of theorem 1.1.5(3). Use Kimball's method, 6.2.5(9).

Problem: Compare the precisions of the estimates.

Reference: Kimball (1956).

7.1.5. The Increase of the Extremes. The initial distributions of type II have a longer tail than those of type I. The same holds for the asymptotic distributions. Consequently a forecast of largest values from the second asymptote exceeds the forecast from the first one. Since, from 7.1.2(6), lg z increases in proportion to lg T, the variable z increases as a power of T, while the variable x in the first asymptote increases only in proportion to lg T.

We now consider the asymptotic distribution II as the initial one, and calculate the averages of the extremes as functions of the sample size N. From the initial probability 7.1.2(6), which is now written $F(z)$ we obtain the probability $F^N(z) = F(z_N)$ of the largest value z_N as

(1) $$F(z_N) = F(zN^{-1/k}),$$

where k remains unchanged. Thus we have a linear transformation. Consequently, the distributions $f(z_N)$ spread with increasing N and contract with increasing k. The averages of the largest values,

(2) $$\check{z}_N = \check{z}N^{1/k}; \quad \tilde{z}_N = \tilde{z}N^{1/k}; \quad v_N = vN^{1/k},$$

increase with v and N and decrease with increasing k. The larger k, the smaller is the increase of the averages of the largest value with N. The corresponding densities of probability,

(3) $$f(\check{z}_N) = f(\check{z})N^{-1/k}; \quad f(\tilde{z}_N) = f(\tilde{z})N^{-1/k}; \quad f(v_N) = f(v)N^{-1/k},$$

decrease with N and increase with k. (In the first asymptote these probabilities remained unchanged.) The occurrence interval $T(\tilde{z}_N)$, 2.1.2, becomes, from (1), (2) and 7.1.3(1),

(4) $$T(\tilde{z}_N) = \frac{1}{1 - \exp[-(k+1)/kN]} \sim \frac{Nk}{k+1} + \frac{1}{2},$$

which is smaller than N (except for the uninteresting case $k = 2N - 1$). This shows again the difference between the two asymptotes. However, the larger k, the nearer is the occurrence interval to the sample size.

The distribution of the largest (smallest) value taken from the second asymptotic distribution of largest (smallest) values is again the second distribution. However, the asymptotic distribution of the largest (smallest) values taken from the second asymptotic distribution of smallest (largest) values is not the second distribution. For the proof it is sufficient to consider the maximum of the smallest value. Since the second asymptotic distribution of smallest values is limited, the asymptotic distribution of the largest values, if it exists, can only be type III. Now Gnedenko's necessary and sufficient condition for the existence of the third asymptote given in 5.1.3 leads for

$$F(x) = 1 - \exp\left[-(v/x)^k\right]$$

to

$$\frac{1-F(cx)}{1-F(x)} = \exp\left[-\left(\frac{v}{cx}\right)^k + \left(\frac{v}{x}\right)^k\right],$$

which does not converge to c^k for $x \to -0$. Therefore, no asymptotic distribution of the largest (smallest) value taken from the asymptotic distribution II of the smallest (largest) values exists.

Exercises: 1) What is the relation of the return period of $2z$ to the return period of z? 2) By what factor must z be multiplied in order to double the return period?

7.1.6. Generalization. In 7.1.1 and 7.1.2 the variate was non-negative. We may equally well assume that $z \gtreqless \varepsilon$, where $\varepsilon \gtreqless 0$. This leads to

(1) $$\Pi(z) = \exp\left[-\left(\frac{v-\varepsilon}{z-\varepsilon}\right)^k\right],$$

where the meaning of v is conserved [see Table 5.1.3(1)]. Since this asymptotic probability is obtained from a stable one by a linear transformation, it must also be stable. For the proof the probability is written in analogy with Jenkinson's procedure (5.1.3)

(2) $$F(z) = \exp\left[-(a+bz)^{-k}\right],$$

where

$$a = -\varepsilon/(v-\varepsilon) \; ; \qquad b = 1/(v-\varepsilon) \, .$$

Then the probability of the largest among n observations,

$$F^n(z) = \exp\left[-(an^{-1/k} + bzn^{-1/k})^{-k}\right]$$

is again of the form (2) with

(3) $$a_n = an^{-1/k} \; ; \qquad b_n = bn^{-1/k} \, .$$

Thus, the probability function (2) is also a solution of Fisher and Tippett's functional equation 5.1.2(1). They consider only two alternatives, either $a_n = 1$; $b_n \neq 0$, or $a_n \neq 1$; $b_n = 0$. The linear transformation between the variate and the largest value is either a translation (first asymptote) or a change in scale (second and third asymptotes). In the present form both operations exist at the same time. The general case resembles the first asymptote, since the probability function is shifted by taking the extremes. At the same time, it resembles the second asymptote, since the scale is enlarged.

The relations given in 7.1.3 hold, except for a linear translation. Those moments that exist and the reciprocal moments are analogous to 7.1.3(6) and (7),

(4) $$\left(\frac{z-\varepsilon}{v-\varepsilon}\right)^l = \Gamma(1-l/k) \; ; \qquad \left(\frac{v-\varepsilon}{z-\varepsilon}\right)^l = \Gamma(1+l/k).$$

The estimation of the parameters v and k along the lines shown in 7.1.4 is complicated by the existence of the lower limit ε. If it is known that $k > 3$ the parameters may be estimated by the method of moments. If the parameter v is known and $k > 2$, two moments are sufficient. If k and v are known, a rough estimate of the lower limit ε is obtained from the plotting position of the smallest value z_1. For N observations the equation,

$$\exp\left[-\left(\frac{v-\varepsilon}{z_1-\varepsilon}\right)^k\right] = \frac{1}{N+1},$$

leads after trivial transformations to the estimate

(5) $$\varepsilon = \frac{z_1[\lg(N+1)]^{1/k} - v}{[\lg(N+1)]^{1/k} - 1}.$$

Combination with one of the previous methods, shown in 7.1.4, leads to successive approximation for the three parameters.

Exercises: 1) Rewrite the formulae of 7.1.3 for the probability (1). 2) Estimate the three parameters for $k > 3$ by the method of moments. 3) Estimate ε and k for $k > 2$, provided that v is known. 4) Show that the mode, the probability and density at the mode and the median, and the density at the median converge for increasing k to the corresponding values for the first asymptote.

Problems: 1) What are the precisions of the different estimates? 2) Construct a systematic estimate of the three parameters. 3) Estimate the three parameters by assuming that the smallest among N observations is the most probable (or the median) smallest value.

7.1.7. Applications. Thom used the second asymptote for the analysis of maximum wind speed, because the speed is a non-negative variate. For the solution of 7.1.4(10) he uses an I.B.M. machine. Only experience can show whether the first or the second asymptote gives a better fit for such data.

On the basis of the criterion shown in 5.2.6, Jenkinson used the first asymptote for the annual maximum vapor pressures, Milan, 1921–1950, the second asymptote for the floods of the Little River, Westfield, Massachusetts, 1910–1928, and the annual maximum hourly (daily, four day) rainfalls, Naples, 1888–1933, Marseille, 1882–1946, and Tripoli 1919–1948, and the third asymptote for the floods of the Connecticut River, Hartford, Connecticut, 1843–1934, the maximum August temperature, Siros, 1898–1929, the annual maximum pressure, Marseille, 1923–1946, the annual (monthly) maximum hourly wind speed, Trieste, 1903–1920, and Tateno, December, January, February, 1950–1952, the annual minimum temperatures, Athens, 1900–1929, and the annual minimum pressures, Tortosa, 1919–1934. Thus he assumes the existence of a lower limit for maximum rainfalls and of an upper limit for wind speeds and temperatures.

Bernier used the second asymptote for the floods of the Rhine at Rheinfelden, the Colorado River at Black Canyon (see Gumbel, 1954b), and the Durance at Archidiacre. In these cases the observed largest floods exceed strongly the theoretical values obtained from the first asymptote. The second asymptote with estimation of the parameters from the geometric mean and standard deviation led to a better fit.

7.1.8. Summary. In Fréchet's derivation, the initial distributions from which the largest values are taken may differ with respect to the medians. This procedure also leads to the asymptotic distribution of the absolute largest deviation from the median. The second asymptotic distribution of the largest value possesses as many moments as the initial one. The smaller the order of the lowest moment which diverges, the larger is the spread of the distribution of the largest value. If it is known that the initial distribution is of Cauchy's type, the order of the first moment which diverges may be estimated. The estimation of the parameters is complicated by the lack of certain moments, but facilitated by the logarithmic relations between the first and second asymptote. A maximum likelihood solution for the estimation of two parameters exists, but requires a prohibitive amount of numerical calculations.

7.2. THE THIRD ASYMPTOTE

7.2.0. Introduction. The third asymptotic distribution of extremes was obtained from the stability postulate in 5.1.2. It holds for limited

7.2.1   SECOND AND THIRD ASYMPTOTES

distributions in contrast to the exponential and the Cauchy types, both of which are unlimited with respect to the corresponding extreme. It may also be obtained from the first asymptote by a logarithmic transformation. Conversely, the first asymptote is reached from the third one by a linear transformation of the variate and a limiting process on one of the parameters. The first and third asymptotes possess all moments, in contrast to the second one. In other respects, the third asymptote resembles the second one more than the first one, since the generalized second and the third asymptotes contain upper (or lower) limits ω (or ε).

As previously, we derive the asymptotic distribution of the largest value. However, we study mainly the smallest value, because it leads to fruitful applications in the analysis of droughts and breaking strength of materials. Here the lower limit $\varepsilon \, (\geq 0)$ turns out to be the most important parameter. In the study of one extreme, no assumption on the behavior of the initial distribution at the other extreme is necessary.

7.2.1. The Von Mises Derivation. For a limited variate x, the probability $F(x)$ reaches unity for a finite value $x = \omega$, and the largest value is also limited by ω. To obtain the distribution of the largest value, R. von Mises makes certain assumptions about the way in which $F(x)$ approaches unity, namely, that the probability $F(x)$ possesses at least k derivatives and that the first $k - 1$ vanish at $x = \omega$,

(1) $$f(\omega) = f'(\omega) = \ldots = f^{(k-2)}(\omega) = 0,$$

while there is a positive value b of dimension x^{-k} defined by

(2) $$f^{(k-1)}(\omega) = F^{(k)}(\omega) = (-1)^{k-1} b.$$

The number k is an integer and exceeds unity. Both assumptions will be dropped later. Furthermore, let the next derivative be finite in a small interval,

(3) $$|f^{(k)}(\omega)| \leq A ; \quad \omega - \delta \leq x \leq \omega.$$

The usual Taylor series about ω becomes, from (2),

(4) $$F(x) = F(\omega) + \frac{(x-\omega)^k}{k!} F^{(k)}(\omega) + \frac{(x-\omega)^{k+1}}{(k+1)!} F^{(k+1)}(\omega) + \ldots$$

Since for $x = \omega$,

$$\lg F(x) = F(x) - 1,$$

we may write from (4) in the neighborhood of ω, for $0 < \Theta < 1$,

(4′) $$\lg F(x) = \frac{(-1)^{k-1}(x-\omega)^k b}{k!} + \frac{(x-\omega)^{k+1} \Theta A}{(k+1)!} = -\frac{(\omega-x)^k b}{k!}$$
$$+ \frac{(-1)^{k+1}(\omega-x)^{k+1} \Theta A}{(k+1)!}.$$

The difference between the limit ω and the largest value x is used to define a reduced largest value z by

(5) $$z = \beta(x - \omega) \leq 0,$$

where the parameter β is defined by

(5') $$\beta^k = \frac{nb}{k!}.$$

This parameter has the dimension $[x^{-1}]$, similar to the parameter α in the first asymptote with the modification that β increases as a root of n. From (5) and (5') the variate becomes

(5") $$x - \omega = z(k!/nb)^{1/k},$$

and the probability $\Phi^{(3)}(z) = \exp[n \lg F(x)]$ of the largest value is, from (4'),

$$\lg \Phi^{(3)}(z) = -(-1)^k z^k + \frac{Cz^{k+1}}{n^{1+1/k}},$$

where C is a constant, independent of z and n. As n increases, the second factor vanishes and the asymptotic probability becomes

(6) $$\Phi^{(3)}(z) = \exp[-(-z)^k] = \exp[-\beta^k(\omega - x)^k];$$
$$\beta > 0; \quad \omega \geq 0; \quad x \leq \omega; \quad k > 1.$$

For both even and odd values of k, ($k = 2l$; $k = 2l + 1$) the corresponding probabilities,

$$\Phi^{(3)}(z) = \exp(-z^{2l}); \qquad \Phi^{(3)}(z) = \exp(z^{2l+1}); \qquad z \leq 0,$$

are zero for $z = -\infty$ and unity for $z = 0$. The probabilities $\Phi^{(3)}(z)$ traced on normal paper in Graph 7.2.1 pass for all values of k through the common point $\Phi^{(3)}(-1) = 1/e$. The distributions $\varphi^{(3)}(z)$ of the reduced largest value and $\varphi^{(3)}(x)$ of the largest value itself are

(7) $$\varphi^{(3)}(z) = k(-z)^{k-1} \exp[-(-z)^k]; \qquad\qquad z < 0$$

$$\varphi^{(3)}(x) = k\beta[\beta(\omega - x)]^{k-1} \exp\{-[\beta(\omega - x)]^k\}; \qquad x \leq \omega.$$

For $k = 2l$ and $k = 2l + 1$, the densities are non-negative, as they ought to be, and contract with increasing k.

By analogy with the procedure used for the second asymptote, we introduce a parameter v (designated by u in the first asymptote) by writing

(8) $$\beta = 1/(\omega - v).$$

7.2.1 SECOND AND THIRD ASYMPTOTES

Then the variate x and the reduced variate z are related by (5), i.e.,

$$(9) \qquad -z = \frac{\omega - x}{\omega - v},$$

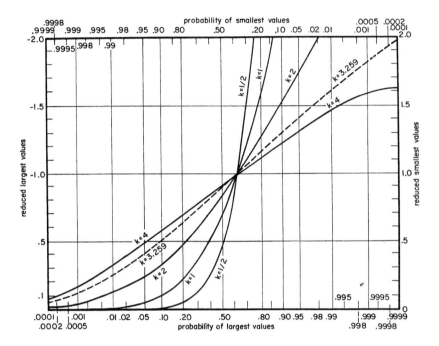

Graph 7.2.1. Probabilities III of Extreme Values, Traced on Normal Paper

and the third asymptotic probability and the corresponding distribution read

(III) (10) $\quad \Phi^{(3)}(x) = \exp\left[-\left(\frac{\omega - x}{\omega - v}\right)^k\right];$

(10′) $\quad \varphi^{(3)}(x) = \frac{k}{\omega - v}\left(\frac{\omega - x}{\omega - v}\right)^{k-1} \Phi^{(3)}(x).$

For the case $\omega = 0$, the variate x is negative and formula III simplifies to

(11) $\quad \Phi_0^{(3)}(x) = \exp\left[-(x/v)^k\right]; \qquad \varphi_0^{(3)}(x) = \frac{k}{-v}\left(\frac{x}{v}\right)^{k-1} \Phi_0^{(3)}(x);$

$$x \leqq 0; \quad v < 0; \quad k > 1,$$

which differs from the second asymptote only with respect to the signs of k and x. The reduced form (6) contains only one parameter, while (11) has two and (10) has three parameters. The initial probability $F(x)$ for which (10) holds is, from (4), (2), (5') and (8), for large values of the variate,

$$F(x) = 1 - \frac{1}{n}\left(\frac{\omega - x}{\omega - v}\right)^k, \tag{12}$$

as stated in Table 5.1.3(1). Here v and k are functions of n, while ω is a fixed value. It is easily verified that Gnedenko's condition 5.1.3(3) is fulfilled and that III holds for a uniform distribution with $\omega = 1, k = 1$.

Exercise: Derive the third asymptote from Fréchet's equation 7.1.1(17) and the conditions $z \geq 0$, $F(0) = 1$.

7.2.2. Other Derivations. In 7.1.2 (10) the second asymptotic probability of the largest value was derived from the first one by a logarithmic transformation. The same method leads to the third one. Let ξ be the variate in the first asymptotic probability,

$$\Phi^{(1)}(\xi) = \exp\left[-e^{-\alpha(\xi - u)}\right],$$

and let a new variate x be defined by

$$\xi - u = -\lg\frac{\omega - x}{\omega - v}; \qquad \alpha = k. \tag{1}$$

Then the probability of x is (III). This transformation is due to B. F. Kimball (1942b), who tried to find a new asymptotic distribution. The relation (1) leads to the construction of a probability paper, Graph 7.2.2(1), where the variate is scaled in logarithms.

The third asymptotic probability of largest values may also be obtained from the generalized second asymptotic probability of smallest values, Table 5.1.3(2). Let z be the variate in this probability,

$$1 - {}_1\Phi^{(2)}(z) = \exp\left[-\left(\frac{\omega - v_1}{\omega - z}\right)^{k_1}\right];$$

then the reciprocal transformation,

$$\frac{\omega - v_1}{\omega - z} = \frac{x - \omega}{u - \omega}; \qquad k_1 = k, \tag{2}$$

leads again to III.

In the derivation 7.2.1, the parameter k was defined as a positive integer. Within the transformations (1) and (2) and Gnedenko's condition

7.2.2 SECOND AND THIRD ASYMPTOTES

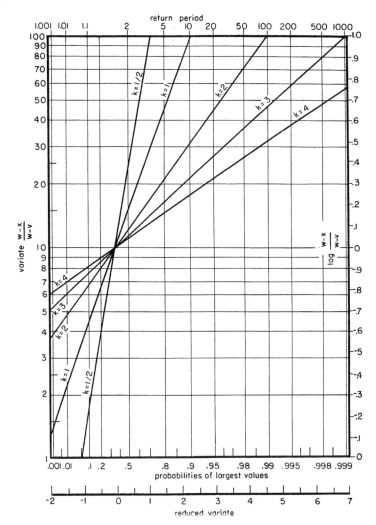

Graph 7.2.2(1). Probabilities III, Traced on Logarithmic Extremal Paper

5.1.3, $k = \alpha$ may be any positive value. This will be assumed throughout in the following.

The third asymptotic probability and distribution of the reduced *smallest* value obtained from 7.2.1(6) and the symmetry principle are

(3) $\quad 1 - {}_1\Phi^{(3)}(z) \equiv {}_1P(z) = \exp(-z^k)\,;\quad {}_1p(z) = kz^{k-1}\exp(-z^k)\,;$
$$z \geqq 0\,.$$

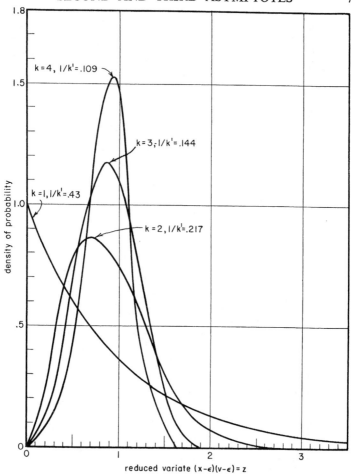

Graph 7.2.2(2). Influence of k on the Shape of the Distribution III of Smallest Values

If we introduce a lower limit ε and the parameter v by writing

(4) $$z = (x - \varepsilon)/(v - \varepsilon),$$

then the probability and distribution for the variate x are

(5) $$P(x) = \exp\left[-\left(\frac{x-\varepsilon}{v-\varepsilon}\right)^k\right]; \quad p(x) = \frac{k}{v-\varepsilon}\left(\frac{x-\varepsilon}{v-\varepsilon}\right)^{k-1} P(x);$$

$$x \geq \varepsilon,$$

with the conditions

(5') $$v > \varepsilon; \quad k > 0; \quad P(\varepsilon) = 1; \quad P(v) = 1/e.$$

7.2.2 SECOND AND THIRD ASYMPTOTES

The distribution (3) is traced in Graph 7.2.2(2). In (3) and (5) the expressions $P(x)$ and $_1P(z)$ designate the *probabilities of a value larger than x or z*. The symmetry relation between largest and smallest values is used in the Graphs 7.2.1 and 7.2.2(1), which show the probabilities for both extremes. In the formulae 7.2.1(6) used by R. A. Fisher (1928), R. von Mises (1936), and B. Gnedenko (1943), and in (3) the reduced variate z is negative for the largest value and positive for the smallest one. This is confusing if we think simultaneously of both extremes, because the largest value cannot be smaller than the smallest one. However, this confusion will be avoided in the following because we deal mainly with the smallest value. The expression (5) is sometimes called the Weibull distribution, since this engineer used it for the first time in the analysis of breaking strengths. (See 7.3.5.)

In 7.1.2(13) it was shown that the first asymptotic distribution may be reached as a limit from the second one, if the parameter k increases. The same holds for the third one. If we introduce into (3) a new variate y by

(6) $$z = 1 + y/k,$$

the corresponding probability,

(7) $$V(y) = 1 - \exp[-(1 + y/k)^k],$$

degenerates for $k = 0$ into $V(y) = 1 - 1/e$ independently of y. For $k = 1$, we obtain the truncated exponential distribution. With increasing k, the probability $V(y)$ converges to the first asymptotic probability of smallest values,

(8) $$_1\Phi^{(1)}(y) = 1 - \exp(-e^y).$$

Therefore, Fisher and Tippett call the third asymptotic probability "*penultimate.*" To illustrate the convergence, consider the median $\breve{y}$ obtained from (7) and the mode $\tilde{y}$, given respectively by

(9) $\breve{y} = 0.36651 \, (-1 + 0.36651/2k + \ldots) \, ; \quad (1 + \tilde{y}/k)^k = 1 - 1/k$.

They decrease as k increases, and reach their smallest values,

(10) $$\breve{y} = -0.36651 \, ; \quad \tilde{y} = 0,$$

identically with the values of the first asymptote at $k = \infty$. The positions of the averages are

(11) $$\breve{y} < 0 \leq \tilde{y} \, ; \quad V(\tilde{y}) \geq V(0) > V(\breve{y}).$$

The probability $V(\tilde{y})$ and the density of probability $v(\tilde{y})$ at the mode also approach the values of the first asymptote as k increases.

Exercise: Analyze the asymptotic behavior of the median and the moments of (7) as functions of k.

7.2.3. Averages and Moments of Smallest Values. We shall now study the averages and the influence of the parameters on the third asymptotic distribution of the *smallest* value 7.2.2(5). The averages depend upon

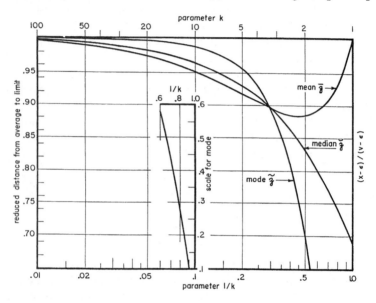

Graph 7.2.3. Averages of the Third Asymptote

the two location parameters v and ε and the scale parameter k, which in the practical applications 7.3.4 and 7.3.5 exceeds unity. The median $\check{x}$ is

(1) $$\check{x} = \varepsilon + (v - \varepsilon)(\lg 2)^{1/k}.$$

The corresponding density,

(2) $$p(\check{x}) = \frac{0.34657k}{v - \varepsilon} e^{0.36651/k}.$$

increases (decreases) with k for $k > 0.36651$ ($k < 0.36651$). The mode $\tilde{x}$ which exists only for $k > 1$,

(3) $$\tilde{x} = \varepsilon + (v - \varepsilon)(1 - 1/k)^{1/k},$$

increases with k. The median and the mode approach v as shown in Graph 7.2.3.

7.2.3 SECOND AND THIRD ASYMPTOTES

The mode $\begin{Bmatrix}\text{precedes}\\ \text{equals}\\ \text{exceeds}\end{Bmatrix}$ the median if $1/k \begin{Bmatrix}>\\ =\\ <\end{Bmatrix} 0.30685$; $k \begin{Bmatrix}<\\ =\\ >\end{Bmatrix} 3.25889$.

The probability at the mode diminishes toward $1 - 1/e$ with increasing k. The density of probability at the mode,

$$p(\tilde{x}) = \frac{k}{v - \varepsilon}\left(\frac{1 - 1/k}{e}\right)^{1-1/k},$$

increases with k for $k > 1$, i.e., whenever a mode exists, while in the first asymptotic distribution of the smallest value, the mode exists independently of the value of the other parameter, precedes the median, and the probability at the mode is fixed.

The reduced moments $\overline{z^l}$ of order l obtained from 1.1.5(1),

$$\overline{\left(\frac{x - \varepsilon}{v - \varepsilon}\right)^l} = -\int_\varepsilon^\infty \left(\frac{x - \varepsilon}{v - \varepsilon}\right)^l d\exp\left[-\left(\frac{x - \varepsilon}{v - \varepsilon}\right)^k\right],$$

become, from 7.2.2(4),

$$-\int_0^\infty z^{l/k} de^{-z} = \frac{l}{k}\Gamma\left(\frac{l}{k}\right),$$

whence

(4) $$\overline{(x - \varepsilon)^l} = (v - \varepsilon)^l \Gamma(1 + l/k).$$

All moments exist. For $l = 1$ the mean,

(5) $$\bar{x} = \varepsilon + (v - \varepsilon)\Gamma(1 + 1/k),$$

first decreases, later increases with k, and converges to v as shown in Graph 7.2.3. From (5) it follows that

(6) $$\bar{x} \gtreqless v \quad \text{if} \quad k \gtreqless 1.$$

The distances from the characteristic value v to the averages are weighted means of ε and v, i.e.,

(7) $v - \tilde{x} = (v - \varepsilon)(1 - 1/k)^{1/k}$; $\quad k > 1$;

(8) $v - \check{x} = (v - \varepsilon)[1 - (\lg 2)^{1/k}]$; $\quad v - \bar{x} = (v - \varepsilon)[1 - \Gamma(1 + 1/k)]$.

The variance obtained from (4) is

(9) $$\sigma^2 = (v - \varepsilon)^2 [\Gamma(1 + 2/k) - \Gamma^2(1 + 1/k)].$$

The standardized difference $B(k)$ from the lower limit to the characteristic value,

(10) $$B(k) \equiv (v - \varepsilon)/\sigma = [\Gamma(1 + 2/k) - \Gamma^2(1 + 1/k)]^{-\frac{1}{2}},$$

and the standardized difference $A(k)$ from the characteristic value to the mean,

(11) $\qquad A(k) \equiv (v - \bar{x})/\sigma = [1 - \Gamma(1 + 1/k)]B(k)$,

are given in Table 7.2.3, columns 2 and 3, as function of k.

Table 7.2.3. *Reduced Averages and Third Moment of the Third Asymptote Multiple of Standard Deviation*

Scale Parameter $1/k$	For Lower Limit ε $B(k)$ [7.2.3(10)]	For Parameter v $A(k)$ [7.2.3(11)]	Reduced 3rd Moment $\sqrt{\beta_1(k)}$ [7.2.3(12)]
.01	78.981714	.448164	−1.081272
.02	39.989044	.446110	−1.024853
.03	26.986212	.443926	−.970702
.04	20.480808	.441603	−.918459
.05	16.574350	.439150	−.867967
.06	13.967343	.436568	−.819101
.07	12.102862	.433863	−.771740
.08	10.702446	.431038	−.725772
.09	9.611395	.428096	−.681102
.10	8.736889	.425043	−.637637
.11	8.019861	.421881	−.595296
.12	7.420934	.418614	−.554002
.13	6.912848	.415245	−.513687
.14	6.476131	.411778	−.474287
.15	6.096505	.408216	−.435743
.16	5.763261	.404563	−.398002
.17	5.468210	.400822	−.361012
.18	5.204984	.396996	−.324729
.19	4.968556	.393087	−.289108
.20	4.754903	.389100	−.254110
.21	4.560770	.385036	−.219697
.22	4.383495	.380900	−.185835
.23	4.220878	.376693	−.152490
.24	4.071085	.372419	−.119634
.25	3.932577	.368079	−.087237
.26	3.804052	.363678	−.055272
.27	3.684400	.359218	−.023715
.28	3.572672	.354700	+.007458
.29	3.468048	.350129	+.038270
.30	3.369818	.345505	+.068742
.31	3.277364	.340832	+.098893
.32	3.190146	.336112	+.128743
.33	3.107688	.331348	+.158308
.34	3.029573	.326541	+.187606
.35	2.955428	.321694	+.216653

Table 7.2.3 (continued)

$1/k$	$B(k)$	$A(k)$	$\sqrt{\beta_1(k)}$
.36	2.884924	.316809	.245465
.37	2.817768	.311889	.274055
.38	2.753697	.306935	.302437
.39	2.692475	.301949	.330625
.40	2.633890	.296935	.358632
.41	2.577752	.291893	.386468
.42	2.523887	.286825	.414147
.43	2.472138	.281734	.441678
.44	2.422364	.276622	.469072
.45	2.374435	.271490	.496340
.46	2.328232	.266340	.523491
.47	2.283647	.261174	.550535
.48	2.240583	.255993	.577481
.49	2.198946	.250801	.604336
.50	2.158655	.245597	.631111
.51	2.119632	.240384	.657812
.52	2.081807	.235163	.684448
.53	2.045114	.229937	.711026
.54	2.009492	.224706	.737553
.55	1.974885	.219472	.764038
.56	1.941242	.214237	.790486
.57	1.908514	.209002	.816904
.58	1.876656	.203768	.843299
.59	1.845626	.198537	.869677
.60	1.815385	.193311	.896045
.61	1.785897	.188090	.922408
.62	1.757128	.182875	.948772
.63	1.729045	.177669	.975143
.64	1.701620	.172473	1.001527
.65	1.674824	.167287	1.027929
.66	1.648631	.162113	1.054354
.67	1.623017	.156951	1.080808
.68	1.597958	.151804	1.107296
.69	1.573432	.146672	1.133822
.70	1.549420	.141557	1.160393
.71	1.525901	.136459	1.187012
.72	1.502857	.131379	1.213685
.73	1.480272	.126318	1.240415
.74	1.458130	.121278	1.267209
.75	1.436413	.116260	1.294070
.76	1.415109	.111263	1.321004
.77	1.394204	.106290	1.348013
.78	1.373683	.101340	1.375104
.79	1.353536	.096416	1.402280
.80	1.333750	.091517	1.429545

Table 7.2.3 (continued)

$1/k$	$B(k)$	$A(k)$	$\sqrt{\beta_1(k)}$
.81	1.314314	.086644	1.456904
.82	1.295217	.081799	1.484362
.83	1.276450	.076982	1.511921
.84	1.258002	.072194	1.539587
.85	1.239865	.067435	1.567363
.86	1.222031	.062706	1.595254
.87	1.204489	.058008	1.623264
.88	1.187234	.053341	1.651396
.89	1.170256	.048707	1.679655
.90	1.153550	.044105	1.708045
.91	1.137107	.039537	1.736570
.92	1.120922	.035002	1.765232
.93	1.104988	.030501	1.794038
.94	1.089299	.026035	1.822990
.95	1.073849	.021605	1.852093
.96	1.058632	.017211	1.881350
.97	1.043644	.012852	1.910765
.98	1.028880	.008531	1.940343
.99	1.014333	.004247	1.970086
1.00	1.0	0.0	2.0
1.0	1.	0.	2.
1.1	.867491	−.040326	2.309348
1.2	.752233	−.076579	2.640035
1.3	.651524	−.108817	2.996146
1.4	.563330	−.136421	3.382013
1.5	.486053	−.160077	3.802311
1.6	.418382	−.179747	4.262142
1.7	.369209	−.195656	4.767125
1.8	.307573	−.208071	5.323478
1.9	.262625	−.217284	5.938118
2.0	.223607	−.223607	6.618761
3.0	.038236	−.191180	19.584859
4.0	.005016	−.115370	60.091733
5.0	.000526	−.062593	190.113240

The third central moment μ_3 is obtained from (4) as

$$\mu_3 = \overline{(x-\varepsilon)^3} - 3\overline{(x-\varepsilon)^2}\,\overline{(x-\varepsilon)} + 2\overline{(x-\varepsilon)}^3.$$

The first moment quotient $\sqrt{\beta_1}$ (different from the parameter β in 7.2.1(5)),

(12) $\quad \sqrt{\beta_1} = [\Gamma(1+3/k) - 3\Gamma(1+2/k)\Gamma(1+1/k)$
$$+ 2\Gamma^3(1+1/k)]B^3(k),$$

depends only upon k. The expression (12) as a function of k is given in Table 7.2.3, column 4. As k increases, the skewness passes from its

7.2.4 SECOND AND THIRD ASYMPTOTES

limiting value -1.1396, valid for the first asymptote, to large positive values and vanishes for $1/k = .27760$. Leme gave a short table for

(13) $\beta_2 = [\Gamma(1 + 4/k) - 4\Gamma(1 + 3/k)\Gamma(1 + 1/k)$
$\qquad + 6\Gamma(1 + 2/k)\Gamma^2(1 + 1/k) - 3\Gamma^4(1 + 1/k)]B^4(k)$

Exercises: 1) Analyze the densities of probability at the averages as function of k. 2) Calculate the inflexion points for $k > 1$. 3) Trace the β_1, β_2 diagram.

Problems: 1) Prove that the differences between mean, mode, median, and the characteristic value reduced by the standard deviation and the Betas converge with increasing k to the corresponding values for the first asymptote. 2) Prove that $B(k)$ and $A(k)$ decrease and $\sqrt{\beta_1(k)}$ increases with k decreasing.

References: A. J. Johnson, Weibull.

7.2.4. Special Cases. The third asymptotic probability is more flexible than the two other functions. For $k = 1$ the probabilities and distributions 7.2.2(3) and 7.2.2(5) are exponential functions. The case $\varepsilon = 0$; $k = 2$, called Rayleigh's function, was considered as an initial distribution and analyzed for its largest value in 6.3.8. Since the variate is limited to the left and unlimited to the right, the distribution 7.2.2(3) is asymmetrical. However, the relative position of the averages and the sign of the third moment change with the parameter k. Therefore, pseudosymmetrical cases exist where the distributions *look* symmetrical. Table 7.2.4 gives the reduced averages 7.2.3(1), (3), (5) for those values of k where two of the three averages coincide and where the third moment 7.2.3(12) vanishes.

Table 7.2.4. *The Four Pseudosymmetrical Cases*

Condition	Parameter		Median	Mode	Mean	Skewness
	k	$1/k$	$\check{z}$	$\tilde{z}$	$\bar{z}$	$\sqrt{\beta_1}$
$\check{z} = \tilde{z}$	3.25889	0.30685	0.89363	0.89363	0.89646	0.09350
$\tilde{z} = \bar{z}$	3.31125	0.30189	0.89525	0.89719	0.89719	0.07447
$\bar{z} = \check{z}$	3.43938	0.29075	0.89892	0.90494	0.89892	0.04057
$\beta_1 = 0$	3.60232	0.27760	0.90326	0.91369	0.90114	0.00000

The corresponding averages $\hat{x}$ are obtained from

(1) $\qquad\qquad\qquad \hat{x} = \hat{z}v + (1 - \hat{z})\varepsilon\,.$

When two of the averages coincide, the third one differs but slightly from the others and the same holds for the three averages when the skewness is zero. The probability points corresponding to $\check{z} = \tilde{z}$ are traced in Graph 7.2.1. The slight curvature explains why some small series of observed extremes have been misinterpreted as normal values. The distribution looks symmetrical if the parameter k is contained in the interval $0.27 < 1/k$

$< 0.31; 3.2 < k < 3.7$. This fact marks an essential difference between the third and the two preceding asymptotic distributions, since the skewness was constant and negative for the first asymptotic distribution of smallest values, and the asymmetry was negative and increasing with k for the second one.

In the case $\varepsilon = 0$ the transformations,

(2) $\qquad \xi = \log x\,; \qquad u = \log v\,; \qquad \alpha' = 2.30259k\,,$

lead from 7.2.2.(5) to the first asymptotic probability of smallest values,

(3) $\qquad {}_1\Phi^{(1)}(y) = 1 - \exp(-e^y)\,; \qquad y = \alpha'(\xi - u)\,.$

The logarithms of observed smallest values plotted on the extremal probability paper designed for the first asymptote should be scattered about the line (3).

It is also interesting to study the properties of the return period. Since the smallest values *decrease* with increasing sample size, we use the definition 1.2.3(8) of the return period ${}_1T$ for values below the median. From 7.2.2(5) this return period is, for $\varepsilon = 0$,

(4) $\qquad {}_1T(x) = 1/(1 - \exp[-(x/v)^k])\,.$

For small values of x the first approximation is

$$ {}_1T(x) \sim (v/x)^k $$

or

(5) $\qquad \lg x \sim \lg v - [\lg {}_1T(x)]/k\,.$

The logarithm of the variate decreases as a linear function of the logarithm of the return period, and

(6) $\qquad \dfrac{d \lg x}{d \lg {}_1T} = -\dfrac{1}{k}$

is the slope of this line. The variate does not vanish for any finite value of the return period. The smallest value $x(2\,{}_1T)$ corresponding to the return period $2\,{}_1T$ is related to the smallest value $x({}_1T)$ by

(7) $\qquad x(2\,{}_1T) \sim x({}_1T) 2^{-1/k}\,,$

while the return period, ${}_1T(x/2)$, is related to ${}_1T(x)$ by

(8) $\qquad {}_1T(x/2) = {}_1T(x) 2^k\,.$

If we double the return period, the variate is divided by $2^{1/k}$. If we take one half the variate the return period is multiplied by 2^k.

Exercises: 1) Calculate the seminvariant generating function for $k = 1$ and $k = 2$. 2) Calculate the distribution for $k = 3.25889$ and the distributions and probabilities for the other three values of k given in Table

7.2.4. 3) Construct formulae corresponding to (4) and (8) for the first asymptote and analyze the differences between the two asymptotes. 4) Misinterpret largest values of the third type as normal. See Graph 5.2.7.(2).

References: Fisher and Tippett, Von Mises (1936), Weibull (1939, 1949).

7.2.5. The Increase of the Extremes. Consider the third asymptotic distribution of the smallest value in the reduced form 7.2.2(3) as the initial one. The probability for the minimum of N observations to exceed $z_1(N) = z$ is, dropping the index,

(1) $$P^N(z) = \exp(-Nz^k),$$

where k remains unchanged in the transition. In the linear transformation of the variate, the distributions of the minima contract with increasing number of extremes. The choice of the minimum simply amounts to a reduction of the scale by $N^{-1/k}$. Therefore, all averages $\hat{z}_1(N)$ of the minimum are equal to the corresponding initial averages $\overset{\circ}{z}$ except for change in scale.

(2) $$\hat{z}_1(N) = \overset{\circ}{z} N^{-1/k}.$$

The quotients of any two of the averages are the same for the minimum and for the original variate. The decrease (or increase) of the averages of the minimum (or maximum) as a function of the number N of smallest (or largest) values is

(3) $$\frac{d \lg \hat{z}_1(N)}{d \lg N} = -\frac{1}{k}; \quad \frac{d \lg \hat{z}_N(N)}{d \lg N} = \frac{1}{k}.$$

The logarithms of the averages of the minima (maxima) traced against the logarithms of the number of extremes are straight lines. The variances of the minimum of the smallest values and of the maximum of the largest values obtained from 7.2.3(9),

(4) $$\sigma_1^2(N) = \sigma_N^2(N) = \frac{\Gamma(1+2/k) - \Gamma^2(1+1/k)}{N^{2/k}} = \frac{\sigma^2}{N^{2/k}},$$

decrease with increasing values of N and $1/k$. If $k < 2$, this variance decreases with increasing number of extremes N more quickly and the precision increases more quickly than that of the mean extremes. The coefficient of variation is independent of N.

The third asymptotic distribution of the smallest (largest) value is stable with respect to the smallest (largest) value. This does not hold for the opposite extremes. Consider the maximum taken from the distribution of the smallest value, as in 6.3.8. The intensity function is

(7) $$\mu(z) = kz^{k-1}; \quad z \geq 0.$$

The logarithmic derivative converges, for increasing z, toward the intensity function so that

(8) $$\frac{-p'(z)/p(z)}{\mu(z)} \gtreqless 1 \quad \text{as} \quad k \gtreqless 1.$$

In consequence of 4.1.4, the distribution of the maximum (minimum) value taken from the third asymptotic distribution of the smallest (largest) value belongs to the exponential type. If the third asymptotic distribution of one extreme is considered as the initial one, the asymptotic distribution of the other extreme is the first asymptote, while the first asymptotic distribution is stable with respect to both extremes.

In consequence of (2) and (7), the logarithms of the average maxima (minima) of the smallest (largest) values decrease (increase) and the average maxima (minima) of the smallest (largest) values increase (decrease) as the logarithms of the numbers of extremes increase.

7.2.6. The 15 Relations among the 3 Asymptotes. The asymptotic probabilities (as far as they exist) for the maxima and minima corresponding to the six asymptotic probabilities of extremes considered as initial probabilities are given in Table 7.2.6.

Table 7.2.6. *Asymptotic Probabilities of Iterated Extremes*

Initial Probability	$\Phi^{(1)}$	$_1\Phi^{(1)}$	$\Phi^{(2)}$	$_1\Phi^{(2)}$	$\Phi^{(3)}$	$_1\Phi^{(3)}$
Asymptotic Probability						
Maxima	$\Phi^{(1)}$	$\Phi^{(1)}$	$\Phi^{(2)}$	—	$\Phi^{(3)}$	$\Phi^{(1)}$
Minima	$_1\Phi^{(1)}$	$_1\Phi^{(1)}$	—	$_1\Phi^{(2)}$	$_1\Phi^{(1)}$	$_1\Phi^{(3)}$

Among the six asymptotic probabilities of reduced extreme values, y, z, $\mathfrak{z}$, where $\mathfrak{z}$ stands for the reduced variable in the third distribution,

$$\Phi^{(1)}(y) = \exp(-e^{-y}); \qquad _1\Phi^{(1)}(y) = 1 - \exp(-e^y)$$

$$\Phi^{(2)}(z) = \exp(-z^{-k}); \quad z > 0, k > 0; \qquad _1\Phi^{(3)}(z) = 1 - \exp(z^{-k});$$
$$z < 0, k > 0$$

$$\Phi^{(3)}(\mathfrak{z}) = \exp[-(-\mathfrak{z})^k]; \quad \mathfrak{z} < 0, k > 0; \qquad _1\Phi^{(3)}(\mathfrak{z}) = 1 - \exp(-\mathfrak{z}^k);$$
$$\mathfrak{z} > 0, k > 0,$$

taken two at a time, exist the following fifteen relations:

$$\Phi^{(2)} = \Phi^{(1)}(k \lg z)$$

$$\Phi^{(3)} = \Phi^{(1)}[-k \lg(-\mathfrak{z})] = \Phi^{(2)}(1/\mathfrak{z})$$

$$_1\Phi^{(1)} = 1 - \Phi^{(1)}(-y) = 1 - \Phi^{(2)}(-k \lg \mathfrak{z}) = 1 - \Phi^{(3)}[k \lg(-\mathfrak{z})]$$

$$_1\Phi^{(2)} = 1 - \Phi^{(1)}(k \lg z) = 1 - \Phi^{(2)}(-z) = 1 - \Phi^{(3)}(-1/\mathfrak{z})$$
$$= {_1\Phi^{(1)}}[-k \lg(-z)]$$

$$_1\Phi^{(3)} = 1 - \Phi^{(1)}(-k \lg \mathfrak{z}) = 1 - \Phi^{(2)}(1/\mathfrak{z}) = 1 - \Phi^{(3)}(-\mathfrak{z})$$
$$= {_1\Phi^{(1)}}(k \lg \mathfrak{z}) = {_1\Phi^{(2)}}(1/\mathfrak{z}).$$

7.3.1 SECOND AND THIRD ASYMPTOTES 289

This scheme contains the transformations between the asymptotic distributions of extremes.

Exercise: Analyze the order statistics taken from the second and third asymptotes.

Problem: Calculate the asymptotic distributions of extreme distances for the second and third type.

7.3. APPLICATIONS OF THE THIRD ASYMPTOTE

7.3.0. Problems. The moments of the third asymptotic distribution of smallest values given in 7.2.3(4) will now be used for the estimation of the parameters. If the lower limit is zero, the logarithmic transformation 7.2.4(2), which links the third to the first asymptote, leads to another simple estimation. Later the third asymptote is used for the distribution of the smallest values taken from the Gamma distribution, for normal extremes and for the analysis of droughts and fatigue failures.

7.3.1. Estimation of the Three Parameters. If we know the initial distribution and the sample size n, we know the parameters in the asymptotic distribution of the extremes. However, in general the initial distribution is unknown. If we know only the observed smallest values and have reason to believe that the initial distribution is limited to the left and subject to Gnedenko's condition 5.1.3(3) transformed for the smallest value, the three parameters k, v, ε, in the probability $\Pi(x)$ of smallest values to exceed x,

(1) $$\Pi(x) = \exp\left[-\left(\frac{x-\varepsilon}{v-\varepsilon}\right)^k\right],$$

have to be estimated.

The parameter k may be estimated from the sample skewness 7.2.3(12) with the help of Table 7.2.3, columns 1 and 4, as the solution $\hat{k}$ of

(2) $$\sqrt{b_1} = \left[\Gamma\left(1+\frac{3}{k}\right) - 3\Gamma\left(1+\frac{2}{k}\right)\Gamma\left(1+\frac{1}{k}\right) + 2\Gamma^3\left(1+\frac{1}{k}\right)\right] / \left[\Gamma\left(1+\frac{2}{k}\right) - \Gamma^2\left(1+\frac{1}{k}\right)\right]^{3/2}.$$

For the estimation of v, we use 7.2.3(11). Consequently, the characteristic value,

(3) $$\hat{v} = \bar{x}_0 + sA(\hat{k}),$$

is estimated from the previous knowledge of $\hat{k}$ with the help of the sample mean and standard deviation. The values of $A(k)$ are given in Table 7.2.3, column 3, as function of k, and traced in Graph 7.3.1(1).

The lower limit, which in practical applications is the most important of the three parameters, because it may be a physical constant, is related to the mean by 7.2.3(5), whence the estimate,

$$\hat{\varepsilon} = \frac{\bar{x}_0 - \hat{v}\,\Gamma(1 + 1/\hat{k})}{1 - \Gamma(1 + 1/\hat{k})}.$$

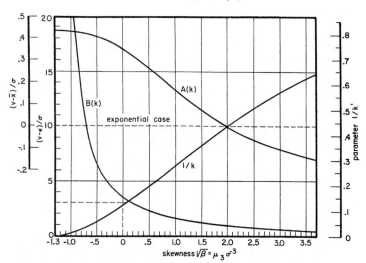

Graph 7.3.1(1). Parameter k and the Standardized Distances $A(k)$ and $B(k)$ as Functions of the Skewness

From (3) and 7.2.3(10),

(4) $$\hat{\varepsilon} = \hat{v} - sB(\hat{k}),$$

where $B(k)$ is given in Table 7.2.3, column 2, as function of k and traced in Graphs 7.3.1(1) and (2). Therefore, ε may be estimated from the previous estimates of $\hat{v}$, $\hat{k}$, and the sample standard deviation s without approximations.

Combination of equations (3) and (4) shows that

(5) $$\hat{\varepsilon} = \bar{x}_0 - s(B(\hat{k}) - A(\hat{k}))$$
$$= \bar{x}_0 - s\Gamma(1 + 1/\hat{k})/[\Gamma(1 + 2/\hat{k}) - \Gamma^2(1 + 1/\hat{k})]^{1/2}$$

is a linear function of the sample mean and standard deviation. The standardized distances $B(k)$ and $B(k) - A(k)$ from the characteristic value v and from the mean $\bar{x}$ to the lower limit ε are traced in Graph 7.3.1(2) against the skewness $\sqrt{\beta_1}$ and the parameter $1/k$.

7.3.1 SECOND AND THIRD ASYMPTOTES

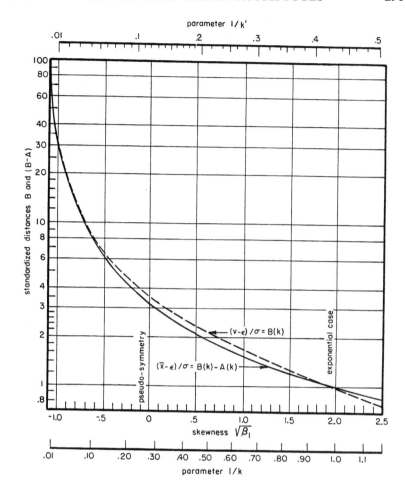

Graph 7.3.1(2). Standardized Distances from Lower Limit

The estimation of k, v, and ε with the help of (2), (3), and (5) requires the calculation of the third moment. This may be avoided if the characteristic value v is estimated from an order statistic. Let the ranks m be counted from the bottom, then the characteristic value is estimated as the m'th value with

(6) $\qquad 1 - m'/(N+1) = 0.36788 \; ; \qquad m' = 0.63212(N+1)\,.$

Hence, it is obtained by interpolation between the mth and $m + 1$th value where
$$m < m' < m + 1\,.$$

If the extremal probability paper is used for the logarithms of x, linear interpolation between $\log x_m$ and $\log x_{m+1}$ is advocated.

This estimate will be reliable if the difference between the consecutive observations x_m and x_{m+1} used to obtain $x_m{}'$ is small, especially when these two observations coincide. Experience has shown that the estimation of v is practically independent of the estimator used. The corresponding estimate of k is obtained from 7.2.3(11) which leads to

(7) $$A(k) = (\hat{v} - \bar{x}_0)/s .$$

Thus $\hat{k}$ is obtained from Table 7.2.3, columns 1 and 3.

The lower limit may be estimated from (4) and (5). However, these estimates cannot be used if they are larger than the smallest observation x_1. This value gives an indication of the lower limit, although the difference $x_1 - \varepsilon$ may be quite large if the series is small. The difference can be evaluated as a function of the sample size N. The most probable smallest value $\tilde{x}_1$, obtained from 7.2.3(3) and 7.2.5(2), is

(8) $$\frac{\tilde{x}_1 - \varepsilon}{v - \varepsilon} = \left(\frac{1 - 1/k}{N}\right)^{1/k} = f(N,k) ,$$

whence the lower limit

(9) $$\varepsilon = \frac{\tilde{x}_1 - f(N,k)}{1 - f(N,k)} .$$

The definition 7.2.3(10) of $B(k)$ leads to

$$B(k) = \frac{v - \tilde{x}_1}{\sigma[1 - f(N,k)]} .$$

If v has been estimated previously, say from (6) as an order statistic, and $\tilde{x}_1$ is replaced by the observed minimum x_1, the parameter k may be estimated from

(10) $$(\hat{v} - x_1)/s = B(k)[1 - f(N,k)]$$

with the help of the table for $B(k)$. The value k thus obtained is introduced in (9) and leads to an estimate of the lower limit. If v has not been estimated independently, the parameter k may be estimated from x_1, $\bar{x}_0$, and s with the help of (8), 7.2.3(5), and 7.2.3(9), which lead after substitution of the sample values to

(11) $$(\bar{x}_0 - x_1)/s = [\Gamma(1 + 1/k) - f(N,k)]B(k).$$

On the left are sample values. The right side can easily be calculated as a function of N and k. Hence, an estimate $\hat{k}$ is obtained. Finally, the remaining parameters v and ε are obtained again from 7.2.3(5) and 7.2.3(9) as

(12) $$\hat{v} = \bar{x}_0 + sA(\hat{k}): \quad \varepsilon = \hat{v} - sB(\hat{k}).$$

7.3.2 SECOND AND THIRD ASYMPTOTES

This procedure avoids the use of the third moment and of the small difference $\hat{v} - \bar{x}_0$, both of which have high errors of estimation, and warrants that the estimate of ε is smaller than the smallest observation x_1. Therefore, this procedure is advocated in those cases where the method of moments leads to an estimate $\hat{\varepsilon} > x_1$.

Exercises: 1) Calculate a table for $f(N,k)$; $N = 10(10)100$, $k = 1(0.5)10$. 2) Replace the most probable smallest observation in (8) by the mean or median smallest observation. 3) Replace the characteristic value in (8) by the mean. 4) Estimate the lower limit from the plotting position of the smallest value. 5) Rewrite the procedure for the corresponding distribution of the largest value. 6) Assuming $1/k$ to be known, calculate the errors of estimation of $\hat{\varepsilon}$ and $\hat{v}$.

Problems: 1) What are the errors of estimation especially for the lower limit? 2) Construct a test for the significance of $x_1 - \hat{\varepsilon}$. 3) Calculate the probability function for $(\bar{x}_0 - x_1)/s$. (See 4.2.4 for the normal distribution.) 4) How do the estimations $\hat{\varepsilon} = 0$; and $\hat{\varepsilon} > 0$ influence the estimation of k?

References: Freudenthal and Gumbel (1953, 1954, 1956), Lieblein, (1955), Weibull (1949).

7.3.2. Estimation of Two Parameters. If the parameter k is known, the two remaining parameters ε and v are linear functions of the mean and the standard deviation and their difference is a multiple of the standard deviation. The formulae 7.3.1(3) and (4) lead to estimates for ε and v. For $k = 1$, we have a truncated exponential distribution with mean and standard deviation,

(1) $$\bar{x} = v; \qquad \sigma = v - \varepsilon.$$

For $k = 1$, 2 and one of the pseudo-symmetrical cases the estimates are, respectively,

$k = 1$	$\hat{\varepsilon} = \bar{x}_0 - s$	$\hat{v} = \bar{x}_0$
$k = 2$	$\hat{\varepsilon} = \bar{x}_0 - 1.91306s$	$\hat{v} = \bar{x}_0 + 0.24560s$
$k = 3.25889$	$\hat{\varepsilon} = \bar{x}_0 - 2.96418s$	$\hat{v} = \bar{x}_0 + 0.34231s$

The composite hypothesis $k = \hat{k}$, $v = \hat{v}$ may be checked by plotting $(x/\hat{v})^{\hat{k}}$ on semilogarithmic paper.

If the lower limit is known, no generality is lost by writing x instead of $x - \varepsilon$. Then the first two moments are, from 7.2.3(4),

(2) $$\bar{x} = v\Gamma(1 + 1/k); \qquad \sigma^2 = v^2[\Gamma(1 + 2/k) - \Gamma^2(1 + 1/k)].$$

The squared coefficient of variation given by

(3) $$1 + V^2 = \Gamma(1 + 2/k)/\Gamma^2(1 + 1/k)$$

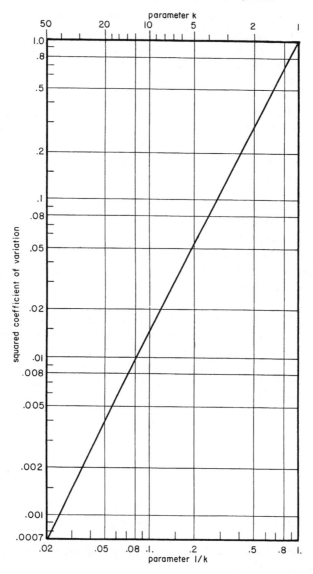

Graph 7.3.2(1).   Estimation of k from the Coefficient of Variation

and traced in Graph 7.3.2(1) depends only on k. Therefore, $1/k$ may be estimated from the sample coefficient of variation, and the characteristic value is estimated as

(4) $$\hat{v} = \bar{x}_0/\Gamma(1 + 1/\hat{k}).$$

7.3.2 SECOND AND THIRD ASYMPTOTES

For $\varepsilon = 0$ the parameters may also be linked to the geometric moments, since the transformations 7.2.4(2) lead from 5.2.4(5) and 5.2.5(1) to the geometric mean and standard deviation,†

(5) $\overline{\log x} = \log v + \gamma/k'$; $\sigma(\log x) = \pi/(\sqrt{6}k')$; $k' = 2.30258k$.

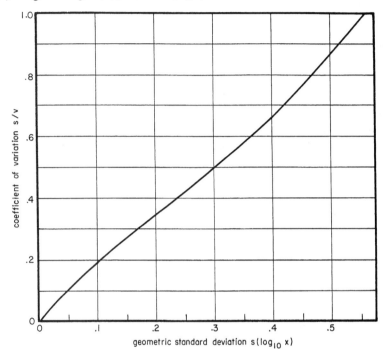

Graph 7.3.2(2). Criterion for Lower Limit Zero

Combination of the second equations in (2) and (5) leads to a relation between $\sigma(\log x)$ and $\sigma(x)/v$ traced in Graph 7.3.2(2), which may serve as a test of the hypothesis $\varepsilon = 0$. For the estimation of the parameters we use

(6) $\dfrac{1}{\hat{k}'} = \dfrac{s(\log x)}{\sigma_N}$; $\log \hat{v} = \overline{\log x} + \bar{y}_N/\hat{k}'$,

which are the formulae 6.2.3(1) for the first asymptote if the variate x is replaced by $\log x$. The factors $\bar{y}_N$ and σ_N are found in Table 6.2.3. This estimation is designed for use with logarithmic extremal probability paper where

(7) $\log x = \log v + y/k'$.

† We designate the decimal logarithm by log, while lg stands for the natural logarithm.

This paper may also be used in case $\varepsilon > 0$, since we may write in 7.2.3(1)

$$e^y = \left(\frac{x-\varepsilon}{v-\varepsilon}\right)^k,$$

whence

(8) $$\lg[-\lg(P(x))] = y = k'\log(x-\varepsilon) - k'\log(v-\varepsilon)$$

and

(9) $$\log(x-\varepsilon) = \log(v-\varepsilon) + y/k'.$$

In contrast to (7), $\log x$ is no longer a linear function of y. Thus, the deviation from the line (7) gives an indication of the existence of a lower limit. In this case, the observed values x corresponding to large probabilities will exceed in a systematic way the theoretical values situated on the line (7). The probability paper cannot be used for $\varepsilon < 0$.

The probability paper may be used for an estimate of the lower limit ε. We choose a plausible value $\hat{\varepsilon}_1$ and plot the "observations" $\log(x - \hat{\varepsilon}_1)$. Then we choose a better $\hat{\varepsilon}_2$ and continue the plotting until an approximate straight line is obtained. Let $\hat{\varepsilon}$ be the final solution. Then the use of (6) applied to the "observations" $\log(x - \hat{\varepsilon})$ leads to the estimates for k and v.

In the case $\varepsilon = 0$, the estimation of the two parameters v and k by the maximum likelihood method goes along the same lines as for the first and second asymptotic distributions and presents the same numerical difficulties. For $k = \alpha$, the distribution 7.2.2(5) may be written

$$\lg p(x) = \lg \alpha + (\alpha - 1)\lg x - \alpha \lg v - (x/v)^\alpha.$$

The two derivatives with respect to α and v are

(10) $$\frac{\partial \lg p(x)}{\partial \alpha} = \frac{1}{\alpha} + \lg x - \lg v - (x/v)^\alpha \lg(x/v),$$

$$\frac{\partial \lg p(x)}{\partial v} = -\frac{\alpha}{v} + \frac{\alpha x^\alpha}{v^{\alpha+1}},$$

whence, after summation and division by N,

(12) $$\frac{1}{\alpha} + \overline{\lg x} = \lg v + \frac{1}{N}\sum_{i=1}^{N}(x_i/v)^\alpha \lg(x_i/v)$$

$$\frac{1}{N}\Sigma(x_i/v)^\alpha = 1$$

and

(13) $$\frac{1}{\alpha} + \overline{\lg x} = \frac{\sum_{1}^{N} x_i^\alpha \lg x_i}{\sum_{1}^{N} x_i^\alpha}.$$

7.3.2 SECOND AND THIRD ASYMPTOTES

This equation contains only the unknown parameter α and the sample values x_i. Therefore it can be solved by successive approximations. If this is done, the other parameter v is obtained from the equation (12) as the solution of

$$(14) \qquad v^{\hat{\alpha}} = \frac{1}{N} \sum_{1}^{N} x_i^{\hat{\alpha}}$$

If α is known, this equation gives an estimation for v. However, in general α is unknown. In order to obtain a first solution of (13), we write it in the form

$$(15) \qquad \frac{1}{\alpha} = -\overline{\lg x} + \frac{1}{N} \sum_{i=1}^{N} (-\lg P_i) \lg (x_i).$$

Now the probability function is replaced by the plotting position and the correction A_N given in 6.2.6(5),

$$-\lg P_i = -A_N \lg (N + 1 - i) + \lg (N + 1) \cdot A_N,$$

whence, after introduction of decimal logarithms with $\alpha' = 2.30258\alpha$,

$$(16) \qquad \frac{1}{\alpha'} = [A_N \lg (N + 1) - 1] \overline{\log x} - \frac{A_N}{N} \sum \lg (N + 1 - i) \log x_i.$$

This equation contains in addition to the unknown parameter $1/\alpha'$ two statistics, the geometric mean and the sum, which can easily be calculated. The result, $\hat{\alpha}'$ introduced into (14), leads to a first approximation $\hat{v}_1$, which, used in (12), leads to a second approximation for $\hat{\alpha}'$ and so on.

Exercises: 1) Calculate the parameters for the other three pseudo-symmetrical cases. 2) Show that for $k = 1$ the asymptotic errors of estimation of $\hat{\varepsilon}$ and $\hat{v}$ are equal. 3) Apply Kimball's method, 6.2.5(10).

Problems: 1) Show the analytic reasons for the approximately linear relation traced in Graph 7.3.2(1). 2) What is the power of the tests for $\varepsilon = 0$? 3) Compare the asymptotic efficiency of the estimates (3) (6) and (16) for $1/k$. 4) What are the errors of estimation involved in (4) and (14)? 5) Estimate the three parameters by maximum likelihood. 6) Estimate the parameters from the r smallest among k observations (See Epstein and Sobel, 1952 for the case $\alpha = 1$).

References: Epstein and Sobel (1954), Freudenthal and Gumbel (1956), Kimball (1946a, 1947b, 1956), Lieblein (1955), Stange, Sukhatme, Weibull (1939a,b, 1949, 1955).

7.3.3. Analytical Examples. Consider the smallest value of the Gamma distribution (4.2.5), where v is replaced by $k \geq 1$. The probability function is

$$1 - F(x) = 1 - \frac{1}{(k-1)!} \int_0^x z^{k-1} e^{-z} \, dz$$

$$= 1 - \frac{1}{(k-1)!} \sum_{v=0}^{\infty} \frac{(-1)^v x^{k+v}}{v! \; k+v}.$$

The development in the neighborhood of the lower limit zero is

$$1 - F(x) = 1 - \frac{x^k}{k!} + \frac{x^{k+1}}{(k-1)!(k+1)} - \cdots$$

If we put, as in 7.2.1(5''),

(3) $$\frac{x^k}{k!} = \frac{z^k}{n},$$

the series becomes

(4) $$1 - F(x) = 1 - \frac{z^k}{n} + \frac{\text{const } z^{k+1}}{n^{1+1/k}},$$

and the probability of the smallest value converges with increasing n to the third asymptotic probability.

For $k = 1$, the distribution of the smallest value taken from the exponential distribution is the exponential distribution itself, with a change of scale as shown in 4.1.2(4).

We return now to normal extremes. Even for samples of $n = 1000$, their probabilities depart considerably from the first asymptote. Since, as shown in 7.2.2, the third asymptotic probability converges for increasing values of the parameter k toward the first one, Fisher and Tippett proposed to use the third function. The parameter v_n in III for $\omega = 0$, i.e.,

(5) $$\Phi^{(3)}(x) = \exp\left[-(x/v_n)^{k_n}\right]$$

is the characteristic largest normal value, previously designated by u_n. Numerical values are given in Table 4.2.2(2) and are traced in Graph 7.3.3(1) against n. For the determination of k_n as function of n, it would seem natural to use 7.2.5(3):

(6) $$\frac{1}{k_n} = \frac{d \lg u_n}{d \lg n}.$$

Fisher and Tippett use instead

(7) $$\frac{1}{k_n} = \frac{d \lg \alpha_n}{d \lg n},$$

7.3.4 SECOND AND THIRD ASYMPTOTES

and replace the right side by its first approximation 4.2.3(4'), so that
$$(8) \quad 1/k_n = (u_n{}^2 - 1)/(u_n{}^2 + 1)^2.$$

The values of α_n and k_n are given in Table 4.2.2(3).

Since the two parameters v_n and k_n in (5) are known, the logarithmic transformation 7.2.2(1) leads to the probability points x corresponding to $y(\Phi^{(3)})$ from

$$\log x = \log v_n + y 0.43429/k_n.$$

The probabilities of the normal extremes for samples of size $n = 10^3$, 10^4,

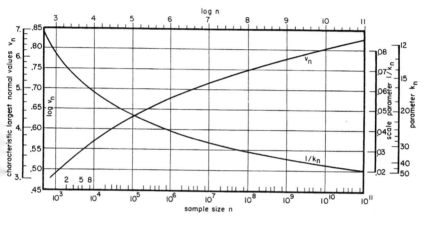

Graph 7.3.3(1). Parameters v_n and k_n for Normal Extremes

10^5, and 10^6 are traced in Graph 7.3.3(2) on logarithmic extremal probability paper, together with the exact values calculated by Tippett for $n = 1000$. The fit for $n = 1000$ is good in the interval $0.10 < \Phi^{(3)} < 0.98$, while the use of (6) leads to a worse result. Of course it is not known how good the other approximations are.

Exercise: Calculate a second approximation to (7).
Problem: Why does (7) lead to a better fit than (6)?

7.3.4. Droughts. In 6.3.1, the daily discharges of a river were considered as statistical variates, unlimited to the right, and their annual maxima, the floods, were analyzed by the first asymptotic distribution of largest values. Since the discharges are limited to the left, the annual minima, henceforth called *droughts*, will now be analyzed by the third asymptotic distribution of the smallest value. The extremal probability paper designed for floods is used for the logarithms of the droughts plotted in decreasing order of magnitude, so that the minima decrease with increasing return periods. If the observations are scattered about a straight line, the minimum of the droughts may be assumed to be zero. In this case, the two parameters may be estimated from 7.3.2(6) and the theoretical droughts are obtained from

$$\log x = \log v + y/\alpha'.$$

If the points ($\log x, y$) fall off less rapidly than a straight line and tend to flatten out, we have to assume that a lower limit exists. In this case, the three parameters are estimated from 7.3.1(2), (3), and (5).

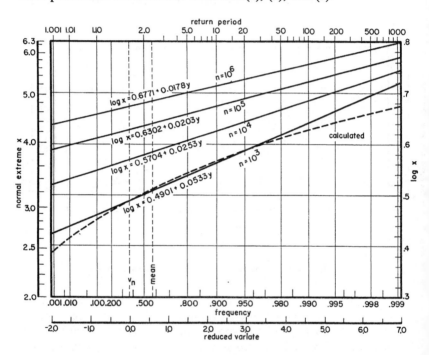

Graph 7.3.3(2). Approximations of the Probabilities of Normal Extremes by the Third Asymptote

Graph 7.3.4 shows the observed and calculated droughts for the Colorado River at Lees Ferry, Arizona, 1922–39, and for the Connecticut River at Sunderland, Massachusetts, 1905–32, taken from Water Supply Papers 879 and 1105 respectively. For the first (second) river we assume $\varepsilon = 0$ ($\varepsilon > 0$); The estimation of the parameter is given in Table 7.3.4. Other examples were given in a previous paper (Gumbel, 1954) and showed close conformance with the theory. If the number of years is large enough so that the sampling errors are small enough, and if it is permissible to

7.3.4 SECOND AND THIRD ASYMPTOTES

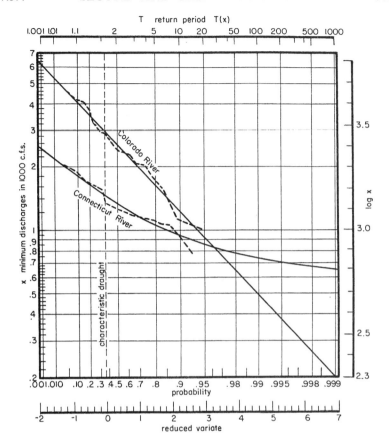

Graph 7.3.4. Droughts: Colorado and Connecticut Rivers

Table 7.3.4. Analysis of Droughts

		Colorado River			Connecticut River
	Symbol			Symbol	
Number of Years	N	18	Number of Years	N	28
Mean Logarithm	$\overline{\log x}$	3.38528	Mean Drought	$\bar{x}_0$	1353.9 c.f.s.
Standard Deviation of $\log x$	$s(\log x)$	0.17712	Standard Deviation	s	332.51 c.f.s.
			Skewness	$\sqrt{b_1}$	0.39915
Scale Parameter	$1/\hat{\alpha}'$	0.16879	Scale Parameter	$1/\hat{\alpha}'$	0.18005
Location Parameter	$\log \hat{v}$	3.47307	Characteristic Drought	$\hat{v}$	1450.2 c.f.s.
Characteristic Drought	$\hat{v}$	2972 c.f.s.	Lower Limit	$\hat{\varepsilon}$	601.3 c.f.s.
Observed Minimum	x_N	1000 c.f.s.	Observed Minimum	x_N	780 c.f.s.
100 Years Drought	x'_{100}	497 c.f.s.	100 Years Drought	x_{100}	724 c.f.s.

assume that the basic conditions prevail, the theoretical curves thus obtained can be used to estimate the most severe droughts to be expected within a given number of years. This procedure may be useful for solving storage and irrigation problems.

References: Goodrich, Gumbel (1954c), Velz (1950a), H. E. Hudson and W. I. Roberts (1955). Illinois droughts, 1952–55, Illinois State Water Survey Division, Bull. 43. C. J. Velz and J. J. Gannon, Drought flow of Michigan streams. Mich Water Ress Comm. (1960).

7.3.5. Fatigue Failures.

"*Gutta cavat lapidem non vi sed saepe cadendo.*"

(*Continuous drops hollow the stone.*)

In 6.3.6, we considered the static breaking strength for rubber. Now we consider the more complicated dynamic case, where a constant stress S (load per unit of surface) is applied during a number of cycles, or different stresses are applied during a constant number of cycles, until the specimen tested breaks. A guidance in the analysis can be found by the following properties of the probabilities of survival: 1) For constant stress S, the probability of survival decreases with increasing number of cycles (fatigue failure). 2) For a constant number of cycles the probability of survival decreases with increasing stress (dynamic breaking strength). It follows that with increasing S a smaller number of cycles is needed to obtain fracture with a given probability. No two survival functions for different values of S, and no two survival functions for different number of cycles, can ever intersect for the same material and the same testing procedure.

These theoretical properties may be used to decide whether sample values are homogeneous. If the observations contradict one of these properties, especially if two observed frequencies of survival intersect, we conclude that the errors of random sampling prevent the comparison and the statistical interpretation of such observations.

W. Weibull derived in numerous writings the third asymptotic distribution of smallest values in an empirical way and applied it to the analysis of dynamic breaking strength. In the following we apply it to fatigue failures. Graph 7.3.5 gives Ravilly's observed numbers of cycles x (in 1000) at failure of nickel specimens under reversed torsion for two stresses $S = \pm 37.5$ kg mm^{-2} and $S = \pm 18$ kg mm^{-2}. For physical reasons which are explained elsewhere (Freudenthal and Gumbel, 1953, 1956) the third asymptotic probability of smallest values,

$$l(x) = \exp\left[-\left(\frac{x-\varepsilon}{v-\varepsilon}\right)^\alpha\right],$$

is chosen for the probability $l(x)$ of survival of x cycles. The location

7.3.5 SECOND AND THIRD ASYMPTOTES

parameters v and ε, which are now numbers of cycles, decrease with increasing stress levels. The parameter v is called the characteristic number at failure, while ε is called the minimum life, the number of cycles before which no failure occurs. It is the most important parameter in the theory of fatigue failures. If it does not vanish, we can state with probability one

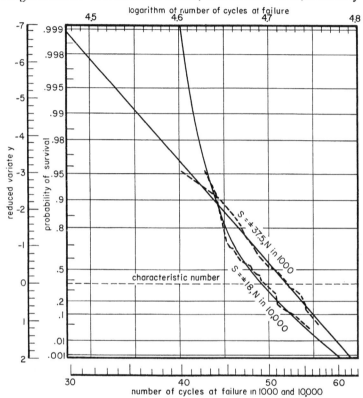

Graph 7.3.5. Fatigue Failures of Nickel

that the specimen will survive this number of cycles or a smaller one for the given stress. Of course statistical methods can only lead to an estimate for this physical constant. Graph 7.3.5 shows a practically linear survivorship function for $S = \pm 37.5$. Thus, it may be assumed that no lower limit for the number of cycles at breakage exists and that failure starts with the first cycle. The calculation of the two parameters with the help of 7.3.2(6) is shown in Table 7.3.5, column 3. The theoretical line,

$$\log x = 4.71793 + 0.03588y,$$

drawn in Graph 7.3.5 gives a very good fit.

The survivorship function for the smaller stress, $S = \pm 18$ kg mm^{-2}, shows for small numbers of cycles a definite upward curvature, which implies the existence of a non-vanishing lower limit. The calculation of the parameters from 7.3.1(2), (3), and (5) is also shown in Table 7.3.5. The curve,

$$\log(x - 396.13) = \log 94.16 + 0.16791 y,$$

traced in Graph 7.3.5 gives a good fit.

Table 7.3.5. *Parameters for Fatigue Failures of Nickel*

Stress in kg mm^{-2}	S	± 37.5	Stress in kg mm^{-2}	S	± 18
Mean Logarithm of Number of Cycles at Failure	$\overline{\log x}$	4.69950	Mean Number of Cycles at Failure	$\bar{x}$	479.75
Geometric Standard Deviation	$s(\log x)$	0.03813	Standard Deviation	s	34.705
Scale Parameter	$1/\alpha'$	0.03588	Skewness	$\sqrt{b_1}$	0.32112
Logarithmic Characteristic Number	$\log v$	4.71793	Scale Parameter	$1/\alpha'$	0.16791
Characteristic Number of Cycles in 1000	v	52.23	Characteristic Number of Cycles in 1000	v	490.29
			Minimum Life in 1000 Cycles	ε	396.13

Further numerical examples are given in the publications of Freudenthal and Gumbel. However numerous problems are still unsolved.

The answers to these questions will depend upon the materials tested and the testing procedures, and can only be found by a combination of physical and statistical considerations. Unfortunately, systematic tests made under statistical control with the same large number of specimens for different stresses and continued for the millions of cycles necessary to check such theories, are practically non-existent, although the expense caused by a single failure may exceed the cost of such experiments.

Problems: 1) Can the static and the dynamic case be treated by the same formulae so that the static case becomes a degeneration of the dynamic one? 2) How does the minimum life decrease with increasing stress? 3) How does α depend upon stress? 4) What is the nature of the probability of survival as function of the stress for constant numbers of cycles? 5) Under what conditions are the assumptions that both survivorship functions are the asymptotic probabilities of the third type compatible? 6) Is there a stress called endurance limit, which is so small that the specimen can withstand it during an infinity of cycles, and how can the endurance limit be estimated? 7) How are the stresses related to the numbers of cycles at failure for given probabilities of survival, in the special case that the minimum life and the endurance limit vanish, and in the general case that both do not vanish? 8) How is the probability of survival as a

7.3.5 SECOND AND THIRD ASYMPTOTES

function of the numbers of cycles transformed if the stresses themselves vary according to an exponential distribution? Can this probability of survival be reproduced by the same function and how are the three parameters related to the parameters valid for the single stresses? 9) What is the influence of the dimensions?

References: Afanassiev, Bastenaire, Davidenkov, Frankel, Freudenthal, Freudenthal and Gumbel, L. G. Johnson, Lieblein (1954b) Lieblein and Zelen, Oding, Prot (1949b, 1950), Ravilly, Reagel and Willis, Stange, Tucker (1941, 1945) Weibull.

Chapter Eight: THE RANGE

Divis manibus magistri.
(*To the memory of L. von Bortkiewicz.*)

8.1. ASYMPTOTIC DISTRIBUTIONS OF RANGE AND MIDRANGE

8.1.0. Problems. In this chapter, we develop the asymptotic theory of the range, the midrange, and connected extremal statistics. As in the preceding analysis, only certain properties of the initial distribution are required, but its form need not be specified. We shall deal mainly with initial distributions of the exponential type, because only for these distributions can relatively simple results be obtained.

M. G. Kendall wrote sadly: "It is not known whether asymptotic distributions of the range exist and what they are." Indeed, it would be very difficult to start from the exact distribution 3.2.2(2) of the range, and to calculate the asymptote toward which it converges with increasing sample sizes. Consequently, we assume at once the existence of a large sample. The size for which the asymptotic considerations hold depends upon the approach of the distributions of the extremes toward their asymptote. Since the two extremes are independent for sufficiently large samples, the existence of an asymptotic distribution of the range can at once be affirmed in some (trivial) cases. If the distribution is limited to the left (or to the right), the distribution of the smallest (largest) value contracts with increasing sample size, and the asymptotic distribution of the range is the asymptotic distribution of the largest (smallest) value, provided that the latter exists. We speak of the span of human life and mean by it the oldest age. More specifically, if the initial distribution is skewed to the right, the largest value will asymptotically outweigh the smallest one, and the asymptotic distribution of the range becomes the asymptotic distribution of the largest value, provided it exists. The same holds if the initial distribution is skewed to the left, because the asymptotic distribution of the smallest value with negative sign is also a distribution of the largest value. Therefore, no asymptotic distribution of the range can be obtained by convolution of the largest value for one type and smallest value of another type. The three asymptotic distributions of the largest value are also asymptotic distributions of the range. Three other asymptotic distributions may be obtained for symmetrical initial distributions by the convolution of extreme values taken from the same type. Then the problem consists in the convolution of two distributions of largest values with the same parameters.

Exercise: Calculate the asymptotic distribution of the range for the uniform distribution.

Problem: Calculate the asymptotic distributions of the range and mid-range for the Cauchy and the limited type.

References: Baker, Carlton, Feller, Gartstein (1948, 1951), Rider, (1950, 1951a, b, 1953), Simon, W. R. Thompson (1938).

8.1.1. The Range of Minima. The range w of minima taken from N samples of large size n has been studied by B. McMillan. We give only the results and refer the reader to the original paper for the proof.

To show the general meaning of the problem, consider the exponential distribution. Since, from 4.1.2(4), the probability of the smallest value is equal to the initial one, except for a change in scale, the same holds for the probability $\Psi_N(w_n)$ of the range w_n of the smallest values, which becomes from 4.1.2(7)

$$\Psi_N(w_n) = [1 - \exp(-nw_n)]^{N-1}.$$

If we introduce the modal range of the smallest values $\widetilde{w}_n$, being the solution of

$$\exp(n\widetilde{w}_n) = N - 1,$$

the probability function

$$\left(1 - \frac{\exp[-n(w_n - \widetilde{w}_n)]}{N-1}\right)^{N-1} \to \exp[-e^{-n(w_n - \widetilde{w}_n)}]$$

converges to the first asymptote and the distribution contracts with n increasing.

If the initial distribution is limited by $x \geq \varepsilon$, so that $F(\varepsilon) = 0$, the minima are also limited and will cluster near ε, so that the distribution of the range contracts to zero as n increases. McMillan proves that this holds also if the tail of the initial distribution is sufficiently small. More generally he proves the following theorems: If

$$\liminf_{x = -\infty} F(x)/F(x + w) = l \leq 1,$$

then the probability $\Psi_n(w)$ of the range is such that

$$\limsup_{n = \infty} \Psi_n(w) \leq (1 - l)^{N-1}.$$

On the other hand, if the initial probability converges to zero for large negative value of x in the same way as an exponential function, the probability function $\Psi'_n(w)$ does not contract with n increasing. If the initial variate is unlimited to the left and

$$\lim_{x=-\infty} \sup F(x)/F(x+w) = L,$$

then for any $\alpha > 0$,

$$\liminf_{n=\infty} \Psi'_n(w) \geqq (e^{-\alpha L} - e^{-\alpha})^N.$$

The two theorems show that for initial distributions of the exponential type the probability function of the range is bounded away from zero for any $w > 0$ and bounded away from unity for w sufficiently small.

Exercise: Calculate the distribution of the range for the double exponential distribution.

Problems: 1) Study the relations between McMillan's and Gnedenko's conditions 5.1.3. 2) Study the range of the maxima. 3) What are the initial distributions such that the probability of the range of the maxima (or minima) is equal to the probability of the range proper except for a linear transformation of the range?

8.1.2. Generating Function of the Range. In the following, we consider mainly initial distributions of the exponential type. For large samples the generating function $G_w(t)$ of the difference $w = x_n - x_1$ of the two extremes is, from 1.1.7(5),

(1) $$G_w(t) = G_n(t) G_1(-t),$$

and, for the exponential type, from 5.2.4(2) and (3),

(2) $$G_w(t) = [\exp(u_n - u_1)t] \Gamma(1 - t/\alpha_n) \Gamma(1 - t/\alpha_1).$$

If we assume symmetry, the generating function becomes

(3) $$G_w(t) = G_n^2(t)$$

(4) $$= \exp[2u_n t] \Gamma^2(1 - t/\alpha_n).$$

The mean range $\bar{w}_n$ and the variance σ_w^2,

(5) $$\bar{w}_n = 2u_n + 2\gamma/\alpha_n; \qquad \sigma_w^2 = \pi^2/3\alpha_n^2,$$

are obviously twice the mean and the variance of the largest value. Of course, this follows also directly from the independence. Since $\bar{w}_n = 2\bar{x}_n$, the mean range increases faster than the mean of the extremes.

The next moments are

(6) $$\mu_3 = 4S_3/\alpha_n^3 \qquad \mu_4 = 7\pi^4/(15\alpha_n^4).$$

The expressions,

(7) $$\beta_1 = 2S_3^2/S_2^3, \quad \beta_2 - 3 = 1.2,$$

are one-half the corresponding characteristics of the largest value.

In order to apply these formulae to a given initial distribution, the characteristic largest value u_n and the parameter $1/\alpha_n$ have to be calculated from their definitions 3.1.4(1) and 3.1.5(1), or estimated from the observed ranges. This procedure will be shown in 8.3.2.

Exercises: 1) Calculate the two asymmetrical distributions of the range for skewed distributions which are of the exponential type in both directions. 2) Calculate the asymptotic standard errors of b_1 and b_2. 3) Construct the moment generating function of the mth range and midrange and calculate the first moments. (See Gumbel, 1944.)

Problem: Connect formulae (5) for the mean range and for the variance to Tippett's formulae, 3.2.3(3) and (4).

Reference: Ruben.

8.1.3. The Reduced Range. Since the variates considered are unlimited, the range increases without limit with the sample size, and a reduction becomes necessary. In the similar case of the largest value, the reduction used was

$$y = \alpha_n(x - u_n).$$

As the range is the sum of two extremes, it is natural now to use $2u_n$ instead of u_n. Therefore, the *reduced range R*, a pure number, is defined by

(1) $$R \equiv \alpha_n(w - 2u_n).$$

The expression $2u_n$, the distance from the characteristic smallest to the characteristic largest value, may be called the characteristic range. Since from 4.1.4(3) the modes of the extremes converge toward the characteristic values, the characteristic range converges toward the range of the modes, which, of course, is not the mode of the range. The reduced range R is thus the range itself minus the range of the modes, divided by a factor proportional to the standard deviation of the range. It is unlimited toward the right, but limited toward the left by

(2) $$R \geq -2\alpha_n u_n.$$

Since from 4.1.7(2) the product $\alpha_n u_n$ increases with n, this condition is no real limitation of the range for sufficiently large sample sizes.

The asymptotic generating function $G_R(t)$ and the seminvariant function $L_R(t)$ of the reduced range R are, from 1.1.7(5),

(3) $$G_R(t) = \Gamma^2(1-t); \qquad L_R(t) = 2\lg \Gamma(1-t) - 2\gamma$$
and
(4) $$L_R(t) = 2 \sum_2^\infty \frac{S_v t^v}{v}.$$

The numerical values of the first S_v are given in 1.1.8. The mean $\bar{R}$, the variance σ_R^2, and the seminvariants λ_v of the reduced range are

(5) $$\bar{R} = \bar{y}_N - \bar{y}_1 = 2\gamma = 1.154431330; \quad \sigma_R^2 = \sigma_N^2 + \sigma_1^2 = \pi^2/3;$$
$$\lambda_v = 2S_v(v-1)! \quad v \geq 2.$$

The reduction (1) may therefore be written

(6) $$R = \frac{\pi}{\sqrt{3}} \frac{w - \bar{w}}{\sigma_w} + 2\gamma,$$

which is a linear function of the customary transformation.

The moments of the reduced range are obtained from (3) and the relations 1.1.8(4) as $\sigma = 1.81380$,

(7) $$\mu_2 = 2S_2 = 3.28986814; \qquad \mu_3 = 4S_3 = 4.80822760.$$
(8) $$\mu_4 = 12(S_4 + S_2^2) = 15\pi^4/30 = 45.4575759$$
$$\mu_5 = 16(3S_5 + 5S_3 S_2) = 207.956762$$
$$\mu_6 = 40(6S_6 + 9S_2 S_4 + 4S_2^3 + 3S_2^4) = \frac{31}{21}\pi^6 + 160S_3^2$$
$$= 1650.38410.$$

The Betas are $\sqrt{\beta_1} = 0.8057814760$;

$$\beta_1 = .649284; \quad \beta_2 = 4.2.; \quad \beta_3 = 8.67186; \quad \beta_4 = 167.730.$$

The mode $\tilde{R}$ of the reduced range is a fixed value 0.50637 derived in 8.1.6. The mode $\tilde{w}_n$ is obtained from the transformation (1) as

(9) $$\tilde{w}_n = 2u_n + \tilde{R}/\alpha_n,$$

while the difference of the modes of the extremes is asymptotically

$$\tilde{x}_n - \tilde{x}_1 = 2u_n.$$

Consequently,

(9') $$\tilde{w}_n = \tilde{x}_n - \tilde{x}_1 + \tilde{R}/\alpha_n.$$

For a symmetrical initial distribution of the exponential type, the mode of the range converges toward the range of the modes of the smallest and of the largest value, provided that α_n increases with n.

Exercises: 1) Calculate the moments of the mth range and midrange. (See Gumbel, 1944.) 2) Interpret the generating function $\Gamma^2(1+t)$.

8.1.4 THE RANGE

Problem: Calculate the asymptotic distribution of the range from the generating function $G_R(t)$ by inversion.

Reference: J. Geffroy (1957). Sur la notion d'independance limité de deux variables aléatoires. Application a l'étendue at au milieu d'un échantillon, Compt. Rend. Ac. Sc., 245: 1291.

8.1.4. Asymptotic Distribution of the Midrange. The asymptotic generating function $G_V(t)$ of the midrange V for initial distributions of the exponential type is
$$G_V(t) = G_n(t) G_1(t),$$
whence, from 5.2.4(2) and (3), for an asymmetrical distribution,

(1) $\qquad G_V(t) = \exp[(u_n + u_1)t] \, \Gamma(1 - t/\alpha_n)\Gamma(1 + t/\alpha_1).$

In the sum $V = x_n + x_1$, one factor becomes very large compared to the other one, and the distribution of the midrange converges to the distribution of the largest (smallest) value if the initial skewness is positive (negative). Hence, as stated in 3.2.6, the midrange is of no interest for skewed distributions. The symmetrical case $u_n = -u_1$; $\alpha_n = \alpha_1$ is interesting because it refutes the widespread opinion that all measures of central tendency converge toward normality. The variance S_V^2 of the midrange, which equals asymptotically the variance of the range, is

(2) $\qquad\qquad\qquad \sigma_V^2 = \pi^2/3\alpha^2.$

The fourth central moment, and the second Beta, are

(3) $\qquad\qquad \mu_{4,V} = 7\pi^4/15\alpha^4; \qquad \beta_{2,V} = 4.2.$

Since the generating function uniquely determines the distribution, it follows from 4.2.1(2) that *the asymptotic distribution $h(v)$ and the probability $H(v)$ of the reduced midrange v are the "logistic expressions,"*

(4) $\quad h(v) = e^{-v}(1 + e^{-v})^{-2}; \quad H(v) = (1 + e^{-v})^{-1}; \quad v = \alpha(V - u).$

For increasing sample sizes, the distribution of $h(V)$ contracts (is invariant, or spreads) if $1/\alpha_n$ increases (is invariant, or decreases) with n. In the two latter cases, estimates from the midrange cannot be improved by increasing the sample size.

The asymptotic probability function of the reduced midrange is traced in Graph 8.1.4, together with the probability functions of the reduced extremes. The asymptotic probability of the midrange is practically identical with the asymptotic probability of the smallest (largest) value, for small (large) values of the midrange. This behavior follows immediately from the three probability functions. The fourth curve in the graph shows the reduced range, and will be derived later.

The logistic distribution is stable with respect to the midrange, as the normal distribution is stable with respect to the mean. However, in the latter case, the precision of the mean increases with increasing sample size, while in the first case this need not hold. The extremal properties of this distribution were studied in 4.2.1.

Exercises: 1) Calculate the generating function $G_v(t)$ from the distribution $h(v)$ and derive the moments directly. 2) Calculate the generating function, the moments, the distributions, and probabilities of the mth

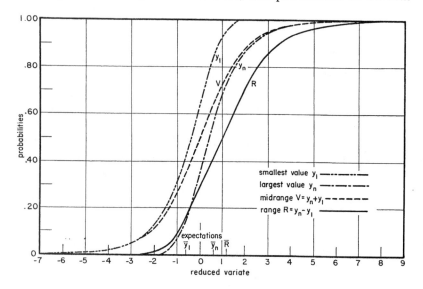

Graph 8.1.4. Asymptotic Probabilities of Extremes, Range and Midrange

midrange, for initial distribution of the exponential type for $m \ll n$. (See Gumbel, 1944.) 3) Prove that the distribution of the mth midrange is of the exponential type, and converges toward normality for m increasing.

Problems: 1) Do other symmetrical distributions exist which are stable with respect to the midrange? 2) What is the distribution of the midrange for the Cauchy type?

8.1.5. A Bivariate Transformation. The asymptotic distribution of the range for initial symmetrical distributions of the exponential type may be obtained from the convolution of the distributions of the extreme values. Instead, we are going to use a transformation of the joint distribution of the extreme 3.2.2(1), which is due to G. Elfving. It leads to four extremal statistics which have different meanings for different initial types. For the exponential type, they are the midrange and an exponential function of the

8.1.5 THE RANGE

range. The interpretation for the symmetrical limited type will be given in the next section, and leads to distributions of two other extremal statistics. This procedure is more general than the method originally used by the author for the derivation of the asymptotic distribution of the range.

Elfving studied the asymptotic distribution of certain probability integral transformations applied to the extremes. He assumes the knowledge of the initial distribution, its parameters, and the sample size. In the following, this knowledge is not required.

The joint distribution 3.2.2(1) of the extremes is written in the form

(1) $\quad \mathfrak{w}_n(x_1, x_n) = n(n-1)f(x_1)[1 - F(x_1) - (1 - F(x_n))]^{n-2} f(x_n)$

where x_1 and x_n are taken in standard units. Two new variates, ξ and ζ, are introduced by the transformation, similar to 3.2.7(2):

(2) $\quad \xi = 2n\sqrt{F(x_1)(1 - F(x_n))}$; $\quad \zeta = \frac{1}{2}\lg[F(x_1)/(1 - F(x_n))]$;

$$\zeta' = 2\zeta .$$

Evidently, ξ is positive, and ζ is unlimited in both directions. From

$$\xi \frac{e^\zeta + e^{-\zeta}}{2} = n(F(x_1) + 1 - F(x_n)),$$

it follows that

(2') $\quad\quad\quad\quad\quad 0 < \xi \cos h\zeta \leq n .$

The joint distribution $\mathfrak{w}_n(\xi, \zeta)$ of the new variates is

(3) $\quad \mathfrak{w}_n(\xi, \zeta) = n(n-1)f(x_1)f(x_n)\left(1 - \frac{\xi}{n} \cos h\zeta\right)^{n-2} \frac{\partial(x_1, x_n)}{\partial(\xi, \zeta)} .$

The Jacobian is calculated from

$$\frac{1}{J} = \begin{vmatrix} nf(x_1)\sqrt{(1 - F(x_n))/F(x_1)} & -nf(x_n)\sqrt{F(x_1)/(1 - F(x_n))} \\ \frac{1}{2}\frac{f(x_1)}{F(x_1)} & \frac{1}{2}\frac{f(x_n)}{1 - F(x_n)} \end{vmatrix}$$

Consequently, from (2) and (2'),

$$1/J = 2n^2 f(x_1) f(x_n)/\xi .$$

From (3), the *asymptotic* bivariate distribution becomes simply

(4) $\quad\quad\quad\quad \mathfrak{w}(\xi, \zeta) = \frac{\xi}{2} e^{-\xi \cos h\zeta} .$

The univariate, asymptotic distributions $h(\zeta)$ and $\varphi(\xi)$ are obtained by integrations. Consider first the variate ζ. Integration over ξ leads to

(5) $$h(\zeta) = 2(e^{\zeta} + e^{-\zeta})^{-2},$$

or, from (2),

(5') $$h^*(\zeta') = e^{-\zeta'}(1 + e^{-\zeta'})^{-2}.$$

The variate ζ' has asymptotically the *logistic distribution* studied in 4.2.1 and 8.1.4. The asymptotic distribution $\varphi(\xi)$ of ξ is obtained in the same way as

(6) $$\varphi(\xi) = \xi \int_0^\infty e^{-\xi \cos h\zeta}\, d\zeta$$
$$= \frac{1}{2}\int_{-\infty}^{+\infty} \exp\left[-e^y - e^{-y}\xi^2/4\right] dy$$

since $\cos h$ is an even function. We introduce

$$\cos h\zeta = t\,;\quad \zeta = 0, t = 1\,;\quad \zeta = \infty, t = \infty.$$

The identity
$$\cos^2 h\zeta - \sin^2 h\zeta = 1$$
leads to
$$d\zeta = dt/\sqrt{t^2 - 1}.$$

Finally, from (6), the asymptotic distribution $\varphi(\xi)$ becomes

(7) $$\varphi(\xi) = \xi \int_1^\infty \frac{e^{-\xi t}}{\sqrt{t^2 - 1}}\, dt,$$

which will now be expressed as a Bessel function. If we write

(7') $$J(\xi) = \int_1^\infty \frac{e^{-\xi t}}{\sqrt{t^2 - 1}}\, dt,$$

it is easily verified that the function $J(\xi)$ is subject to the differential equation of Bessel's type,

(8) $$J''(\xi) + \frac{1}{\xi}J'(\xi) - J(\xi) = 0.$$

Its solution is (see G. N. Watson)

(9) $$J(\xi) = K_0(\xi),$$

where $K_0(\xi)$, in the notation of the *British Mathematical Tables*, is the modified Bessel function of the second kind and of order zero, defined by

(10) $$K_0(\xi) = -(\gamma - \lg 2 + \lg \xi)\sum_0^\infty \left(\frac{\xi}{2}\right)^{2v}\frac{1}{v!v!} + \sum_1^\infty \left(\frac{\xi}{2}\right)^{2v}\frac{S_v}{v!v!}.$$

8.1.5 THE RANGE

The asymptotic distribution $\varphi(\xi)$ and the probability $\Phi(\xi)$ are, from (9), (7'), and (6),

(11) $$\varphi(\xi) = \xi\, K_0(\xi)\,;\qquad \Phi(\xi) = 1 - \xi K_1(\xi),$$

where $K_1(\xi)$ is the modified Bessel function of the second kind and of order 1, defined by

(12) $$K_1(\xi) = (\gamma - \lg 2 + \lg \xi) \sum_0^\infty \frac{1}{v!(v+1)!} \left(\frac{\xi}{2}\right)^{2v+1}$$
$$+ \frac{1}{\xi} - \sum_1^\infty \frac{1}{(v-1)!v!} \left(\frac{\xi}{2}\right)^{2v-1} \left(\sum_{\lambda=1}^v \frac{1}{\lambda} - \frac{1}{2v}\right)$$

One of its integral forms is

(12') $$\xi K_1(\xi) = \int_{-\infty}^{+\infty} \exp\left[y - e^y - e^{-y}\xi^2/4\right] dy$$

The two Bessel functions are related by

(13) $$\xi K'_1(\xi) = -K_1(\xi) - \xi K_0(\xi)\,;\qquad K_1(\xi) = -K'_0(\xi).$$

The second expression in (11) is immediately verified by differentiation. Numerical values of the distribution $\varphi(\xi)$ and the probability $\Phi(\xi)$ obtained from the *British Mathematical Tables* are given in Table 8.1.5. The mode $\breve{\xi}$, the median $\tilde{\xi}$, and the probability at the mode $\Phi(\breve{\xi})$ are found from the *British Mathematical Tables* to be

(14) $$\breve{\xi} = 0.595 < \tilde{\xi} = 1.257\,;\qquad \Phi(\breve{\xi}) = 0.214.$$

The moments may be obtained from the formula,

$$\int_0^\infty \xi^{l-1} K_0(\xi)\, d\xi = 2^{l-2} \Gamma^2(l/2),$$

given by Watson (p. 388). From (11) this formula may be written, after substitution of l for $l - 2$,

(15) $$\overline{\xi^l} = 2^l \Gamma^2(1 + l/2).$$

Therefore, the mean and standard deviation are

(16) $$\bar{\xi} = \pi/2\,;\qquad \sigma_\xi = 1.23798.$$

The probability at the mean obtained from the *British Mathematical Tables* is

(17) $$\Phi(\bar{\xi}) = .606.$$

The interval $\bar{\xi} \pm \sigma_\xi$ has a probability of about .58.

Exercises: 1) Verify the relation (13). 2) Calculate the first Betas.
Reference: Gumbel (1947b).

Table 8.1.5. Asymptotic Distribution and Probability of ξ

Variate ξ	Distribution $\varphi(\xi)$	Probability $\Phi(\xi)$
0.01	.04721	.00026
0.05	.15571	.00452
0.1	.24271	.01462
0.2	.35054	.04481
0.3	.41174	.08320
0.4	.44581	.12626
0.5	.46221	.17178
0.6	.46651	.21830
0.7	.46236	.26480
0.8	.45228	.31057
0.9	.43806	.35512
1.0	.42102	.39809
1.2	.38221	.47849
1.4	.34112	.55083
1.6	.30073	.61449
1.8	.26268	.67128
2.0	.22779	.72027
2.5	.15587	.81527
3.0	.10422	.87953
3.5	.06860	.92216
4.0	.04464	.95007
4.5	.02880	.96815
5	.01846	.97978
6	.00746	.99194
7	.00297	.99682
8	.00117	.99876
9	.00046	.99952
10	.00018	.99981

8.1.6. Asymptotic Distribution of the Range. The derivation of the asymptotic distributions 8.1.5(5) and 8.1.5(11) for ζ and ξ was distribution-free. Now the initial distribution is specified. We assume that it is symmetrical and of the exponential type. (Other types will be considered in 8.2.) In this case, from Table 5.1.3(1),

(1)
$$\lg F(x_1) = \alpha_n(x_1 + u_n) - \lg n ;$$
$$\lg (1 - F(x_n)) = -\alpha_n(x_n - u_n) - \lg n .$$

Therefore, the variate ζ' defined by 8.1.5(2),

$$\zeta' = \alpha_n(x_1 + x_n) ,$$

is the reduced midrange. From 8.1.5(5) it has the logistic distribution, as shown in 8.1.4.

Elfving has proven that the variate ξ defined by 8.1.5(2) converges, for the normal probability $F(x)$, toward

$$\xi^* = 2n\sqrt{1 - F(w/2)} .$$

8.1.6 THE RANGE 317

The variate ξ^* is thus a function of the probability transformation of the normal range. Since it depends upon the sample size, its distribution is not truly asymptotic, and since the transformation is not linear, this result cannot be used directly for observed normal ranges.

Instead of using Elfving's method, we construct the general meaning of the variate ξ for an initial symmetrical distribution of the exponential type. Then, from (1) and 8.1.5(2)

$$\xi^2 = 4e^{\alpha(x_1 - x_n + 2u_n)}.$$

The exponent is the reduced range R defined in 8.1.3(1). Therefore,

(2) $$\xi = 2e^{-R/2}.$$

If we designate the asymptotic probability and distribution functions of the reduced range by $\Psi(R)$ and $\psi(R)$, we obtain from 8.1.5(11) and the transformation (2), taking into account that ξ decreases with R,

(3) $$\Psi(R) = 2e^{-R/2} K_1(2e^{-R/2}) ; \qquad \psi(R) = 2e^{-R} K_0(2e^{-R/2}).$$

For clarity's sake these expressions are also written as sums of exponential functions of the reduced range. From the definition 8.1.5(10) and (12) of the Bessel Function,

(4) $$1 - \Psi(R) = \sum_{v=1}^{\infty} \frac{\exp(-Rv)}{v!(v-1)!} (R - 2\gamma + 2S_v - 1/v)$$

and

(5) $$\psi(R) = \sum_{v=0}^{\infty} \frac{\exp(-(v+1)R)}{v!v!} (R - 2\gamma + 2S_v).$$

with $S_v = \sum_1^v 1/\lambda$; $S_0 = 0$. From 8.15(10') and 8.15(12') it follows that

(3') $$\Psi(R) = \int_{-\infty}^{+\infty} \exp[y - e^y - e^{-y-R}] dy,$$

(3'') $$\psi(R) = e^{-R} \int_{-\infty}^{+\infty} \exp[-e^y - e^{-y-R}] dy$$

are the asymptotic probability and distribution functions of the range for symmetrical distributions of the exponential type.

(6) $\Psi(R) = \xi K_1(\xi) ; \qquad \Psi'(R) = \xi^2 K_0(\xi)/2$ and, from (2),

(7) $d\xi/dR = -\xi/2$

the next derivative is $\Psi''(R) = -\dfrac{\xi^2}{2} K_0(\xi) - \dfrac{3}{4} K_0'(\xi)$.

Therefore, $\Psi'''(R) + \Psi'(R) - e^{-R}\Psi(R) = -\dfrac{\xi^3}{4} (K_0'(\xi) + K_1(\xi))$.

The right side vanishes from 8.1.5(13). The resulting differential equation,

(8) $$\Psi'''(R) + \Psi'(R) - e^{-R}\Psi(R) = 0,$$

with the initial conditions,

$$\Psi(-\infty) = 0 \,; \quad \Psi(\infty) = 1 \,; \quad \Psi''(-\infty) = 0 \,; \quad \Psi''(\infty) = 0.$$

which is also of interest in the theory of wave propagation, was used by the Calculation and Ballistics Department of the Naval Proving Ground,

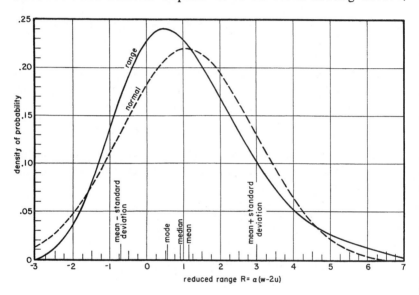

Graph 8.1.6(1). Asymptotic Distribution of the Range

Dahlgren, Virginia, to calculate the probability function $\Psi(R)$ and the distribution $\psi(R)$ with the help of the Special Relay Calculator of the IBM Corporation. The tables were published in 1947, and at greater length in 1949, and were recalculated by the National Bureau of Standards. The mode $\tilde{R}$ and the median $\check{R}$ reduced ranges are

$$\tilde{R} = 0.506366440 \,; \quad \check{R} = 0.928597642.$$

The distribution $\psi(R)$ is compared to the normal one with mean 2γ and standard deviation $\pi/\sqrt{3}$ in Graph 8.1.6(1). The probability function $\Psi(R)$ was shown in Graph 8.1.4, together with the probabilities of extremes. However, the different variates in this graph were not brought into a common scale. This is done in Graph 8.1.6(2) where all variates are standardized and normal probability paper is used.

8.1.6 THE RANGE

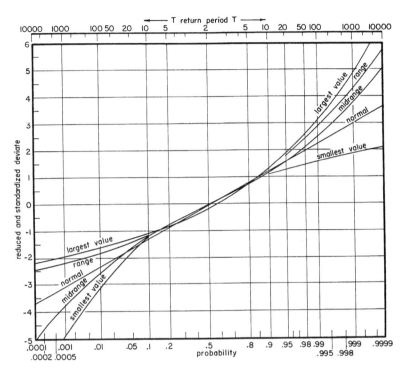

Graph 8.1.6(2). Asymptotic Probabilities of Standardized Extremes, Range and Midrange

Table 8.1.6. Characteristics of Reduced Extremes, Range, and Midrange

Characteristic	Largest Value	Smallest Value	Midrange	Range
Mode	0	0	0	0.50637
Expectation	$\gamma = 0.57722$	-0.57722	0	$2\gamma = 1.15443$
Median	$-\lg \lg 2 = 0.36651$	-0.36651	0	0.92860
Generating Function	$\Gamma(1-t)$	$\Gamma(1+t)$	$\Gamma(1-t)\Gamma(1+t)$	$\Gamma^2(1-t)$
Variance	$\pi^2/6$	$= 1.64493$	$\pi^2/3$	$= 3.28986$
Moment Quotient	$\beta_1 = 1.29857$	-1.29857	0	0.64928
	$\beta_2 = 5.4$	5.4	4.2	4.2
95% Probability Point	2.97020	1.09718	2.94444	4.4644
99% Probability Point	4.60015	1.52718	4.59512	6.4452
$F(\bar{x}+\sigma) - F(\bar{x}-\sigma)$	0.72373	0.72373	0.71963	0.7040
$F(\bar{x}+2\sigma) - F(\bar{x}-2\sigma)$	0.95705	0.95705	0.94821	0.9574

Table 8.1.6 summarizes the connection between reduced extremes, range, and midrange. The last two lines give the probabilities corresponding to the intervals about the respective means and one (two) respective standard deviations; the first probability is slightly larger than for the

normal distribution, where it is 0.68628; the second probability is about the same as in the normal case 0.95449.

Exercises: Calculate Elfving's distribution of ξ and ζ' for the normal and the logistic distributions. 2) Apply the method to the mth range and mth midrange. 3) Construct the distribution of the range for symmetrical distribution of the exponential type by convolution of the initial distribution of the largest and the smallest value. 4) Show that (5) leads to the generating function, 8.1.3(3). 5) Derive the differential equation (8) from the generating function, 8.1.2(3), (communicated by H. von Schelling). 6) Derive the differential equation (8) from the differential equation 8.1.5.(8). 7) Calculate the distribution of the mth reduced range. 8) Verify that the differential equation (8) is fulfilled by the explicit formulae (4) and (5). 9) Prove that the moments exist and are equal to the values given in 8.1.2. 10) Prove that the distribution of the midrange of the double exponential distribution for $n = 2$ is the asymptotic distribution of the range.

Problems: 1) Construct the convolution of three and, generally, n double exponential distributions. 2) Construct the convolution of two, or more, double exponential distributions, each truncated at zero. 3) Does an asymptotic distribution of the range exist for symmetrical distributions of the Cauchy type? 4) Find the order statistic with minimum variance. 5) Reduce the distribution of the mth range to the distribution of the range.

References: D. R. Cox (1948, 1949, 1954), Feller, Gartstein (1951), Gumbel (1949a), N. L. Johnson, Tukey (1955).

8.1.7. Boundary Conditions. We now investigate the analytic behavior and the order of magnitude of the probability $\Psi(R)$ and the distribution $\psi(R)$ for large negative and large positive reduced ranges, i.e., for large and small values of the positive variate ξ. If ξ is so large and R so large in the negative domain that

(1) $$\xi^{-3} = \tfrac{1}{8}e^{3R/2} \ll 1,$$

the expressions for $K_1(\xi)$ and $K_0(\xi)$ become (*British Mathematical Tables*, p. 271)

(2) $$K_1(\xi) = \sqrt{\frac{\pi}{2\xi}}\, e^{-\xi}\left(1 + \frac{3}{8\xi} - \frac{15}{128\xi^2}\right)$$

(3) $$K_0(\xi) = \sqrt{\frac{\pi}{2\xi}}\, e^{-\xi}\left(1 - \frac{1}{8\xi} + \frac{9}{128\xi^2}\right).$$

The probability $\Psi(R)$ becomes, from 8.1.6(6),

(4) $$\Psi(R) = \sqrt{\pi}\left(\exp\left[-\frac{R}{4} - 2e^{-R/2}\right]\right)\left(1 + \frac{3}{16}e^{R/2} - \frac{15}{512}e^{R}\right).$$

Condition (1) holds, say, for $R = -4$. In the same way, the distribution $\psi(R)$ becomes, for large negative reduced ranges,

(5) $$\psi(R) = \sqrt{\pi}\left(\exp\left[-\frac{3R}{4} - 2e^{-R/2}\right]\right)\left(1 - \frac{e^{R/2}}{16} + \frac{9\,e^R}{512}\right).$$

The probability $\Psi(R)$ and the distribution $\psi(R)$ practically vanish for $R = -4$. This removes the importance of the lower limit $R = -2\alpha u$ stated in 8.1.3(2). If $\alpha u \geq 2$, the distribution of the range may be dealt with as if it were unlimited toward the left.

We are interested to know how the probability $\Psi(R)$ converges to unity for large positive values of R. If R is so large that

(6) $$e^{-R} \ll 1,$$

we obtain, from 8.1.6(4) simply

(7) $$1 - \Psi(R) = e^{-R}(R - 2\gamma + 1).$$

Semilogarithmic paper shows a rapid approach of the probability functions $\Psi(R)$ and $1 - \Psi(R)$ toward exponential functions of R.

The distribution $\psi(R)$ for large reduced ranges obtained from (7) is

(8) $$\psi(R) = e^{-R}(R - 2\gamma).$$

From formulae (4) and (5), valid for large negative values of R, and from formulae (7) and (8), valid for large positive values of R, follow the boundary conditions,

(9) $$\lim_{R=-\infty} \frac{\psi(R)}{\Psi(R)} = e^{-R/2}; \quad \lim_{R=\infty} \frac{\psi(R)}{1 - \Psi(R)} = 1 - (R - 2\gamma + 1)^{-1}.$$

For increasing values of the reduced range, the intensity function increases toward unity, and the distribution of the largest range converges to the first asymptotic distribution. The range of the ranges is not a range, but a largest value.

Problem: Does the asymptotic distribution of the smallest range converge to the third type?

8.1.8. Extreme Ranges. We study now the extreme values of the range as a function of the number N of observed ranges, different from n, the sample sizes from which the ranges are taken and use the procedures introduced in 3.1.4 and 3.1.6.

We define two reduced ranges, R_N and R_1, by the equations

$$\Psi(R_N) = 1 - 1/N; \quad \Psi(R_1) = 1/N$$

or

(1) $$N(1 - \Psi(R_N)) = 1; \quad N\Psi(R_1) = 1.$$

Among N ranges, we may expect one range to be equal to or larger (smaller) than R_N (or R_1). The solutions of (1) are the characteristic largest (smallest) ranges corresponding to a given number N of samples, provided that each sample size n is sufficiently large to warrant the use of the asymptotic distribution, and that the initial distribution is symmetrical and of the exponential type. The functions of R_N and R_1 and other averages derived later are traced against the number N in Graph 8.1.8(1), which thus shows the characteristic largest and smallest ranges in N large samples. Since the

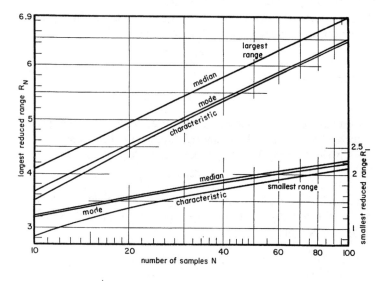

Graph 8.1.8(1). **Averages of Extreme Reduced Ranges**

distribution of the range is asymmetrical, and the skewness is positive, the characteristic largest range increases more quickly than the absolute amount of the characteristic smallest range.

The medians $\breve{R}_N$ and $\breve{R}_1$ of the largest and smallest ranges in N samples are the solutions of

$$\Psi(\breve{R}_N) = \exp[-(\lg 2)/N] = 1 - \Psi(\breve{R}_1),$$

whence

(2) $$\frac{1}{N} = \frac{-\lg \Psi(\breve{R}_N)}{\lg 2} = \frac{-\lg(1 - \Psi(\breve{R}_1))}{\lg 2}.$$

The results are also traced in Graph 8.1.8(1).

Finally we calculate the most probable extreme ranges. The probability

8.1.8   THE RANGE

at the most probable smallest range $\tilde{R}_1$ is, from 3.1.6(1), omitting the argument, the solution of

$$\frac{N-1}{1-\Psi}\psi = -\frac{\psi'}{\psi},$$

whence, from the differential equation 8.1.6(8),

(3) $$N - 1 = \frac{1-\Psi}{\psi}\left(\frac{\Psi e^{-R}}{\psi} - 1\right).$$

The most probable largest reduced range $\tilde{R}_N$ is obtained in the same way from 3.1.6 as the solution of

$$\frac{N-1}{\Psi}\psi = -\frac{\psi'}{\psi},$$

whence, from the differential equation 8.1.6(8),

(4) $$N - 1 = \frac{\Psi}{\psi}\left(1 - e^{-R}\frac{\Psi}{\psi}\right).$$

The results of equations (3) and (4) are given in Table 8.1.8. The modes of the largest (smallest) reduced ranges are also traced in Graph 8.1.8(1). The three averages of the extreme ranges as functions of the logarithm of the number of samples are nearly parallel straight lines.

Table 8.1.8. Modal Extreme Reduced Ranges

Reduced Range $\tilde{R}_1, \tilde{R}_N$	Probability $\Psi(R)$	Number of Ranges N
−2.25	.00695	117.1
−2.00	.01355	58.03
−1.75	.02431	31.13
−1.50	.04048	17.59
−1.25	.06315	10.88
4.0	.92867	12.81
4.5	.95136	19.12
5.0	.96721	28.75
5.5	.97810	43.50
6.0	.98549	66.19
6.5	.99045	100.2
7.0	.99375	155.6

Since the characteristic largest and smallest ranges increase in absolute value, the same holds for their differences, the range of the ranges, which is traced in Graph 8.1.8(2) against the number of ranges N. The largest range which approximately increases as the logarithm of N, increases much more quickly than the smallest range decreases. Therefore, the range of the ranges increases approximately as the logarithm of N.

Problem: Calculate the mean extreme ranges as a function of N.

8.1.9. Summary. For initial symmetrical distributions of the exponential type, the asymptotic distribution of the midrange is logistic and the asymptotic distribution of the reduced range is a Bessel function. The

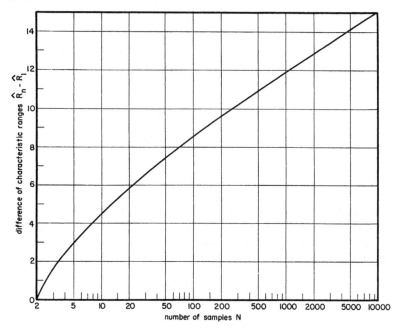

Graph 8.1.8(2). The Range of the Range

even moments of the range and midrange are equal. The distribution of the reduced range is independent of the initial distribution, and its derivation is not based on the assumption of normality. The initial distribution and the sample size influence the position and the shape of the distribution of the range in the same way as they influence the distribution of the largest value.

8.2. EXTREMAL QUOTIENT AND GEOMETRIC RANGE

8.2.0. Problems. For the exponential type, the variate ζ' in 8.1.5 is the reduced midrange and ξ is an exponential function of the reduced range. Both variates possess all moments. For the limited and for the Cauchy types, ζ' and ξ will lead to the extremal quotient $q = x_n/(-x_1)$ considered in 3.2.8 and to the geometric range $\rho = \sqrt{x_n(-x_1)}$. In both cases, only symmetrical initial distributions and asymptotic behavior will be considered. The extremal quotient for small samples has been studied in a previous publication (Gumbel and Herbach).

8.2.1. Definitions. If the initial symmetrical distribution is of the second (third) type, the initial probabilities of the extremes for large samples are, from Table 5.1.3(1), respectively,

(1) $\quad 1 - F(x_n) = \dfrac{1}{n}\left(\dfrac{v}{x_n}\right)^k; \qquad F(x_1) = \dfrac{1}{n}\left(\dfrac{v}{-x_1}\right)^k$

(1') $\quad 1 - F(x_n) = \dfrac{1}{n}\left(\dfrac{\omega - x_n}{\omega - v}\right)^k; \qquad F(x_1) = \dfrac{1}{n}\left(\dfrac{\omega + x_1}{\omega - v}\right)^k,$

where v stands for the characteristic extreme value and k denotes for the second type the order of the lowest moment which diverges. Since we deal with large samples, we may assume from 3.1.1(2) that $x_n > 0$; $x_1 < 0$. The two variates ζ' and ξ, defined in 8.1.5(2), become, respectively, for the second and third type of initial distributions,

(2) $\quad \zeta'_{(2)} = k \lg(-x_n/x_1); \qquad \xi_{(2)} = 2v^k(\sqrt{x_n(-x_1)})^k$

(2') $\quad \zeta'_{(3)} = k \lg \dfrac{\omega + x_1}{\omega - x_n}; \qquad \xi_{(3)} = 2\sqrt{(\omega - x_n)^k(\omega + x_1)^k/(\omega - v)^{2k}}.$

We now introduce two statistics,

(3) $\quad\quad\quad q \equiv x_n/(-x_1) \text{ and } \rho \equiv \sqrt{x_n(-x_1)}.$

The first, the *extremal quotient*, has no dimension. The second, called the *geometric range*, is a symmetrical function of the extremes, and has the same dimension as the variate. They may be obtained from the extremes of N samples, each consisting of a large number n of observations. From (3) the variates ζ' and ξ become for the second type

(4) $\quad\quad\quad \zeta'_{(2)} = -k \lg q; \qquad \xi_{(2)} = 2v^k/\rho^k,$

and for the third type, if the variate is counted from the limit ω,

(4') $\quad\quad\quad \zeta'_{(3)} = k \lg q = -\zeta'_{(2)}; \qquad \xi_{(3)} = 2\rho^k/v^k = 4/\xi_{(2)}.$

They are related to the variates for the second type by a reciprocal transformation and, for ξ, a change in scale. The reduced variates ζ' contain one parameter. The reduced variates ξ contain two parameters.

By virtue of 8.1.5(5), the logarithm of the extremal quotient for initial distributions of the Cauchy type (where no moments of an order equal to or larger than k exist), has the same logistic distribution as the midrange has for distributions of the exponential type (where all moments exist). The reason for this equality is the following: The logarithmic transformation, applied to the asymptotic distribution of the largest value for the exponential type, leads to the asymptotic distribution of the largest value of the Cauchy type.

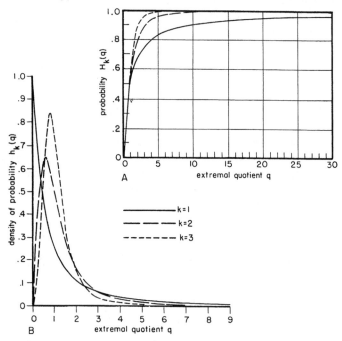

Graph 8.2.2. Asymptotic Probabilities and Distributions of the Extremal Quotient for the Cauchy Type

The analogy between $\zeta'_{(2)}$ and the reduced midrange v is

(5) $\qquad \zeta'_{(2)} = k(\lg z_n - \lg(-z_1)); \quad v = \alpha(x_n - (-x_1))$.

8.2.2. The Extremal Quotient. The distribution $h^{(2)}(q)$ and the probability $H^{(2)}(q)$ for the Cauchy type become from 8.1.5(5) after trivial transformations,

(1) $\qquad h^{(2)}(q) = kq^{k-1}(1 + q^k)^{-2}; \qquad H^{(2)}(q) = q^k(1 + q^k)^{-1}$.

This distribution is invariant under a reciprocal transformation. (See 1.1.2(4).) Therefore, it holds also for the third type, and the index 2 may be omitted. Graph 8.2.2 shows the distribution for $k = 1$, 2, and 3.

For $k \leq 1$, no mode exists, and the distribution diminishes with increasing q. For $k > 1$, the mode of the extremal quotient is, from (1),

(2) $$\tilde{q}^k = (k-1)/(k+1) < 1.$$

The probability density at the mode increases with k. The probability function has the symmetry

(3) $$H(1/q) = 1 - H(q).$$

The probabilities decrease with increasing k for $q < 1$ and increase for $q > 1$. Therefore, the distribution contracts with increasing values of k. The more moments that exist in the initial distribution, the more concentrated is the distribution of the extremal quotient and the smaller becomes the tail.

The median of the extremal quotient is unity. The density of probability,

(4) $$h(1) = k/4,$$

at the median increases with k. The larger k, the smaller is the distance from the median to the mode. The probability at the mode approaches $1/2$. The distribution $h(q)$ belongs to the Pareto Type and has no moments of an order equal to or greater than k.

For an estimate of the unique parameter k, we return to the logistic distribution. Its variance, from 8.1.4(2), is

(5) $$\sigma^2(\lg q) = \pi^2/3k^2.$$

This population value is replaced by the sample value, and k is estimated as

(6) $$\hat{k} = \pi/\sqrt{3} \cdot s(\lg q).$$

Exercises: 1) Calculate the characteristic largest extremal quotients for N samples. 2) Construct the distribution of the extremal quotient for the exponential type. (See Gumbel and Keeney.)

References: Curtiss, Krishna, Norris, Rietz (1936), Sakamoto.

8.2.3. The Geometric Range. For initial distributions of the third type, the geometric range ρ is, from 8.2.1(4'),

(1) $$\rho^k = v^k \xi_{(3)}/2: \qquad \rho = v(\xi_{(3)}/2)^{1/k}; \qquad \xi_{(3)} = 2(\rho/v)^k.$$

The probability $\Psi^{(3)}(\rho)$ and the distribution $\psi^{(3)}(\rho)$ obtained from 8.1.5(11),

(2) $$\Psi^{(3)}(\rho) = 1 - 2\rho^k K_1(2\rho^k/v^k)/v^k; \qquad \psi^{(3)}(\rho) = 4\rho^{2k-1} k K_0(2\rho^k/v^k)/v^{2k}.$$

contain two parameters v and k. The formulae 8.1.5(14) to (17) give the averages and moments for $k = 1$ and $v = 2$. The numerical values of

$1 - \Psi'^{(3)}(\rho)$ are given in Table 8.1.5 as functions of ξ. The corresponding values of ρ are obtained from (1). The probability is traced in Graph 8.2.3(1). For $\rho = v$, $\xi = 2$, the probability becomes $\Psi'^{(3)}(v) = .72027$. Therefore, v may be estimated as an order statistic. For the systematic

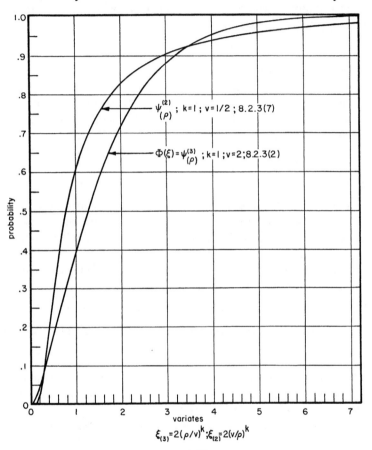

Graph 8.2.3(1). Asymptotic Probabilities $\Psi'^{(2)}(\xi)$ and $\psi^{(3)}(\xi)$ for the Geometric Range

estimation of the two parameters, we calculate the moments from (2). All moments exist. After trivial simplifications, we obtain from 8.1.5(15)

(3) $$\overline{\rho^l} = v^l \Gamma^2(1 + l/2k),$$

whence, for $l = 1$ and $l = 2$,

(4) $$\bar{\rho} = v\Gamma^2(1 + 1/2k); \qquad \overline{\rho^2} = v^2\Gamma^2(1 + 1/k).$$

8.2.3 THE RANGE

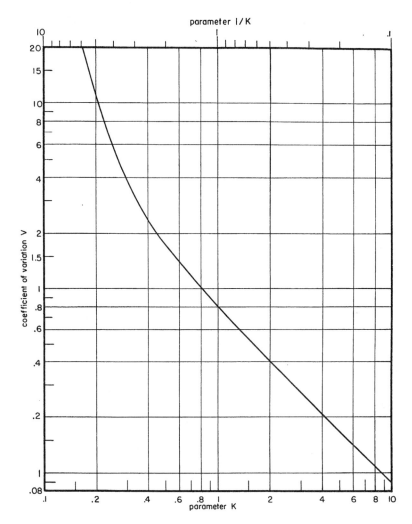

Graph 8.2.3(2). Estimation of k from the Coefficient of Variation of the Geometric Range

Therefore we may estimate the parameter k from the sample coefficient of variation with the help of

(5) $$1 + V^2 = \Gamma^2(1 + 1/k)/\Gamma^4(1 + 1/2k)$$

and v from the sample mean. Graph 8.2.3(2) gives V as a function of k on a double logarithmic scale.

The simplest way to compare observed geometric ranges with the theory is the use of a probability paper based on the linear relation,

(6) $$\lg \rho = \lg v - 1/k \cdot \lg 2 + 1/k \cdot \lg \xi,$$

obtained from (1). On such a paper ρ is plotted on the abscissa in logarithmic scale, and the corresponding values of $\Psi'^{(3)}(\rho)$, given in Table 8.1.5, are plotted on the ordinate.

For initial distributions of the Cauchy type, the geometric range ρ is, from (6),

(7) $$\rho^k = v^k 2/\xi_{(2)}; \quad \rho = v(2/\xi_{(2)})^{1/k}; \quad \xi_{(2)} = 2(v/\rho)^k.$$

The probability $\Psi'^{(2)}(\rho)$ and the distribution obtained from 8.1.5(11) after a reciprocal transforation,

(8) $$\Psi'^{(2)}(\rho) = 2(v/\rho)^k K_1(2(v/\rho)^k): \quad \psi^{(2)}(\rho) = 4kv^{2k}K_0(2(v/\rho)^k)/\rho^{2k+1},$$

contain two parameters v and k. For $\rho = v$, $\xi_{(2)} = 2$, the probability becomes $\Psi'^{(2)}(v) = .27973$. Therefore, v may be estimated as an order statistic. For a systematic estimation of the two parameters we calculate the reciprocal moments $\overline{\rho^{-1}}$ and $\overline{\rho^{-2}}$. Then v and k are obtained from (5) and (6) by substituting the inverse sample moments. The substitution of the inverse geometric range can also be used on probability paper.

The numerical values of ρ corresponding to the probabilities given in Table 8.1.5 are obtained after estimation of the parameters.

The transformations of the extremes x_n and x_1 introduced in 8.1.5(2) and the four extremal statistics R, V, ρ, q to which they lead for the three types of initial distributions are summarized in Table 8.2.3.

Table 8.2.3. *Transformations of Extremes to Extremal Statistics*

Transformations	Exponential Type	Distribution	Cauchy and Limited Type	Distribution
$\xi = 2n\sqrt{F_1(1-F_n)}$	Range R	Bessel Function	Geometric Range ρ	Bessel Function
$\zeta' = \lg [F_1/(1-F_n)]$	Midrange V	Logistic	Extremal Quotient q	8.2.2.(1)

Exercise: Calculate the first two Betas.

Problems: 1) Construct the generating function. 2) Construct the distribution of the geometric range for initial distributions of the exponential type. 3) Analyse the distribution (8). 4) Can general properties of the geometric range be obtained by the procedure used in 3.2.8?

References: Camp, Curtiss, Gumbel and Keeney, Gurland (1948), Rietz (1936).

8.3. APPLICATIONS

8.3.0. Problems. The use of the asymptotic distributions of the midrange and the range requires estimation of the parameters appropriate to the respective probability papers. Since the normal range has widely been studied for small samples, we are interested in knowing for what sample sizes the asymptotic distribution may be used.

8.3.1. The Midrange. As shown in 8.1.4(4), the midrange V taken from a symmetrical distribution of the exponential type has the logistic probability function. The linear relation to be traced on probability paper is

(1) $$v = \alpha(V - u),$$

where v is the reduced midrange. The methods shown in 1.2.8 lead to the estimations

(2) $$\hat{u} = \bar{v}: \quad 1/\hat{\alpha} = s/\sigma(v_N).$$

The population standard deviation $\sigma(v_N)$ as a function of the number of samples N given in Table 8.3.1(1) is obtained from

(3) $$\sigma^2(v_N) = \frac{1}{N} \sum_{m=1}^{N} v_m^2,$$

where the reduced values v_m are obtained from the plotting positions as

(4) $$v_m = \lg[m/(N - m + 1)].$$

Table 8.3.1(1). *Population Standard Deviation of the Midrange*

Sample Size N	Characteristic Largest Midrange $v_N = \lg N$	Standard Deviation $\sigma(v_N)\sqrt{3}/\pi$	Corresponding Reduced Value $\hat{v}_{2,N}$
9	2.19722	.71800	0.9346
19	2.94444	.81857	1.5067
24	3.17805	.84300	1.6876
39	3.66356	.88487	2.0395

Column 4 gives the probability points corresponding to the third column. As for the extreme values, $\hat{v}_{2,N}$ is approximately a linear function of v_N. (See Graph 8.3.1(1).) Therefore $\hat{v}_{2,N}$ may be obtained for other values of N by linear interpolation in Table 8.3.1(1). The values $\sigma(v_N)$ are obtained after multiplication of $\sigma(v_N)\sqrt{3}/\pi$ by $\pi/\sqrt{3} = 1.81380$.

Purcell observed $N = 32$ sets of $n = 5$ bulbs and measured the maxima and minima of their lives. Addition of these values leads to the midranges given in Table 8.3.1(2).

332 THE RANGE 8.3.1

*Table 8.3.1(2). Observed Midranges of Lives of Bulbs (in Ten Hours)**

136	241	260	307	355
153	244	266	311	364
183	246	270	323	386
212	247(2)	292	326	421
219	248	294(2)	334	425
220	252	295	353	471

* Parenthetical figures represent frequencies, different from unity.

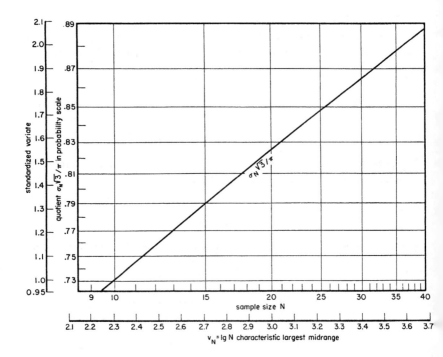

Graph 8.3.1(1). Population Standard Deviation of Midrange

The observations are traced in Graph 8.3.1(2) on logistic probability paper. The sample mean and standard deviations are $\overline{V}_0 = 287.34$; $s_V = 76.44$. For $N = 32$, linear interpolations in columns 2 and 3 of Table 8.3.1 lead to $\sigma_{32}\sqrt{3}/\pi = .8678$; $\sigma_{32} = 1.5740$. The line

$$V = 287.34 + 48.56v$$

gives a very good fit.

Exercises: 1) Calculate a table of $\sigma(v_N)\sqrt{3}/\pi$ for $N = 10(1)100$. 2) Construct the control curves as described in 2.1.6 and 6.1.6.

8.3.2 THE RANGE

Problems: 1) Why is $\hat{v}_{2,N}$ a linear function of v_N? (See the similar relations in 1.2.9.) 2) For what type of initial distributions do such linear relations hold?

Reference: Berkson (1944, 1951, 1957).

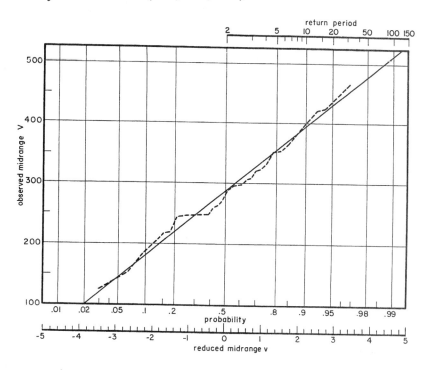

Graph 8.3.1(2). **Midranges of Lives of Bulbs Traced on Logistic Paper**

8.3.2. The Parameters in the Distribution of Range. The asymptotic distribution of the reduced range R is, of course, independent of the sample size, and parameter-free, while the distribution of the range w proper contains the parameters α_n and u_n which depend on the initial distribution. They influence the distribution of the range in the same way as the distribution of the extremes. An increase of the sample size increases the parameters u_n and shifts the distribution toward the right. If α_n increases (decreases) with n, the distribution of the range shrinks (expands) with increasing sample size. If α_n is independent of n, an increase of the sample size does not change the shape of the distribution. Only in the first case may we increase the precision by increasing the sample size from which the ranges were taken.

If the initial distribution and the sample size n are known, the two parameters u_n and $1/\alpha_n$ may be obtained from the definitions. However, in many practical applications the initial distribution, or the parameters it contains, are unknown, or the initial distribution might be known, but the number of observations is insufficient to warrant this procedure, because the most probable largest value still differs from the characteristic value u_n. In these cases the parameters now written α and u have to be estimated from the observed ranges alone.

For the comparison between theory and observations, and the estimation of the parameters, we may use a probability paper, constructed from the National Bureau of Standards *Probability Tables* (*Table* 4, *Range*). The observed ranges w are plotted as ordinate and the reduced ranges R as abscissa respectively, both in linear scales. The abscissa shows the probabilities $\Psi(R)$ corresponding to reduced ranges R. If there is a constant interval of time between the observations, this interval may be used as a unit for the return period $T(R)$ and $_1T(R)$ of the ranges above (and below) the median, traced on a parallel to the probability scale. The probability paper does not assume normality, but only symmetry of the initial distribution of the exponential type. [See Graphs 8.3.3(1) and 8.3.5.]

For the estimation of the parameters α and u in the line

(1) $$w = 2u + R/\alpha,$$

we use the procedure explained in 1.2.8, which leads to

(2) $$1/\alpha = s_w/\sigma(R_N); \quad 2u = \bar{w}_0 - \bar{R}_N/\alpha.$$

Here $\bar{w}_0$ and s_w stand for the sample mean and standard deviation. The population values $\bar{R}_N$ and $\sigma(R_N)$ calculated from the Bureau of Standards *Table*, are given for some values of N in Table 8.3.2.

Table 8.3.2. Population Mean and Standard Deviation of the Range

N	$\bar{R}_N$	$\tilde{R}_N$	$\tilde{R}_N/\bar{R}_N$	$\hat{R}_{1,N}$	$\sigma(R_N)$	$\sigma(R_N)/\sigma_R$	$\hat{R}_{2,N}$
9	3.5540	1.0627	0.92042	3.8541	1.3342	.74687	2.1731
19	4.4644	1.0962	0.94958	4.4540	1.5270	.84218	2.8976
24	4.7496	1.1045	0.95678	4.6501	1.5679	.86474	3.1211
49	5.6112	1.1241	0.97371	5.2754	1.6626	.91699	3.7741

In this table, R_N, $\hat{R}_{1N}$, and $\hat{R}_{2N}$ stand for the characteristic largest value and the solutions of

$$\Psi(R_N) = 1 - 1/N; \quad \Psi(\hat{R}_{1,N}) = \bar{R}_N/2\gamma; \quad \Psi(R_{2,N}) = \sigma(R_N)/\sigma_R.$$

In Graph 8.3.2, we plot R_N on the abscissa and write N at its corresponding place. The ordinate gives $\bar{R}_N/2\gamma$ and $\sigma(R_N)3/\sqrt{\pi}$, which have to be multiplied by 1.15443 and 1.81380, respectively, to obtain $\bar{R}_N$ and $\sigma(R_N)$.

8.3.2 THE RANGE

Since the relations between $\hat{R}_{1,N}$, $\hat{R}_{2,N}$, and R_N are approximately linear, values of $\bar{R}_N$ and $\sigma(R_N)$ for other values of N than those given in the table may be obtained by linear interpolation.

If the observed ranges plotted on the probability paper are scattered

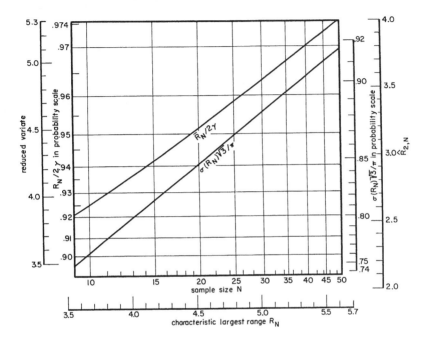

Graph 8.3.2. Population Mean Range and Standard Deviation as Function of Sample Size N

about the straight line (1), they conform to a certain pattern of regularity. If not, we may either pick up those ranges which seem doubtful, or we may conclude that the ranges do not come from a symmetrical distribution of the exponential type, or that the inital standard deviation or the initial mean have undergone a change, or that the sample size n was insufficient, or that it did not remain constant. Which of these reasons hold can only be decided from the way in which the observation actually arose.

Exercises: 1) Prove that the variance of the mean is smaller than the variance of the median. 2) Estimate the parameter u from the median. 3) Show that the linear relations of Graph 8.3.2. hold approximately from $N \geq 3$ onward. 4) Calculate the order statistic with minimum variance.

Problems: 1) Calculate $\bar{R}_N$ and $\sigma(R_N)$, as functions of N for $N = 10(1)$, 100. 2) What are the standard errors of u estimated from the mean or the median? 3) Express $1/\alpha$ by the sample mean deviation. What is the variance of this estimate compared to the variance of the estimate of $1/\alpha$ from s_w? 4) Estimate the two parameters $1/\alpha$ and u by the method of maximum likelihood. 5) Why are $\bar{R}_n/2\gamma$ and $\sigma(R_N)\sqrt{3}/\pi$ plotted on the probability scale approximately linear functions of the characteristic largest range R_N? (See the similar problems in 1.2.9, 6.3.2, and 8.3.1.)

8.3.3. Normal Ranges. The probability function of the normal range was calculated by E. S. Pearson and H. O. Hartley for $n = 2(1)20$. To check how close they are to the asymptotic values, the probability functions for $n = 5, 10, 15$, and 20 are traced on the probability paper in Graph 8.3.3(1). The curves are flat, practically straight within the interval $.1 < \Psi < .9$.

To obtain the asymptotic probability of normal ranges, we have to estimate the parameters. The mean range $\bar{w}_n$ as a function of the sample size n is obtained from 8.3.2(1) as

(1) $$\bar{w}_n = 2u_n + \bar{R}_n/\alpha_n.$$

If u_n and α_n are taken from Table 4.2.2(1), the asymptotic values of $\bar{w}_n$ are for $n = 10, 100, 1000$, as follows: 3.1583, 5.0448, 6.4906, while the exact values calculated by Tippett are 3.0775, 5.0152, 6.4829. Therefore, (1) can certainly be used for these and larger values of n. However, the asymptotic standard deviation of the normal range,

(2) $$\sigma(w_n) = \pi/(\sqrt{3}\alpha_n),$$

obtained from the values of α_n in Table 4.2.3(1) exceeds by far the calculated values as shown in Graph 8.3.3(2). Therefore, the parameters are estimated from the calculated values $\bar{w}_n$ and $\sigma(w_N)$ as

(3) $$1/\hat{\alpha}_n = 0.5553\,\sigma(w_n); \quad 2\hat{u}_n = \bar{w}_n - 1.15443/\hat{\alpha}_n.$$

These values are given in Table 8.3.4, columns 5 and 6. The estimate $2\hat{u}_n$ is practically identical with the values given in Table 4.2.2(1).

The theoretical probabilities obtained from

(4) $$w_n = 2u_n + R/\alpha_n$$

are traced for $n = 20, 50, 100, 200, 500, 1000$ in Graph 8.3.3(1). The fit for $n = 20$ is satisfactory within the interval $.10 < \Psi < .95$. For $n = 200$ (Harley and Pearson) the asymptotic theory holds for a wider interval $.05 < \Psi < .97$. Consequently, the asymptotic theory can safely be used for a small number of observed normal ranges.

The median normal range,

(5) $$\breve{w}_n = 2u_n + 0.9286/\alpha_n,$$

8.3.3 THE RANGE

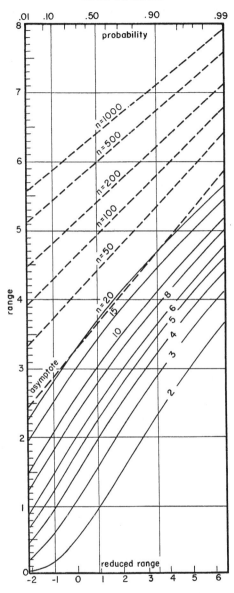

Graph 8.3.3(1). Probabilities of Normal Ranges

as function of the sample size n is given in Table 8.3.4, column 7, and traced in Graph 8.3.3(2). The asymptotic values given by (5) differ only slightly from the exact values, which may be obtained by interpolation

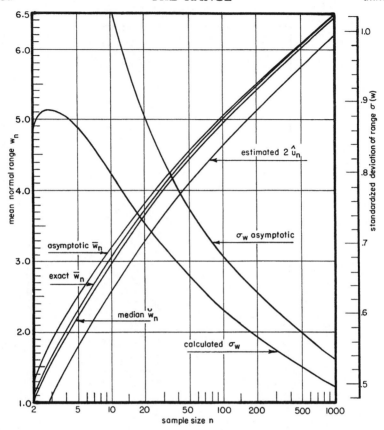

Graph 8.3.3(2). **Characteristics of the Normal Range**

from Pearson and Hartley's table. For $n = 20$, the difference is about 1%. The mth normal ranges have been studied by Cadwell (1953b).

Problem: How does the characteristic largest normal range for samples of size $n = 3, 5, 7, 10$ increase with the number of samples?

References: Bannerjee, Belz and Hooke (1954), Cadwell (1953a), Cyffers and Vessereau, David (1953), Harley and Pearson, Hartley and Pearson (1951), Lavin, E. S. Pearson (1926, 1932), Pearson and Hartley (1942, 1943), Shimada, "Student," Tukey (1955). *Consult also:* Tukey (1956). A rejection criterion based upon the range, Biometrika, 43: 418.

8.3.4. Estimation of Initial Standard Deviation. Observed ranges are often used to estimate the unknown initial standard deviation. This procedure requires the knowledge of the initial distribution and of the sample

8.3.4   THE RANGE

size. In quality control the sample size n is small and the initial distribution is (in many cases wrongly) assumed to be normal. The observed range W_n taken from a sample of size n, or the mean $\bar{W}_n$ of N observed ranges W_v, each taken from samples of size n, are divided by the corresponding theoretical mean $\bar{w}_n$ of normal ranges valid for unit standard deviation. These quotients give consistent estimates of the initial standard deviations,

(1) $$s = W_n/w_n, \qquad s' = \bar{W}_n/w_n.$$

Since the expectation of W_n is $\bar{w}_n \sigma$, the standard errors of (1) are

(2) $$\sigma(s) = \frac{\sigma(w_n)}{\bar{w}_n} \sigma : \qquad \sigma(s') = \frac{\sigma(w_n)}{\bar{w}_n} \frac{\sigma}{\sqrt{N}}.$$

The mean normal range was calculated by Tippett for $n = 2(1)1000$, and the standard deviation was calculated by Hartley and Pearson for selected values, given in Table 8.3.4.

The second expression in (2) is a function of $n \cdot N$, the total number of observations. If this value is kept constant, the question arises whether it is better to take N large and n small, or vice versa. Since $\alpha_n u_n$ increases only as a function of $\lg n$, it is asymptotically preferable to increase the number of samples N, and to take samples of small size n. For the normal distribution, the standard errors $\sigma(s')$, obtained from Table 8.3.4, are basic for the choice of N. The combination of n and N which minimizes the standard error for given values of the product nN, leads to the following rules of action: Up to twelve observations and, of course, for all prime numbers, take one sample. For 12, 14, 16, 20, 22 observations take $N = 2$ samples. For 15, 18, 21, and 24 observations take $N = 3$ samples. Finally, for 25 observations take 5 samples.

The standard error of the estimate (2) of σ from $\bar{W}_n$ becomes asymptotically

(3) $$\sigma(s') = \frac{\pi}{2\alpha_n u_n \sqrt{3N}}.$$

Since $\alpha_n u_n$ increases with n (see 4.1.7), the standard error of the estimate of the standard deviation from the range converges toward zero for increasing n and N. For the normal distribution (3) becomes, from 4.2.3(6) and 4.2.3 (11),

(4) $$\sigma(s') = \frac{\pi}{4\sqrt{3N} \lg(n/\sqrt{2\pi})}.$$

Since this expression decreases with N for constant values of $n \cdot N$, we have to take many samples of small size n in order to minimize the standard error of the estimate of the initial standard deviation for a fixed number of observations. This asymptotic result had been stated by Tippett in 1925.

References: Baker, Cohan, Daly, Davies, Deming, Grubbs and Weaver, Hartley (1942) A. E. Jones, Lord (1947, 1950), E. S. Pearson (1932), Winston.

Table 8.3.4. Characteristics for Normal Ranges

1 Sample Size	2 Mean	3 Standard Deviation	4 Coefficient of Variation	5 Parameters	6	7 Median
n	$\bar{w}_n$	$\sigma_{w,n}$	$\sigma_{w,n}/\bar{w}_n$	$1/\alpha_n$	$2\hat{u}_n$	$\breve{w}_n$
2	1.12838	.8525	.7555	.4702	0.5856	1.022
3	1.69257	.8884	.5247	.4890	1.128	1.582
4	2.05875	.8798	.4273	.4852	1.499	1.950
5	2.32593	.8641	.3715	.4766	1.776	2.219
6	2.53441	.8480	.3346	.4677	1.994	2.428
7	2.70436	.8332	.3081	.4595	2.174	2.600
8	2.84720	.8198	.2880	.4521	2.325	2.745
9	2.97003	.8078	.2720	.4455	2.455	2.869
10	3.07751	.7971	.2589	.4396	2.571	2.979
11	3.17287	.7873	.2481	.4342	2.672	3.075
12	3.25846	.7785	.2389	.4294	2.763	3.161
13	3.35598	.7704	.2296	.4249	2.865	3.260
14	3.40676	.7630	.2240	.4208	2.921	3.312
15	3.47183	.7562	.2178	.4171	2.990	3.377
16	3.53198	.7499	.2123	.4136	3.055	3.439
17	3.58788	.7441	.2074	.4104	3.114	3.495
18	3.64006	.7386	.2029	.4074	3.170	3.548
19	3.68896	.7335	.1988	.4045	3.222	3.598
20	3.73495	.7287	.1951	.4019	3.271	3.644
50	4.49815	.653	.145	.360	4.082	4.416
60	4.63856	.639	.138	.352	4.233	4.560
100	5.01519	.605	.121	.334	4.620	4.939
200	5.49209	.566	.103	.312	5.132	5.422
500	6.07340	.524	.0863	.289	5.739	6.007
1000	6.48287	.497	.0767	.274	6.167	6.421

8.3.5. Climatological Examples. In the following, we give some climatological examples. The temperature at which the air is saturated by containing the maximum quantity of vapor is called the dew point. The frequencies of the range of dew points in January and July in Washington, D.C., 1905–45, are given in Table 8.3.5(1), and plotted in Graph 8.3.5(1), together with the theoretical values obtained from Table 8.3.5(4). The increase in time may be seen from the return period scale. The range of the dew points for the ten months not given here show the same good fit as for the two months traced in Graph 8.3.5(1).

B. J. Birkeland and G. Frogner compared the largest annual ranges of temperatures observed in Bergen, Norway, within N years with theoretical values obtained from the assumption that the corresponding initial distribution is normal. Since the asymptotic distribution of normal ranges was unknown at that time, the authors introduced artificial devices to overcome

8.3.5 THE RANGE

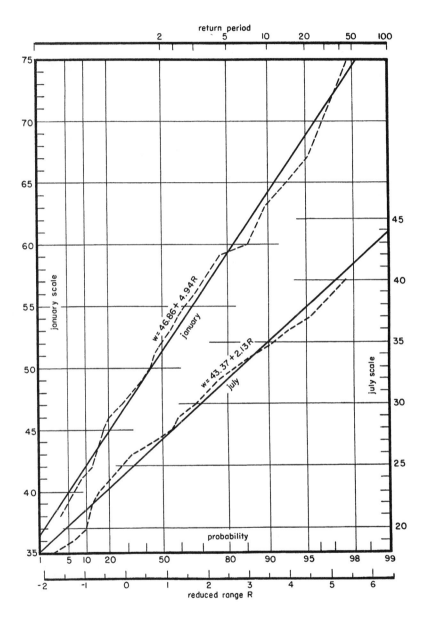

Graph 8.3.5(1). Range of Dew Points: Washington, D.C., 1905–45

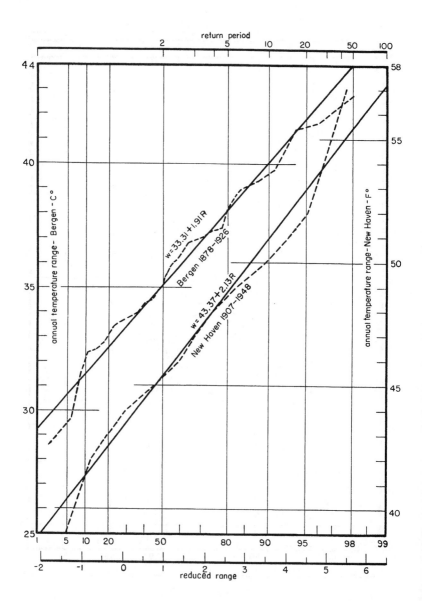

Graph 8.3.5(2). Annual Range of Temperatures: Bergen, Norway, and New Haven, Connecticut

8.3.5 THE RANGE

this lack. The values for the median of the range as a function of the number of samples (i.e., the years) given by the authors, do not differ too much from the values obtained from the asymptotic theory traced in Graph 8.3.3(2). The $N = 49$ observed yearly ranges, the differences of the mean monthly temperatures of the warmest and of the coldest month of a year in Bergen, 1878–1926, given in Table 8.3.5(2) are traced in Graph 8.3.5(2), together with the theoretical values obtained from Table 8.3.5(4).

H. L. Landsberg gives (Table 8.3.5(3)) the annual temperature range in °F for New Haven, Connecticut, traced in Graph 8.3.5(2). The fit of the theoretical values is better than could be expected, since the temperatures at the observed places are not symmetrically distributed over a year. From a comparison with older records, Landsberg concludes that the mean range has decreased, i.e., that the climate has become less continental.

8.3.5(1). Frequencies of Dew Points, Washington, D.C., 1905–1945

	July				January			
F°	Frequency	F°	Frequency	F°	Frequency	F°	Frequency	F°	Frequency
18	1	28	1	38	2	51	1	60	1
19	2	29	2	41	2	52	2	63	2
20	1	30	5	42	2	53	2	65	1
22	1	32	5	45	1	54	1	67	1
23	2	34	2	46	2	55	3	75	1
24	2	35	1	47	3	56	1		
25	3	36	1	48	2	57	2		
26	2	37	1	49	3	58	3		
27	8	40	1	50	1	59	2		

Table 8.3.5(2). Annual Range of Temperature in C°, Bergen, Norway, 1878–1926*

28.6	33.1	34.5	36.4(2)	38.8(2)
29.3	33.4(2)	34.7	36.8	39.0
29.7	33.6	34.9	36.9	39.2
31.7(2)	33.7(2)	35.0	37.0(2)	39.7(2)
32.4	33.8	35.4	37.1	41.4
32.5	33.9(2)	35.8	37.3(2)	41.6
32.6	34.1	36.0	37.4	42.8
32.8	34.4(2)	36.1	38.2	

* Parenthetical figures represent frequency.

Table 8.3.5(3). Annual Range of Temperature in F°, New Haven, Connecticut, 1907–1948

Range	Frequency	Range	Frequency
39	2	47	3
40	1	48	3
42	3	49	4
43	3	50	2
44	6	51	1
45	8	52	1
46	4	57	1

Table 8.3.5(4). Estimation of the Parameters for the Four Ranges

	July Dew Points, Washington, D.C.	Jan. Dew Points, Washington, D.C.	Temperature, New Haven, Connecticut	Temperature, Bergen, Norway
Table	8.3.5(1)	8.3.5(1)	8.3.5(3)	8.3.5(2)
N	41	41	42	49
R_N	5.3948	5.3948	5.4266	5.6112
$\bar{R}_N/R_N$	.9693	.9693	.9699	.9737
$\bar{R}_N$	1.1190	1.1190	1.1197	1.1241
$\sigma(R_N)/\sigma(R)$	.9034	.9034	.9052	.9170
$\sigma(R_N)$	1.6388	1.6386	1.6419	1.6626
$\bar{w}_0$	27.976	52.390	45.762	35.455
s_w	4.997	8.096	3.505	3.172
$1/\hat{\alpha}$	3.049	4.939	2.134	1.908
$2\hat{u}$	24.563	46.862	43.372	33.310

SUMMARY

"A book should either have intelligibility or correctness. To combine the two is impossible."

THE DISTRIBUTIONS of extreme values are derived here either from a specific initial distribution or from a certain type. This genetic procedure is opposed to the empirical methods used by many engineers and practical statisticians, who are inclined to believe that, after all, nearly everything should be normal, and whatever turns out not to be so can be made normal by a logarithmic transformation. This is neither practical nor true. However, some extreme values show certain similarities to these distributions. With the noteworthy exception of the cases considered in 4.2.7, the normal distribution plays no significant role within this theory.

The different transitions from the exact to the asymptotic distributions of exceedances, extremes, and extremal statistics show certain parallelisms and differences with the classical transitions. For the number of exceedances, the rank of the variate plays a role similar to the probability p in the binomial distribution. The normal and Poisson's distributions are reached under conditions similar to those in the binomial case.

For the extremes, three asymptotic stable distributions exist. This implies the existence of three asymptotic distributions of mth extremes. The parallelism to the binomial case is established by the fact that the asymptotic distributions of the mth extremes of the exponential type converge, with increasing m, toward normality.

The third group of transitions concerns the sum and the differences of extremes. This sum, taken from a symmetrical distribution, converges toward the logistic, not the normal, distribution. The difference of the extremes leads to new asymptotes. The convergence toward normality holds for the exponential type if, instead of the sum (or the difference) of the extremes, the sum (or the difference) of the mth extremes is taken, and m becomes large.

In contrast to usual statistics, where the mean plays a dominant role, the theory of extreme values is centered about the characteristic extreme values. In the asymptotic theory, the mode is the most important average, since it enters the distribution directly. Some asymptotic distributions of the extremes, mth extremes, and extremal statistics can be written in a form which is parameter-free. As soon as this theory is applied to observations, two parameters enter for unlimited initial distributions. They are functions of the initial distribution and the sample size n from which the extremes were taken, and depend exclusively upon the way in which the initial probability approaches unity. The parameters are thus independent of the rest of the distribution. The first parameter is the characteristic largest value. The second one is the intensity function at this value for the exponential type, or the order of moments that diverge for the Cauchy type. For limited initial distributions, the limit itself is one of the parameters in the asymptotic theory.

Estimation of the parameters is studied for the three asymptotic distributions of the extremes, and four asymptotic distributions of extremal statistics. Maximum likelihood estimation leads to a lengthy procedure requiring successive approximations. For the time being, it was necessary and sufficient to show that, and how, the parameters can be estimated. The author hopes that further researches will improve his methods. Probability papers were developed which give a rapid check as to whether a series of observations can be interpreted from one of the theories here expounded.

No theory can ever hope to explain all observations to which it is allegedly related. In the acceptance or rejection of a statistical theory as an explanation of observed phenomena, there is always an arbitrary probability judgment involved. Such well-known limitations need not be stressed. However, it is necessary to point out the specific limitations of the asymptotic theories:

1) The observations from which the extreme values are drawn ought to be independent. This condition may be met in an experimental setup. However, it is seldom met in natural observations. Still, the asymptotic theory gives a very good fit for such observations, because it is only the asymptotic behaviour of the initial distribution which counts.

2) The observations must be reliable and be made under identical conditions. The initial distribution and the parameter it contains must be the same for each sample. This condition is not necessary in Fréchet's procedure. It may be partly eliminated by considering not the extremes themselves, but their differences from the average of each sample.

3) The number of observations n from which the extremes are taken must be large. How large n must be depends upon the initial distribution, and the degree of precision we seek. Unfortunately, we are not always free in the choice of the sample size. In meteorology, for instance, the day and the year are natural units of periodicity, and the choice of $n = 365$ is imposed upon us.

4) The initial distribution from which the extreme values are taken must belong to one of the three described types. In reality, the analytic properties of the initial distribution are rarely known, except in artificial

examples. However, the distinctions among the three types of initial distributions are so clear that an assumption in this respect can be made.

The conditions for the use of the asymptotic theory of extreme values are not always met by the observations. This fact, and lack of knowledge about the initial distribution, must lead to certain discrepancies between the theory and the available observations, which cannot be considered as a failure of the theory.

BIBLIOGRAPHY

Afanassiev, N. N. 1940. Statistical theory of the endurance of fatigued metals, J. Tech. Phys. (USSR), 10:1553 (Russian).

Altman, Landy B., jr., and Emil H. Jebe. 1955. Load characteristics of southeastern Iowa farms using electric ranges, Agric. Exp. Station, Iowa State College, Res. Bull. 420. Ames, Iowa.

American Society of Civil Engineers. 1949. Hydrology Handbook (Manual No. 28). New York.

———— 1953. Review of flood frequency methods. Trans., paper 2574.

Anderson, T. W., and D. A. Darling. 1952. Asymptotic theory of certain "Goodness of Fit" criteria based on stochastic processes, Ann. Math. Stats., 23:193.

———— 1954. A test of goodness of fit, J. Am. Stat. Assn., 49:765.

Anis, A. A. 1956. On the moments of the maximum of partial sums of a finite number of independent normal variates, Biometrika, 43:79.

Anonymous. 1934. Study of weather records fails to reveal long-range cycles, Eng. News-Record, 113:46.

Anscombe, F. J. 1956. On estimating binomial response relations, Biometrika, 43:461.

Arsdel, W. B. Van. 1922. A method for the calculation of normal frost data from short temperature records, Monthly Weather Rev., 50:297.

Bachelier, L. 1912. Calcul des probabilités. Paris, Gauthier-Villars.

———— 1937. Les Lois des grands nombres du calcul des probabilités. Paris, Gauthier-Villars.

———— 1938. La Spéculation et le calcul des probabilités. Paris, Gauthier-Villars. p. 43.

———— 1939. Les Nouvelles Méthodes du calcul des probabilités. Paris, Gauthier-Villars. p. 42.

Baird, Ralph W., and Wm. D. Potter. 1950. Rates and Amounts of Runoff for the Blacklands of Texas. U.S. Dept. Agric., Tech. Bull. 1022.

Baker, G. A. 1946. Distribution of the ratio of sample range to sample standard deviation for normal and combinations of normal distributions, Ann. Math. Stats., 17:366.

Balaca, A. P. 1953. La determinacion del coefficiente de seguridad en las distentas obras. Madrid, Instituto tecnico de la construccion y del cemento.

Bannerjee, D. P. 1952. On the distribution of the range of variation of the ordered variates in samples of n from normal universe, Proc. Indian Acad. Sci., A, 35:24.

Barnard, G. A. 1944. An Analogue of Tchebycheff's Inequality in Terms of Range. British Ministry of Supply Advisory Service on Statistical Method and Quality Control, Tech. Rept. QC/R/11.

Barricelli, N. A. 1943. Les plus grands et les plus petits maxima ou minima annuels d'une variable climatique, Arkif for Mathematik og Naturvidenskab, Oslo, 46 (No. 6).

Barrow, H. K. 1948. Floods. New York, McGraw-Hill.

Bartholomew, D. J. 1954. Note on the use of Sherman's statistic as a test for randomness, Biometrika, 41:556.

———— 1957a. Testing for departure from the exponential distribution, Biometrika, 44:253.

———— 1957b. A problem in life testing, J. Am. Stat. Assn., 52:350.

Barton, D. E., and F. N. David. 1956. Tests for randomness of points on a line, Biometrika, 43:104.

Bastenaire, F. 1956. Sur une propriété des distributions statistiques des durées de vie à la fatigue, Compt. Rend. Ac. des Sc., 243:1270.

—— 1957. Distributions statistiques des durées de vie à la fatigue et forme de la courbe de Wöhler, Compt. Rend. Ac. des Sc., 245:136.
Baur, F. 1944. Ueber die grundsätzliche Möglichkeit langfristiger Witterungsvorhersagen, Ann. Hydrogr., 72:15.
Beard, L. R. 1942. Statistical analysis in hydrology, Proc. Am. Soc. Civ. Eng., 68:1077.
—— 1951. Flood Frequency Analyses. Office of Chief of Engineers, Civil Works Bull. 51.
Bednarski, E. J. 1943. Discussion on statistical analysis in hydrology, Proc. Am. Civ. Eng. Soc., 69:299.
Belz, M. H., and R. Hooke. 1953. Approximate distribution of extreme values of the range, Ann. Math. Stats., 24:143.
—— 1954. Approximate distribution of the range in the neighborhood of low percentage points, J. Am. Stat. Assn., 49:620.
Benard, A., and E. C. Bos-Levenbach. 1953. Het uitzetten van waarnemingen op waarschijnlijkheidspapier, Statistica, 7:163.
Bendersky, A. M. 1952. On the distribution of the absolute value of the maximum deviation from the mean in a series of observations, Doklady Akad. Nauk SSSR (N.S.), 85:5 (Russian).
Benham, A. D. 1950. The estimation of extreme flood discharges by statistical methods. Proc., New Zealand Institution of Engineers, 36:119.
Bennett, B. M. 1955. On the joint distribution of the mean and standard deviation, Ann. Inst. Stat. Math. (Tokyo) 7:63.
Benson, M. A. 1950. Use of historic data in flood frequency analysis, Trans. Am. Geophys. Union, 31:419.
Berkson, Joseph. 1944. Application of the logistic function to bio-assay, J. Am. Stat. Assn., 39:357.
—— 1951. Why I prefer logits to probits, Biometrics, 7:327.
—— 1957. Tables for the maximum likelihood estimate of the logistic function, Biometrics, 13:28.
Bernier, J. 1956. Sur l'application des diverses lois limites des valeurs extrêmes au problème des débits de crue, Houille Blanche, No. 5:718.
Bertrand, I. 1907. Calcul des probabilités. Paris, Gauthiers-Villars.
Birkeland, B. J., and E. Frogner. 1935. Die extreme Variabilitaet der Lufttemperatur, Meteorol. Zeitschrift, 52:349.
Birnbaum, Z. W. 1952. Numerical tabulation of the distribution of Kolmogorov's statistics for finite sample size, J. Amer. Stat. Assn., 47:425.
Birnbaum, Z. W., and F. H. Tingey. 1951. One-sided confidence contours in probability distribution functions, Ann. Math. Stats., 22:592.
Birnbaum, Z. W., and H. S. Zuckerman. 1949. A graphical determination of sample size for Wilks' tolerance limits, Ann. Math. Stats., 20:313.
Bliss, C. L., W. G. Cochran, and J. W. Tukey. 1956. A rejection criterion based upon the range, Biometrika, 43:418.
Boehm, Carl. 1953. Zwei Sterblichkeitsuntersuchungen. In: Festschrift: 100 Jahre Victoria-Versicherung, Berlin.
Bonnet, M. 1953. Étude expérimentale de la qualité du béton, Ann. Ponts et Chaussées, 123:185; 287.
Borel, E. 1952. Valeur pratique et philosophie des probabilités. Paris, Gauthiers-Villars.
Bortkiewicz, L. von. 1898. Das Gesetz der kleinen Zahlen. Leipzig, Teubner.
—— 1922a. Variationsbreite und mittlerer Fehler, Sitzungsberichte d. Berliner Math. Ges., 21:3.
—— 1922b. Die Variationsbreite beim Gauss'schen Fehlergesetz, Nordisk Statistik Tidskrift, 1 (No. 1):11; (No. 2):13.
Botts, Ralph R. 1957. "Extreme-value" methods simplified, Agric. Econ. Research, U.S. Dept. Agric., 19:88.
Boyer, M. 1952. Essai d'étude statistique des débits du Rhin à Strasbourg (1881–1951), Houille Blanche, 7:130.

BIBLIOGRAPHY 351

Brakensick, D. L., and A. W. Zingg. 1957. Application of the extreme value statistical distribution to annual precipitation and crop yields, ARS 41–13, U.S. Dept. Agric., Washington.

Breny, H. 1952. Quelques considérations sur la théorie statistique des faisceaux de fibres, Math. Centrum Amsterdam, Rapport S.96, p. 11.

British Association for the Advancement of Science. 1937. Mathematical Tables vol. 6, Bessel Functions, part 1: Functions of Order Zero and Unity. London, Cambridge Univ. Press.

Broadbent, S. R. 1956. Lognormal approximation to products and quotients, Biometrika, 43:404.

Brookner, R. J. 1941. A note on the mean as a poor estimate of central tendency, J. Am. Stat. Assn., 36:410.

Brooks, C. E. P., and N. Carruthers. 1953. Handbook of Statistical Methods in Meteorology. London, H.M.S.O.

Bruges, W. E. 1948. The curve of error in which the maximum error is defined, Phil. Mag., ser. 7, 39:394.

Burr, Irving W. 1952. Distribution of ranges from an arbitrary discrete population, Ann. Math. Stats., 23:145.

———— 1955. Calculation of exact sampling distribution of ranges from a discrete population, Ann. Math. Stats., 26:530.

Cadwell, J. H. 1953a. The distribution of quasi-ranges in samples from a normal population, Ann. Math. Stats., 24:603.

———— 1953b. Approximation to the distributions of a measure of dispersion of a power of χ^2, Biometrika, 40:336.

———— 1954a. The probability integral of range for samples from a symmetrical unimodal population, Ann. Math. Stats., 25:803.

———— 1954b. The statistical treatment of mean deviation, Biometrika, 41:12.

Camp, B. H. 1938. Notes on the distribution of the geometric mean, Ann. Math. Stats., 9:221.

Carlson, P. G. 1956. A least squares interpretation of the bivariate line of organic correlation. Skand. Akt. Tidsk., vol. 1956:7.

Carlton, A. G. 1946. Estimating the parameters of a rectangular distribution, Ann. Math. Stats., 17:355.

Carter, R. W. 1951. Floods in Georgia, Frequency and Magnitude. U.S. Geol. Survey Circ. 100.

Cartwright, D. E., and M. S. Longuet-Higgins. 1956. The statistical distribution of the maxima of a random function, Proc. Roy. Soc., A, 237:212.

Chamayou, H. 1950. Recherche d'une loi périodique sur l'hydraulicité, Houille Blanche, 5:529.

Chandler, K. M. 1952. The distribution and frequency of record values, J. Roy. Stat. Soc., B, 14:220.

Chapin, W. S. 1880. The relation between the tensile strengths of long and short bars, Van Nostrand's Engineering Magazine, 23:441.

Charlier, C. V. L. 1920. Vorlesungen über die Grundzüge der mathematischen Statistik. Lund, Scientia. Trans. by J. A. Greenwood, Cambridge (Mass.). 1947.

Chauvenet, W. 1876. A Manual of Spherical and Practical Astronomy. Philadelphia, Lippincott. Vol. 2, pp. 558–596.

Chernoff, H., and G. J. Lieberman. 1954. Use of normal probability paper, J. Am. Stat. Assn., 49:778.

———— 1956. The use of generalized probability paper for continuous distributions, Ann. Math. Stats., 27:806.

Chow, Ven Te. 1951. Discussion of long-term storage capacity of reservoirs, Proc. Am. Soc. Civ. Eng., 77:1.

———— 1951–52. A general formula for hydrologic frequency analysis. Trans. Am. Geophys. Union, 32:231; Discussion, 33:277.

―――― 1952. Design charts for finding rainfall intensity frequency, Concrete Pipe News, vol. 4, no. 6.
―――― 1953. Frequency analysis of hydrologic data with special application to rainfall intensities. Univ. of Illinois Bulletin No. 414.
―――― 1954. The log-probability law and its engineering applications, Proc. Am. Soc. Civ. Eng. vol. 80, sep. no. 536.
Chu, J. T. 1957. Some uses of quasi-ranges, Ann. Math. Stats., 28:173.
Chung, Kai-Lai. 1947. On the maximum partial sum of independent random variables, Proc. Nat. Acad. Sc., 33:132.
Chung, Kai-Lai, and Paul Erdös. 1947. On the lower limit of sums of independent random variables, Ann. Math., 48:1003.
Cochran, W. G. 1941. The distribution of the largest of a set of estimated variances as a fraction of their total, Ann. Eugenics, 11:47.
―――― 1952. The Chi-square test of goodness of fit, Ann. Math. Stats., 23:315.
Cohan, G. 1947. New uses for the average range, Ind. Qual. Control, 3:17.
Cohen, A. C., jr. 1951. Estimating parameters of logarithmic-normal distributions by maximum likelihood, J. Am. Stat. Assn., 46:206.
Conrad, V. 1940. Investigations into Periodicity of the Annual Range of Air Temperature at State College, Pennsylvania. Pennsylvania State College Studies, No. 8.
Court, Arnold. 1951. Coldest temperatures in the United States, Weatherwise, 4:136.
―――― 1952. Some new statistical techniques in geophysics. In: Advances in Geophysics, vol. 1 (edited by H. E. Landsberg). New York, Academic Press, Inc.
―――― 1953a. Temperature extremes in the United States, Geographical Review, 43:39.
―――― 1953b. Wind extremes as design factors, J. Franklin Institute, 256:39.
Coutagne, Aimé. 1930. Etude analytique des débits de crue. Comité national français de l'Union internationale de Géodésie et de Géophysique, Congrès de Stockholm.
―――― 1934. Les usines hydroéléctriques à barrages-reservoirs, La Téchnique moderne, vol. 26, nos. 14, 15.
―――― 1937. Etude statistique des débits de crue, Revue générale de l'Hydraulique, Paris.
―――― 1938. Etude statistique et analytique des crues du Rhône à Lyon. Compt. Rend., Congrès pour l'Utilisation des Eaux, Lyon.
―――― 1951. Méthodes pour déterminer le débit de crue maximum qu'il est possible de prévoir pour un barrage et pour lequel le barrage doit être établi. Congrès des Grands Barrages, New Delhi.
―――― 1952a. Variabilité des débits des cours d'eau d'après Lane et Kai Lei, Houille Blanche, 7:129.
―――― 1952b. Initiation mathématique à l'hydrologie fluviale, Houille Blanche, 7:245.
―――― 1952c. Commentaires récents sur les travaux de M. Gumbel, Houille Blanche, 7:130.
Cox, D. R. 1948. A note on the asymptotic distribution of range, Biometrika, 35:310.
―――― 1949. The use of the range in sequential analysis, J. Roy. Stat. Soc., ser. B11, p. 101.
―――― 1954. The mean and coefficient of variation of range in small samples from non-normal populations, Biometrika, 41:469.
Cox, S. M. 1948. Kinetic approach to the theory of strength of glass. Nature, 162:947.
Cragwell, J. S. 1952. Floods in Louisiana, Magnitude and Frequency. Washington, D. C., U. S. Geol. Survey and State of Louisiana Dept. of Highways.
Craig, A. T. 1932. On the distribution of certain statistics, Am. Math. J., 54:353.
Cramér, H. 1946. Mathematical methods of statistics. Princeton, N. J., Princeton Univ. Press.
Creager, Wm. P. 1939. Possible and probable future floods, Civ. Eng., 9:668.
Creager, Wm. P., and Julian Hinds. 1945. Engineering for Dams. New York, Wiley.
Creager, Wm. P., and Joel D. Justin. 1927. Hydroelectric Handbook. New York, Wiley.

BIBLIOGRAPHY 353

―――― 1950. Same. 2d ed.
Cross, W. P. 1946. Floods in Ohio, magnitude and frequency. Ohio Water Resources Board, Bull. no. 7.
Curtiss, J. H. 1941. On the distribution of the quotient of two chance variables, Ann. Math. Stats., 12:409.
Cyffers, B., and A. Vessereau. 1955. Mesure et contrôle de la dispersion à partir de groupes de deux, Revue de stat. appl., 3:45.
Czuber, E. 1891. Theorie der Beobachtungsfehler. Leipzig, Teubner.
Dalcher, A. 1955. Statistische Schätzungen mit Quantilen, Mitt. Verein. schweizer. Versicherungsmath., 55:475.
Dalrymple, T. 1946. Use of stream-flow records in design of bridge waterways, Proc. Highway Research Board, vol. 26.
―――― 1950. Regional flood frequency, Proc. Highway Research Board. Rept. 11B.
Daly, J. F. 1946. On the use of the sample range as an analogue of "Student's" t-test, Ann. Math. Stats., 17:71.
Daniel, Cuthbert, and Nicholas Heereman. 1950. Design of experiments for most precise slope estimation or linear extrapolation, J. Am. Stat. Assn., 45:546.
Daniels, H. E. 1941. A property of the distribution of extremes, Biometrika, 32:194.
―――― 1945. The statistical theory of the strength of bundles of threads, Proc. Roy. Soc., A, 183:405.
―――― 1952. The covering circle of a sample from a circular normal distribution, Biometrika, 39:137.
Dantzig, D. van. 1956. Economic decision problems for flood prevention, Econometrica, 24:276.
Darling, D. A. 1952a. On a test for homogeneity and extreme values, Ann. Math. Stats., 23:450. [Correction, 24:135 (1953)].
―――― 1952b. The influence of the maximum term in the addition of independent random variables, Trans. Am. Math. Soc., 73:95.
―――― 1956. The maximum of sums of stable random variables, Trans. Amer. Math Soc., 83:164.
Darling, D. A., and P. Erdös. 1956. A limit theorem for the maximum of normalized sums of independent random variables, Duke Math. J., 23:143.
Darwin, J. H. 1957. The difference between consecutive members of a series of random variables arranged in order of size, Biometrika, 44:211.
David, H. A. 1951. Further applications of range to the analysis of variance, Biometrika, 38:393.
―――― 1953. The power function of some tests based on range, Biometrika, 40:347.
―――― 1954. The distribution of range in certain non-normal populations, Biometrika, 41:463.
―――― 1955a. A note on moving ranges, Biometrika, 42:512.
―――― 1955b. Moments of negative order and ratio-statistics, J. Roy. Stat. Soc., B, 17:121.
―――― 1956a. On the application to statistics of an elementary theorem in probability, Biometrika, 43:85.
―――― 1956b. Revised upper percentage points of the extreme studentized deviate from the sample mean, Biometrika, 43:449.
―――― 1957. Estimation of means of normal populations from observed minima, Biometrika, 44:282.
David, H. A., H. O. Hartley, and E. S. Pearson. 1954. The distribution of the ratio, in a single normal sample, of range to standard deviation, Biometrika, 41:482.
Davidenkov, N., E. Shevadin, and F. Wittman. 1947. The influence of size on the brittle strength of steel, J. Appl. Mech., 14:63.
Davies, O. L., and E. S. Pearson. 1934. Methods of estimating from samples the population standard deviations, J. Roy. Stat. Soc. Suppl. I:76.
Davis, D. J. 1952. An analysis of some failure data, J. Am. Stat. Assn., 47:113.

Dederick, L. S. 1928. Relation Between the Probable Error of a Single Shot and the Bracket of a Salvo or Series. Aberdeen Proving Ground, Md.

Deemer, Walter L., jr., and David F. Votaw, jr. 1955. Estimation of parameters of truncated or censored exponential distributions, Ann. Math. Stats., 26:498.

Deming, W. E. 1944. Some principles of the Shewhart methods of quality control, Mech. Eng., 66:173.

Dick, I. D., and J. H. Darwin. 1954. The prediction of floods, New Zealand Engineering, 8:99.

Dinsmore, J. S., jr. 1953. A Sample Criterion for Rejecting Distal Values. Record of Research, Institute of Statistics, Consolidated Univ. North Carolina, vol. 2.

Dixon, W. J. 1940. A criterion for testing the hypothesis that two samples are from the same population, Ann. Math. Stats., 11:199.

——— 1950. Analysis of extreme values, Ann. Math. Stats., 21:488.

——— 1951. Ratios involving extreme values, Ann. Math. Stats., 22:68.

——— 1953. Processing data for outliers, Biometrics, 9:47.

Dixon, W. J., and A. M. Mood. 1946. The statistical sign test, J. Am. Stat. Assn., 41:557.

Dodd, E. L. 1923. The greatest and the least variate under general laws of error, Trans. Am. Math. Soc., 25:525.

Doornbos, R. 1956. Significance of the smallest of a set of estimated normal variances, Statistica Neerlandica, 10:117.

Dwass, Meyer. 1954. On the asymptotic normality of certain rank order statistics Ann. Math. Stats., 24:303.

Eberhard, O. von. 1938. Über das Fehlergesetz des grössten Fehlers einer Serie und das Gesetz der Salvenausdehnung, Zeitschrift für angew. Math. und Mech., 18:128.

Edgeworth, F. Y. 1887. On discordant observations, Phil. Mag., Series 5, 23:364.

Eggenberger, Florian. 1950. Wahrscheinlichkeitstheoretische Analyse der Wasserführung einiger Flüsse der Schweiz (Thesis). Zürich, Eidgen. Techn. Hochschule.

Egudin, G. I. 1947. Certain relations between the moments of the distribution of extreme values in random samples, Doklady Akad. Nauk SSSR (N.S.), 58:1581 (Russian).

Eisenhart, Ch. 1948. Significance of the Largest of a Set of Sample Estimates of Variance. In: Selected Techniques of Statistical Analysis, ed. by Eisenhart, Hastay, and Wallis. New York, McGraw Hill.

Elfving, G. 1947. The asymptotical distribution of range in samples from a normal population, Biometrika, 34:111.

Epstein, B. 1948a. Statistical aspects of fracture problems, J. Appl. Phys., 19:140.

——— 1948b. Application of the theory of extreme values in fracture problems, J. Am. Stat. Assn., 43:403.

——— 1949a. A modified extreme value problem, Ann. Math. Stats., 20:99.

——— 1949b. The distribution of extreme values in samples whose members are subject to a Markoff chain condition, Ann. Math. Stats., 20:590.

——— 1954. Tables for the distribution of the number of exceedances, Ann. Math. Stats., 25:762.

——— 1955. Comparison of some non-parametric tests against normal alternatives with an application to life testing, J. Amer. Stat. Assn., 50:894.

Epstein, B., and H. Brooks. 1948. The theory of extreme values and its implications in the study of the dielectric strength of paper capacitors, J. Appl. Phys., 19:544.

Epstein, B., and Milton Sobel. 1952. Some tests based on the first r ordered observations drawn from an exponential distribution, Ann. Math. Stats., 23:143.

——— 1953. Life testing, J. Am. Stat. Assn., 48:486.

——— 1954. Some theorems relevant to life testing from an exponential distribution, Ann. Math. Stats., 25:373.

Epstein, B., and Chia Kuei Tsao. 1942. Some tests based on ordered observations from two exponential populations, Ann. Math. Stats., 14:458.

BIBLIOGRAPHY 355

Evans, W. Duane. 1942. The standard error of percentiles, J. Am. Stat. Assn., 37:367.
Eyraud, H. 1934. Sur la valeur la plus précise d'une distribution, Compt. Rend. Ac. des Sc., 199:817.
―――― 1935. Valori osservati di una variabile casuale è loro perequazione, Giorn. Ist. Ital. Att., 6:243.
Feller, W. 1950. An Introduction to Probability Theory and Its Application. New York, Wiley.
―――― 1951. The asymptotic distribution of the range of sums of independent random variables, Ann. Math. Stats., 22:427.
Finetti, B. de. 1932. Sulla legge di probabilità degli estremi, Metron, 9:127.
Finkelstein, B. V. 1953. On the limiting distributions of the extreme terms of a variational series of a two-dimensional random quantity, Doklady Akad. Nauk SSSR (N.S.), 91:209 (Russian).
Fisher, J. C., and J. H. Hollomon. 1947. A Statistical Theory of Fracture. Am. Inst. of Mining and Metallurgical Engineers, Tech. Pub. no. 2218, Metals Technology.
Fisher, R. A. 1922. On the mathematical foundations of theoretical statistics, Phil. Trans. Roy. Soc., 222:309.
―――― 1935. The mathematical distributions used in the common tests of significance, Econometrika, 3:353.
Fisher, R. A., and L. H. C. Tippett. 1928. Limiting forms of the frequency distribution of the largest or smallest member of a sample, Proc. Cambridge Phil. Soc., 24:180.
Fogelson, S. 1930. Sur la détermination de la médiane, Revue trimestrielle de statistique (Warsaw), 7:1.
Foster, Edgar E. 1942. Evaluation of flood losses and benefits, Trans. Am. Soc. Civ. Eng., 107:871.
―――― 1948. Rainfall and Run-off. New York, Macmillan.
Foster, F. G., and A. Stuart. 1954. Distribution-free tests in time-series based on the breaking of records, J. Roy. Stat. Soc., B, 16:1.
Foster, F. G., and D. Teichroew. 1955. A sampling experiment on the powers of the records tests for trend in a time series, J. Roy. Stat. Soc., B, 17:115.
Foster, H. A. 1924. Theoretical frequency curves and their applications to engineering problems, Trans. Am. Soc. Civ. Eng., 87:142.
―――― 1933. Duration curves, Proc. Am. Soc. Civ. Eng., 59:1223.
―――― 1952. Report on Floods and Flood Damage. New York, Parsons, Brinckerhoff, Hall and Macdonald.
Frank, Philipp. 1918. Über die Fortpflanzungsgeschwindigkeit, Physik. Zeitschrift, 19:516.
Frankel, J. P. 1948. Relative strengths of Portland cement mortar in bending under various loading conditions, Proc. Am. Concrete Inst., 45:21.
Fréchet, M. 1927. Sur la loi de probabilité de l'écart maximum, Ann. de la Soc. polonaise de Math. (Cracow), 6:93.
―――― 1954. Interdépendance du centre et du rayon empirique de variation de n observations indépendantes. In: Studies in Mathematics and Mechanics Presented to R. von Mises. New York, Academic Press.
Frenkel, J. I., and T. A. Kontorova. 1943. A statistical theory of the brittle strength of real crystals, J. Phys. (USSR), 7:108 (Russian).
Freudenberg, Karl. 1934. Die Gesetzmässigkeiten der menschlichen Lebensdauer, Ergebnisse der Hygiene, etc., 15:437.
―――― 1955. Die Sterblichkeit in hohen Lebensaltern, Schweiz. Zeitschrift f. Volksw. u. Stat., 91:438.
Freudenthal, A. M. 1946. The statistical aspect of fatigue of materials, Proc. Roy. Soc., A, 187:416.
―――― 1948. Reflections on standard specifications for structural design, Trans. Am. Soc. Civ. Eng., 113:269.
―――― 1951. Planning and Interpretation of Fatigue Tests. Spec. Tech. Publ. no. 121, Am. Soc. Testing Materials.

——— 1953. A random fatigue testing procedure and machine, Am. Soc. Testing Materials, 53:896.
——— 1954. Safety and the probability of structural failure, Proc. Am. Soc. Civ. Eng., vol. 80, sep. no. 468.
Freudenthal, A. M., and E. J. Gumbel. 1953. On the statistical interpretation of fatigue tests, Proc. Roy. Soc., A, 216:309.
——— 1954a. Minimum life in fatigue, J. Am. Stat. Assn., 49:575.
——— 1954b. Failure and permanent survival in fatigue, J. Appl. Phys., 25:1435.
——— 1956a. Physical and statistical aspects of fatigue, Advances in Applied Mechanics, 4:117.
——— 1956b. Distribution functions for the prediction of fatigue life and fatigue strength. Intern. Conf. on Fatigue of Metals, British Inst. of Mechanical Engineers.
Fuller, Weston E. 1914. Flood flows, Trans. Am. Soc. Civ. Eng., 77:564.
Fung, Y. C. 1952. Statistical Aspects of Dynamic Loads. New York, Institute of Aeronautical Sciences, preprint 376.
Gaede, K. 1942. Anwendung statistischer Untersuchungen auf die Pruefung von Baustoffen, Bauingenieur, 23:291.
Galton, F. 1875. Statistics by intercomparison with remarks on the law of frequency of error, Phil. Mag., 4th ser., 49:33.
——— 1879. The geometric mean in vital and social statistics, Proc. Roy. Soc., 29:365.
——— 1902. The most suitable proportions between the values of first and second prizes, Biometrika, 1:385.
Gartstein, B. N. 1948. On certain limit laws for the range, Doklady Akad. Nauk, SSSR, 60:1119 (Ukrainian).
——— 1951. On the limiting distribution of the extreme and mixed ranges of a variational series, Dopovidi Akad. Nauk Ukrain. RSR, no. 1, p. 24 (Ukrainian).
Geary, R. C. 1930. The frequency distribution of the quotient of two normal variates, J. Roy. Stat. Soc., 93:442.
——— 1943. Minimum range for quasi-normal distributions, Biometrika, 33:100.
Gibrat, R. 1931. Les inégalites économiques. Paris.
——— 1932. Amenagement hydroélectrique des cours d'eau: Statistique mathématique et calcul des probabilités, Revue générale de l'Electricité, vol. 32, nos. 15, 16.
——— 1934. L'art de l'ingenieur et la statistique, La statistique mécanique, vol. 1, nos. 18, 19.
Gilbert, F. C. 1954. Range distribution of a particle following the decay of Li and B, Phys. Rev., 93:499.
Gillette, H. P. 1935. Application of rainfall cycles to problems of water supply, Roads and Streets, 78:360.
——— 1937. Compound weather and climatic cycles, Water Works and Sewage, 84:25.
Glaisher, J. W. L. 1872–73. On the rejection of discordant observations, Roy. Astron. Soc., Monthly Notices, 23:391.
Gnedenko, B. V. 1941. Limit theorems for the maximal term of a variational series, Doklady Akad. Nauk SSSR (Moscow), 32:37 (Russian).
——— 1943. Sur la distribution limite du terme maximum d'une serie aléatoire, Ann. Math., 44:423.
——— 1953. On the role of the maximal summand in the summation of independent random variables, Ukrain. Mat. Zurnal, 5:291.
Godwin, H. J. 1949. Some low moments of order statistics, Ann. Math. Stats., 20:279.
Goodman, L. A. 1953. Methods of measuring useful life of equipment under operational conditions, J. Am. Stat. Assn., 48:503.
——— 1954. Some practical techniques in serial number analysis, J. Am. Stat. Assn., 49:97.
Goodrich, R. D. 1927. Straight line plotting of skew frequency data, Trans. Am. Soc. Civ. Eng., 91:1.

Gordon, R. D. 1941. Values of Mill's ratio of area to bounding ordinate and of the normal probability integral for large values of the argument, Ann. Math. Stats., 12:364.
Gould, B. A., jr. 1854. On Peirce's criterion for the rejection of doubtful observations, with tables for facilitating its application, Astron. J., 4:81.
Graf, O., and F. Weise. 1938. Über die Pruefung des Betons in Betonstrassen durch Ermittlung der Druckfestigkeit von Wuerfeln und Bohrproben, Forschungsarbeiten aus dem Strassenwesen, 6:22.
Graf, Ulrich, and Rolf Wartmann. 1954. Die Extremwertkarte bei der laufenden Fabrikationskontrolle, Mitteilungsblatt für Mathematische Statistik (Würzburg), 6 (no. 2):121; (no. 3):188.
Grant, E. L. 1940. The probability viewpoint in hydrology, Trans. Am. Geophys. Union, 21:7.
Grassberger, H. 1932. Die Anwendung der Wahrscheinlichkeitsrechnung auf die Wasserführung der Gewaesser, Wasserwirtschaft, 25:55.
―――― 1936. Die Anwendung der Wahrscheinlichkeitsrechnung auf Hochwasserfragen, Deutsche Wasserwirtschaft, Nos. 9, 10.
Greville, T. N. E. 1946. U. S. Life Tables, 1939–41. Washington, D. C., U. S. Gov. Print. Off.
Griffith, A. A. 1920. The phenomena of rupture and flow in solids, Phil. Trans. Roy. Soc., A, 221:163.
―――― 1924. The theory of rupture, Proc. First Internatl. Congress of Appl. Math., 1:55.
Grubbs, F. E. 1950. Sample criteria for testing outlying observations, Ann. Math. Stats., 21:26.
Grubbs, F. E., and C. L. Weaver. 1947. The best unbiased estimate of population standard deviation based on group ranges, J. Am. Stat. Assn., 42:224.
Gumbel, E. J. 1926. Ueber ein Verteilungsgesetz, Z. Physik., 37:468.
―――― 1935. Les valeurs extrêmes des distributions statistiques, Ann. Inst. Henri Poincaré, 4:115.
―――― 1937a. La Durée extrême de la vie humaine. Paris, Hermann et Cie.
―――― 1937b. Les Intervalles extrêmes entre les émissions radioactives, J. de Physique (series VII), VIII (no. 8), 321; (no. 11), 446.
―――― 1940. Les Crues du Rhône, Annales de l'Université de Lyon (serie 3), no. 3, Section A, p. 39.
―――― 1941. Probability interpretation of the observed return period of floods, Trans. Am. Geophys. Union, p. 836.
―――― 1942a. Simple tests for given hypotheses, Biometrika 32:317.
―――― 1942b. On the frequency distribution of extreme values in meteorological data, Bull. Am. Meteorol. Soc., 23:95.
―――― 1943. On the reliability of the classical Chi square test, Ann. Math. Stats., 14:253.
―――― 1944. Ranges and midranges, Ann. Math. Stats., 15:414.
―――― 1945a. Floods estimated by probability methods, Eng. News-Record, 134:97.
―――― 1945b. Simplified plotting of statistical observations, Trans. Am. Geophys. Union, 26:(Pt. I), p. 69.
―――― 1945c. Théorie statistique des débits de crue, La Renaissance, 2–3:122.
―――― 1947a. Discussion of B. F. Kimball, "Assignment of frequencies to a completely ordered set of sample data," Trans. Am. Geophys. Union, 28:951.
―――― 1947b. The distribution of the range, Ann. Math. Stats., 18:384.
―――― 1949a. Probability tables for the range, Biometrika, 36:142.
―――― 1949b. The Statistical Forecast of Floods. Ohio Water Resources Board, Bull. 15.
―――― 1952. Discussion of Ven te Chow, "A general formula for hydrologic frequency analysis," Trans. Am. Geophys. Union, 33:277.
―――― 1953. La Définition de l'âge limite, Giorn. Ist. Ital. Att., 16:88.

―――― 1954a. The maxima of the mean largest value and of the range, Ann. Math. Stats., 25:76.
―――― 1954b. Statistical Theory of Extreme Values and Some Practical Applications. Natl. Bureau of Stand., Appl. Math. Ser. No. 33.
―――― 1954c. Statistical theory of droughts, Proc. Am. Soc. Civ. Eng., 80, sep. no. 439.
―――― 1955. Die Bedeutung der Parameter in der Gompertz-Makehamschen Formel, Blätter der D. Ges. Versicherungsmath., 2:193.
―――― 1956a. Statistische Theorie der Ermüdungserscheinungen bei Metallen, Mitt. für Math. Stat., 8:97.
―――― 1956b. Discussion of Weibull's paper, Internat. Conf. on Fatigue in Aircraft Structures. New York, Assoc. Press. P. 143.
―――― 1956c. Méthodes graphiques pour l'analyse des débits de crue, Houille Blanche, No. 5:709.
Gumbel, E. J., and Phillip G. Carlson. 1954. Extreme values in aeronautics, J. Aeronaut. Sci., 21:389.
―――― 1956. On the covariance of the sample mean and standard deviation, Metron, 18:113.
Gumbel, E. J., and A. Court. 1951. Evaluation of Climatic Extremes. Off. Quartermaster General Environmental Protection Section, Report No. 175. (Available from the Library of Congress, Washington, D. C., photo-duplication service, PB no. 106360.)
Gumbel, E. J., and J. A. Greenwood. 1951. Table of the asymptotic distribution of the second extreme, Ann. Math. Stats., 22:143.
Gumbel, E. J., and L. H. Herbach, 1951. The exact distribution of the extremal quotient, Ann. Math. Stats., 22:418.
Gumbel, E. J., and R. D. Keeney. 1950. The extremal quotient, Ann. Math. Stats., 21:523.
Gumbel, E. J., and J. Lieblein. 1954. Some applications of extreme value methods. Am. Statistician, 8:14.
Gumbel, E. J., and H. von Schelling. 1950. The distribution of the number of exceedances, Ann. Math. Stats., 21:247.
Gurland, John. 1948. Asymptotically Normal Estimates. Inversion Formulae for the Distribution of Ratios. (Thesis.) University of California.
―――― 1955. Distribution of the maximum of the arithmetic mean of correlated random variables, Ann. Math. Stats., 26:294.
Haag, J. 1924. Sur la combinaison des résultats d'observations, Compt. Rend. Ac. des Sc., 179:1388.
Haag, J., and E. Borel. 1926. Applications au tir. Traité du calcul des probabilités et de ses applications. Vol. 4. Paris, Gauthier-Villars.
Hagstroem, K. G. 1956. Variables fondamentales du hasard, Gior. Ist. Ital. Attuari, 19:84.
Hald, A. 1952. Statistical Theory with Engineering Applications. New York, Wiley.
Haldane, J. B. S. 1945. On a method of estimating frequencies, Biometrika, 33:222.
Halperin, Max, S. W. Greenhouse, J. Cornfield, and J. Zalokar. 1955. Tables of percentage points for the Studentized maximum absolute deviate in normal samples, J. Am. Stat. Assn., 50:185.
Hammer, Hans-Karl. 1938. Zu einer Theorie der Versuchszahlen. Doctoral Dissertation. Munich, Borna-Leipzig.
Harley, B. I., and E. S. Pearson. 1957. The distribution of range in normal samples with $n = 200$, Biometrika, 44:257.
Harris, Lee B. 1952. On a limiting case for the distribution of exceedances, with an application to life-testing, Ann. Math. Stats., 23:295.
Hartley, H. O. 1942. The range in random samples, Biometrika, 32:334.
―――― 1944. Studentization or the elimination of the standard deviation of the present population from the random sample-distribution of statistics, Biometrika, 33:173.

BIBLIOGRAPHY

—— 1950. The maximum F ratio as a short-cut test for heterogeneity of variance, Biometrika, 37:308.
Hartley, H. O., and H. A. David. 1954. Universal bounds for mean range and extreme observation, Ann. Math. Stats., 25:88.
Hartley, H. O., and E. S. Pearson. 1951. Moment constants for the distribution of range in normal samples, Biometrika, 38:463.
Hastings, C., jr., F. Mosteller, J. W. Tukey, and C. P. Winsor. 1947. Low moments for small samples: A comparative study of order statistics, Ann. Math. Stats., 18:413.
Hazen, A. 1930. Flood Flows. New York, Wiley.
Helmert, R. 1877. Über den Maximalfehler einer Beobachtung, Ztschr. f. Vermessungswesen, 6:131.
Hesselberg, T., and B. J. Birkeland. 1940. Saekulare Schwankungen des Klimas von Norwegen. Die Lufttemperatur. Geofys. Publikasj., vol. 14, No. 4.
—— 1941. Der Niederschlag. Geofys. Publikasj., vol. 14, no. 5.
—— 1943. Luftdruck und Wind. Geofys. Publikasj., vol. 14, no. 6.
Himsworth, F. R. 1954. The variability of concrete and its effect on mix design, Proc. Inst. Civ. Engrs., London, Vol. 3, pt. 1, p. 163.
Hoeffding, Wassily. 1955. The extrema of the expected value of a function of independent random variables, Ann. Math. Stats., 26:268.
Homma, T. 1951. On the asymptotic independence of order statistics, Rep. Stat. Appl. Res. JUSE (Japan), 1:1.
—— 1952. On the limit distributions of some ranges, Rep. Stat. Appl. Res. JUSE, 1:15.
Hooker, P. F., and L. H. Longley-Cook. 1953. Life and Other Contingencies. London, Cambridge Univ. Press. Vol. 1.
Horton, Robert E. 1936 Hydrologic conditions as affecting the results of the applications of methods of frequency analysis to flood records. U.S. Geol. Survey, Water Supply Paper 771.
Hotelling, H. 1930. The consistency and ultimate distribution of optimum statistics, Trans. Am. Math. Soc., 32:847.
Howell, J. M. 1949. Control chart for largest and smallest values, Ann. Math. Stats., 20:305.
—— 1950. Errata to "Control chart for largest and smallest values," Ann. Math. Stats., 21:615.
Hudimoto, Hirosi. 1956. Note on fitting a straight line when both variables are subject to error and some applications, Ann. Inst. Stat. Math. (Japan), 7:159.
Hurst, H. E. 1950. Long-term storage capacity of reservoirs, Proc. Am. Soc. Civ. Eng., 76:1.
Huss, P. O. 1946. Relation Between Gusts and Average Winds for Housing Load Determination. Daniel Guggenheim Airship Institute, Akron, Ohio. Report No. 140.
Ilyrcnius, H. 1953. On the use of ranges, cross-ranges and extremes in comparing small samples, J. Am. Stat. Assn., 48:534.
Insolera, F. 1937–38. Au sujet de l'age limite, Aktuarske Vedy (Prague), 7:49.
Irwin, J. O. 1925a. On a criterion for the rejection of outlying observations, Biometrika, 17:238.
—— 1925b. The further theory of Francis Galton's individual difference problem, Biometrika, 17:100.
Iwai, Shigehisa. 1950. Duration curves of logarithmic normal distribution type and their applications, Memoirs of Engineering, Kyoto University, 12:1.
Izzard, Carl F. 1953. Peak discharge for highway drainage design, Proc. Am. Soc. Civ. Eng., 79:320.
Jarvis, Clarence S., et al. 1936. Floods in the United States, Magnitude and Frequency, U. S. Geol. Survey, Water Supply Paper 771.
Jasper, N. H. 1956. Statistical distribution patterns of ocean waves and of wave-induced ship stresses and motions with engineering applications. Society of Naval Architects and Marine Engineers. New York Meeting, preprint No. 6.

Jeffreys, H. 1932. An alternative to the rejection of observations, Proc. Roy. Soc., A, 137:78.
Jenkinson, A. F. 1955. The frequency distribution of the annual maximum (or minimum) values of meteorological elements, Q. J. Roy. Meteor. Soc., 87: 158.
Johnson, A. I. 1953. Strength, Safety and Economical Dimensions of Structures. Stockholm, Bull. of the Div. of Bldg. Stat. and Structural Engineering, Roy. Inst. Tech., No. 12.
Johnson, J. W. 1943. Effect of height of test specimens on compressive strength of concrete, Am. Soc. Testing Materials Bull. 120:19; 123:48.
Johnson, L. G. 1954. An axiomatic derivation of a general S–N equation, Industrial Math., 4:1.
Johnson, N. L. 1952. Approximation to the probability integral of the distribution of range, Biometrika, 39:417.
Jones, A. E. 1946. A useful method for the routine estimation of dispersion from large samples, Biometrika, 33:274.
Jones, Howard L. 1948. Exact lower moments of order statistics in small samples from a normal distribution, Ann. Math. Stats., 19:270.
——— 1953. Approximating the mode from weighted sample values, J. Am. Stat. Assn., 48:113.
Jordan, C. 1927. Statistique mathématique. Paris, Gauthier-Villars.
Juncosa, M. L. 1948. On the distribution of the minimum of a sequence of mutually independent random variables, Duke Math. J., 16:609.
Kalinske, A. A. 1946. On the logarithmic-probability law, Trans. Amer. Geophys. Union, 27:709.
Kaplansky, I. 1945. A common error concerning kurtosis, J. Am. Stat. Assn., 40:259.
Kapteyn, J. C. 1903. Skew Frequency Curves in Biology and Statistics. Groningen.
——— 1916. Same. 2d ed.
Kase, Shigeo. 1953. A theoretical analysis of the distribution of tensile strength of vulcanized rubber, J. Polymer. Science, 11:425.
——— 1954. How to treat tensile data of rubber. A computational method, J. Polymer Science, 14:497.
——— 1955a. Nomographs for evaluation of tensile data of rubber, Canad. J. of Tech., 33:409.
——— 1955b. The specification for tensile strength of rubber, J. Appl. Chem., 5:323.
Kawata, Tatsuo. 1951. Limit distributions of single order statistics, Rep. Stat. Appl. Res., JUSE (Japan), 1:4.
Kendall, M. G. 1940–41. Note on the distribution of quantiles for large samples, J. Roy. Stat. Soc., Suppl. no. 7:83.
——— 1946. The Advanced Theory of Statistics. London, J. B. Lippincott. Vol. 1.
Kennedy, E. 1942. Analyzing the degree of randomness in weather data, Civ. Eng., 12:34.
Kimball, B. F. 1938–42a. Probability-distribution-curve for flood-control studies, Trans. Am. Geophys. Union, 19:460; Discussion, 23:501.
——— 1942b. Limited type of primary probability distribution applied to annual maximum flood flows, Ann. Math. Stats., 13:318.
——— 1946a. Sufficient statistical estimation functions for the parameters of the distribution of maximum values, Ann. Math. Stats., 17:299.
——— 1946b–47a. Assignment of frequencies to a completely ordered set of sample data, Trans. Am. Geophys. Union, 27:843; Discussion, 28:952.
——— 1947b. Some basic theorems for developing tests of fit for the case of non-parametric probability distribution functions, Ann. Math. Stats., 18:540.
——— 1949. An approximation to the sampling variances of an estimated maximum value of given frequency based on fit of doubly exponential distribution of maximum values, Ann. Math. Stats., 20:110.

——— 1955. Practical applications of the theory of extreme values, J. Am. Stat. Assn., 50:517.
——— 1956. The bias in certain estimates of the parameters of the extreme-value distribution, Ann. Math. Stats., 27:758.
Kincer, J. B. 1946. Our changing climate, Trans. Am. Geophys. Union, 27:342.
King, E. P. 1953. On some procedures for the rejection of suspected data, J. Am. Stat. Assn., 48:531.
Kinnison, H. A., and B. R. Colby. 1945. Flood formulas based on drainage basin characteristics, Trans. Am. Soc. Civ. Eng., 110:849.
Kinnison, H. A., L. F. Conover, and B. L. Bigwood. 1938. Stages and Flood Discharges of the Connecticut River at Hartford, Conn., U. S. Geol. Survey, Water Supply Paper 836a.
Kolmogorov, A. N., and A. Ya. Hinshin. 1951. The work of N. V. Smirnov on the investigation of properties of variational series and on nonparametric problems of mathematical statistics, Uspekhi Matematicheskich Nauk N.S. 6 (44), p. 190 (Russian).
Komatu, Y. 1955. Elementary inequalities for Mills ratio, Rep. Stat. Appl. Res. JUSE (Japan), 4:69.
Kondo, Takayuki. 1954. Evaluation of some ω_n^2 distributions, J. Gakugei, Tokushima Univ., Japan (Nat. Sci. Ser.), 4:45.
Kontorova, T. A. 1940. Statistical theory of endurance, J. Tech. Phys. (USSR), 10:886 (Russian).
Kozelka, R. M. 1956. Approximate upper percentage points for extreme values in multinomial sampling, Ann. Math. Stats., 27:507.
Kresge, Ralph F., and Tor J. Nordenson. 1954. Use of Flood Forecasting Procedures in the Derivation of Flood Frequencies. Washington, D. C., U. S. Dept. of Commerce, Weather Bureau.
Krishna Sastry, K. V. 1948. On a Bessel function of the second kind and Wilks' Z-distribution, Proc. Indian Acad. Science, A, 28:532.
Krumbein, W. C., and J. Lieblein. 1956. Geological application of extreme-value methods to interpretation of cobbles and boulders in gravel deposits, Trans. Am. Geophys. Union, 37:313.
Kudô, Akio. 1955. On the confidence interval of the extreme value of a second sample from a normal universe, Bull. Math. Stat. Japan, 6:51.
Landsberg, H. 1949. Climatic trends in the series of temperature observations at New Haven, Connecticut. Glaciers and Climate, Geografiska Annaler 1949, p. 125, Stockholm.
Lane, E. W., and Kai Lei. 1950. Stream flow variability, Trans. Am. Soc. Civ. Eng., 115:1084.
Langbein, W. B. 1949. Annual floods and the partial-duration flood series, Trans. Am. Geophys. Union, 30:879.
——— 1953. Flood insurance, Land Economics, 29:323.
Lavin, M. 1946. Inspection efficiency and sampling inspection plans, J. Am. Stat. Assn., 41:432.
Leme, R. A. Da Silva. 1954. Os extremos de amostras ocasionais e suas applicações à engenharia. (Thesis, University of São Paulo) (Portuguese).
Levert, C. 1953. La durée de retour et la durée d'absence de la plus grande valeur, Compt. Rend. Ac. des Sc., 237:374.
——— 1954a. Quelques détails mathématiques sur le théorème de Gumbel appliqué aux evênements rares, Compt. Rend. Ac. des Sc., 239:149.
——— 1954b. On the statistical theory of return and absence periods of rare phenomena (large daily rainfalls, heavy rains) and on considerations of risk concerning practical application, Compt. Rend., Tenth Congress U.G.G.I., Rome.
——— 1954c. Return periods and periods of absence of rare events, Verhandelingen, no. 59:353. The Hague, Nederlands Meteorol. Instituut (Dutch).
——— 1956. On the testing of outlying observations, Sankhyā, 17:67.

Lévi, R. 1949. Calculs probabilistes de la sécurite des constructions, Ann. Ponts et Chaussées, 119:493.

Lévy, Paul. 1948. Processus stochastiques et mouvement Brownien. Paris, Gauthier-Villars.

L'Hermite, R. 1939. Étude statistique des bétons de chantier, Ann. de l'Inst. Tech. Bâtim. et Trav. Publ., 4:66.

Lieblein, Julius. 1953. On the exact evaluation of the variances and covariances of order statistics in samples from the extreme-value distribution, Ann. Math. Stats., 24:282.

——— 1954a. A New Method of Analyzing Extreme-value Data. Washington, D. C. Natl. Advisory Committee for Aeronautics, Tech. Note 3053.

——— 1954b. A historical note on the relation between extreme values and tensile strength, Ann. Math. Stats., 25:172.

——— 1955. On moments of order statistics from the Weibull distribution, Ann. Math. Stats., 26:330.

Lieblein, Julius, and M. Zelen. 1956. Statistical investigation of the fatigue life of deep groove ball bearings, J. Research Natl. Bur. Standards, 57:273. Paper 2719.

Linsley, R. H., Max A. Kohler, and J. H. L. Paulhus. 1949. Applied Hydrology. New York, McGraw-Hill. Chapter 20.

Longuet-Higgins, M. S. 1952. On the statistical distribution of the heights of sea waves, Sears Foundation J. of Marine Res., 11:245.

Lord, E. 1947. The use of range in place of standard deviation in the t-test, Biometrika, 34:41.

——— 1950. Power of the modified t-test (u-test) based on range, Biometrika, 37:64.

McAllister, D. 1879. The law of the geometric mean, Proc. Roy. Soc. A., 29:367.

McKay, A. T. 1935. The distribution of the difference between the extreme observation and the sample mean in samples of n from a normal universe, Biometrika, 27:466.

McKay, A. T., and E. S. Pearson. 1933. A note on the distribution of range in samples of n, Biometrika, 25:415.

McMillan, Brockway. 1949. Spread of minima in large samples, Ann. Math. Stats., 20:444.

Malmquist, S. 1950. On a property of order statistics from a rectangular distribution, Skand. Akt. Tidsk., 33:214.

Mann, H. B., and A. Wald. 1942. On the choice of the number of intervals in the application of the Chi square test, Ann. Math. Stats., 13:306.

Marinescu, G. 1951. Functia de distributie a maximului modulului en variabile statistice, Comunicarile Acad. Rep. Pop. Romane, I:309 (Rumanian).

Massey, Frank J., jr. 1951a. The Kolmogorov-Smirnov test for goodness of fit, J. Am. Stat. Assn., 46:68.

——— 1951b. The distribution of the maximum deviation between two sample cumulative step functions, Ann. Math. Stats., 22:125.

Masuyama, Motosaburo. 1955. Table of two-sided 5% and 1% control limits for individual observations of the rth order, Sankhyā, 15:291.

Mather, K. 1949. The analysis of extinction time data in bioassay, Biometrics, 5:127.

May, Joyce M. 1952. Extended and corrected tables of the upper percentage points of the "Studentized" range, Biometrika, 39:192.

Meizler, D. G. 1949. On a problem of B. V. Gnedenko, Ukrain. Mat. Zurnal, 1:67 (Ukrainian).

——— 1955. On partial limit distributions for the maximal term of a variational series, L'vov. Politehn. Inst. Nauk Zap. 30, Ser. Fiz.-Mat. No. 1, p. 24 (Ukrainian).

Miklowitz, J. 1949. Influence of dimensional factors on mode of yielding and fracture in medium-carbon steel, Trans. Am. Soc. Mech. Engr., 72:159.

Mises, R. von. 1923. Über die Variationsbreite einer Beobachtungsreihe, Sitzungsber. d. Berliner Math. Ges., 22:3.

——— 1931. Wahrscheinlichkeitsrechnung und ihre Anwendung in der Statistik und theoretischen Physik. Leipzig, Deuticke.

——— 1934. Problème de deux races, Recueil Math. (Moscow), 41:359.
——— 1936. La distribution de la plus grande de n valeurs, Revue math. de l'Union Interbalkanique (Athens), 1:1.
Mitra, S. K. 1957. Tables for tolerance limits for a normal population based on sample mean and range or mean range, J. Am. Stat. Assn., 52:88.
Mitchell, William D. 1954. Floods in Illinois: Magnitude and Frequency. U. S. Geol. Survey and Div. of Waterways, State of Illinois.
Molina, E. C. 1942. Poisson's Exponential Binominal Limit. New York, Van Nostrand.
Moran, P. A. P. 1954. A probability theory of dams and storage systems, Australian Journal of Applied Science, 5(No. 2):116.
——— 1957. The statistical treatment of flood flows, Trans. Am. Geophys. Union, 18:518.
Moriguti, Sigeiti. 1951. Extremal properties of extreme value distributions, Ann. Math. Stats., 22:523.
——— 1954. Bounds for second moments of the sample range, Rep. Stat. Appl. Res. JUSE (Japan), 3:1
Morlat, Billiet, and Bernier. 1956. Les crues de la haute Durance et la théorie statistique des valeurs extrêmes, Pub. No. 42 de l'Assoc. Intern. d'Hydrologie (de l'U.G.G.F.), Symposia Darcy, p. 99, Dijon.
Moshman, Jack. 1952. Testing a straggler mean in a two-way classification using the range, Ann. Math. Stats., 23:126.
——— 1953. Critical values of the log-normal distribution, J. Am. Stat. Assn., 48:600.
Mosteller, F. 1946. On some useful "inefficient" statistics, Ann. Math. Stats., 17:377.
——— 1948. A k-sample slippage test for an extreme population, Ann. Math. Stats., 19:58.
Muniruzzaman, A. N. M. 1951. On some distributions in connection with Pareto's law, Proc. of the first Pakistan Statistical Conference. Lahore, Punjab Univ. Press.
Murty, V. N. 1955. The distribution of the quotient of maximum values in samples from a rectangular distribution, J. Amer. Stat. Assn., 50:1136.
Nag, S. K., and N. C. Dutta. 1951. Determination of maximum likely one, two, three, four, and five day rainfall in a river catchment, Bull. Internatl. Stat. Inst. (Calcutta, India), 33(Part 5):241.
Nair, K. R. 1940. The median in tests by randomization, Sankhyā, 4:543.
——— 1948a. The distribution of the extreme deviate from the sample mean and its studentized form, Biometrika, 35:118.
——— 1948b. Studentized form of the extreme mean square test in the analysis of variance, Biometrika, 35:16.
Nair, K. R., and H. A. David. 1952. Tables of percentage points of the extreme "Studentized" deviate from the sample mean, Biometrika, 39:190.
National Bureau of Standards. 1953. Probability Tables for the Analysis of Extreme Value Data (preface by E. J. Gumbel), Appl. Math. Ser. No. 22.
——— 1954. Extreme value methods for engineering problems, Tech. News Bull, 38:29.
Newman, D. 1939. The distribution of range in samples from a normal population, expressed in terms of an independent estimate of standard deviation, Biometrika, 31:20.
Newman, F. W. 1892. The Higher Trigonometry; Superrationals of Second Order. Cambridge, MacMillan and Bowes.
Neyman, J. 1923. Sur les valeurs théoriques de la plus grande de n erreurs, Revue Math. et Phys. (Warsaw), 33:1.
——— 1949. Contribution to the theory of the Chi square test, Proc. of the Berkeley Symposium on Math. Stat. and Probability. Berkeley, Univ. of Calif. Press. P. 239.
Neyman, J., and E. S. Pearson. 1928. On the use and interpretation of certain test criteria for purposes of statistical inference, Biometrika, 20A:175.

Noether, Gottfried E. 1948. On confidence limits for quantiles, Ann. Math. Stats., 19:416.
——— 1955. Use of the range instead of the standard deviation, J. Am. Stat. Assn., 50:1040.
Norris, N. 1940. The standard errors of the geometric means and their application to index numbers, Ann. Math. Stats., 11:445.
Oding, J. A. 1955. Dislocation theory for the fatigue of metals, Dokl. Ak. Nauk, SSSR (N.S.), 105:1238 (Russian).
Ogawa, Junjiro. 1951. Contributions to the theory of systematic statistics, I, Osaka Math. J., 3:175.
Ogawara, M., et al. 1955. Stochastic limits for the maximum possible amount of precipitation, Papers in Meteorology and Geophysics (Japan), 5:8.
Ogawara, M., and K. Tomatsu. 1955. A prediction for the next maximum of sunspot numbers, Papers in Meteorology and Geophysics (Japan), 5(Nos. 3–4):212.
Olds, E. G. 1935. Distribution of greatest variates, least variates, and intervals of variation in samples from a rectangular universe, Bull. Am. Math. Soc., 41:297.
Olmstead, P. S. 1940. Note on theoretical and observed distributions of repetitive occurrences, Ann. Math. Stats., 11:363.
Patnaik, P. B. 1950. The use of mean range as an estimator of variance in statistical tests, Biometrika, 37:78.
Paulson, Edward. 1943. A note on tolerance limits, Ann. Math. Stats., 14:90.
Pearl, R. 1940. Medical Biometry and Statistics. 3d ed. Philadelphia and London, Saunders.
Pearson, E. S. 1926. A further note on the distribution of range in samples taken from a normal population, Biometrika, 18:173.
——— 1932. The percentage limits for the distribution of range in samples from a normal population ($n \leq 100$), Biometrika, 24:404.
——— 1950. Some notes on the use of range, Biometrika, 37:88.
——— 1952. Comparison of two approximations to the distribution of the range in small samples from normal populations, Biometrika, 39:130.
Pearson, E. S., and N. P. Adyanthaya. 1928. The distribution of frequency constants in small samples from non-normal symmetrical and skewed distributions, Biometrika, 20:356.
Pearson, E. S., and J. Haines. 1935. The use of range in place of standard deviation in small samples, J. Roy. Stat. Soc. Suppl. II, p. 83.
Pearson, E. S., and H. O. Hartley. 1935. The Application of Statistical Methods to Industrial Standardization and Quality Control. British Standards Institution, Brochure no. 600.
——— 1942. The probability integral of the range in samples of n observations from a normal population, Biometrika, 32:301.
——— 1943. Tables of the probability integral of the "studentized" range, Biometrika, 33:89.
——— 1954. Biometrika Tables for Statisticians, vol. I. London, Cambridge Univ. Press.
Pearson, E. S., and C. Chandra Sekar. 1936. The efficiency of statistical tools and a criterion for the rejection of outlying observations, Biometrika, 28:308.
Pearson, Karl. 1902. Note on Francis Galton's difference problem, Biometrika, 1:390.
——— 1920. On the probable errors of frequency constants, Biometrika, 13:113.
——— 1931a. Tables for Statisticians and Biometricians. Vol. II. London, Cambridge Univ. Press.
——— 1931b–32. On the mean character and variance of a ranked individual and on the mean and variance of the intervals between ranked individuals, Biometrika, 23:364; 24:203.
——— 1934. Tables of the Incomplete Beta Function. London, Cambridge Univ. Press.

BIBLIOGRAPHY

——— 1934. Tables of the Incomplete Gamma Function. London, Cambridge Univ. Press.
Peirce, B. 1852. Criterion for the rejection of doubtful observations, Astron. J., 2:161.
Peiser, A. M. 1946. An Analysis of the Airspeeds and Normal Accclerations of Douglas DC-3 Airplanes in Commercial Transport Operation. Washington, D. C., Natl. Adv. Comm. for Aeronautics. Tech. Note 1142.
Peiser, A. M., and W. G. Walker. 1946. An Analysis of the Airspeeds and Normal Accelerations of Boeing S-307 Airplanes in Commercial Transport Operation. Washington, D. C., Natl. Adv. Comm. for Aeronautics. Tech. Note 1141.
Peiser, A. M., and M. Wilkerson. 1945. A Method of Analysis of V-G Records from Transport Operations. Washington, D. C., Natl. Adv. Comm. for Aeronautics. Rept. 807.
Pierce, F. T. 1926. The weakest link, J. Textile Inst., Trans. 17:355.
Pierce, L. B. 1954. Floods in Alabama, Magnitude and Frequency. U. S. Geol. Survey, Circ. 342.
Pillai, K. C. S. 1950. On the distribution of midrange and semi-range in samples from a normal population, Ann. Math. Stats., 21:100.
——— 1951. Some notes on ordered samples from a normal population, Sankhyā, 11:23.
Plackett, R. L. 1947. Limits of the ratio of mean range to standard deviation, Biometrika, 34:120.
Plum, N. M. 1953. Quality control of concrete, Proc. Inst. Civ. Engrs., Vol. 2, part 1, p. 311.
Potter, W. D. 1949a. Normalcy Tests of Precipitation and Frequency Studies of Runoff in Small Watersheds. U. S. Dept. of Agric., Tech. Bull. No. 985, p. 1.
——— 1949b. Simplification of the Gumbel Method for Computing Probability Curves. Soil Conservation Service Research Tech. Paper No. 78.
——— 1949c. Effect of rainfall on magnitude and frequency of peak rates of surface runoff, Trans. Am. Geophys. Union, 30:735.
——— 1950. Surface Runoff from Agricultural Watersheds. Highway Research Board Report No. 11-b.
——— 1957. The effect of nonrepresentative sampling on linear regressions as applied to runoff, Trans. Am. Geophys. Union, 38:333.
Powell, R. W. 1942. A simple method of estimating flood frequencies, Civ. Eng., 14:105.
President's Water Resources Policy Commission. 1950. Water Policy for the American People. Report. Washington, D. C. Vol. 1.
Press, H. 1949, 1950. The Application of the Statistical Theory of Extreme Values to Gustload Problems. Washington, D.C., Natl. Adv. Comm. for Aeronautics, Tech. Note 1926; Rep. 991.
——— 1955. Time series problems in aeronautics, J. Amer. Stat. Assn., 50:1022.
Price, Reginald C. 1943. The Gumbel method of estimating flood frequencies, Civ. Eng., 13:285.
Price, W. H. 1951. Factors influencing concrete strength, J. Am. Concrete Inst., 47:417.
Proschan, Frank. 1953. Rejection of outlying observations, Amer. J. Phys., 21:520.
Prot, M. 1936. Note sur la notion de coéfficient de sécurité, Ann. Ponts et Chaussées, 106:5.
——— 1949a. La sécurité, Ann. Ponts et Chaussées, 119:19.
——— 1949b. Statistique et sécurité, Revue de Métallurgie, Mémoires et Extraits, 46:716.
——— 1950a. Vues nouvelles sur la sécurité des constructions, Soc. Ingénieurs civils de France, 103:50.
——— 1950b. Méthodes modernes d'essai des matériaux, Ann. de l'Inst. Tech. Bâtim. et Trav. Publ., 156:1.
Purcell, W. G. 1947. Saving time in testing life, Ind. Qual. Control, 3:15.

Quensel, C. E. 1945. Studies of the logarithmic normal curve, Skand. Akt., 28:141.
Rajalakashma, D. V. 1943. On the interval between the ranked individuals of samples taken from a rectangular population, J. of the Madras Univ., Section B15, p. 31.
Rantz, S. E., and H. C. Riggs. 1949. Magnitude and Frequency of Floods in the Columbia River Basin. U. S. Geol. Survey, Water Supply Paper 1808.
Ravilly, E. 1938. Contributions à l'étude de la rupture des fils métalliques soumis â des torsions alternées, Publ. scient. et tech. du Ministère de l'Air, no. 120, p. 52.
Rayleigh, J. W. Strutt, 3d Baron. 1880. On the resultant of a large number of vibrations of the same pitch and of arbitrary phase, Phil. Mag., 10–73.
Reagel, F. V., and T. F. Willis. 1931. The effect of the dimensions of test specimens on the flexural strength of concrete, Public Roads, 12:37.
Reiersol, Olav. 1944. Measures of departures from symmetry, Skand. Akt.Tidsk., 27:229.
Renyi, Alfred. 1953. On the theory of order statistics, Acta Mathematica (Budapest), IV:191.
Rider, P. R. 1929. On the distribution of the ratio of mean to standard deviation in small samples from non-normal universes, Biometrika, 21:124.
——— 1933. Criteria for Rejection of Observations. Washington Univ. Studies (New Series), Science and Technology, No. 8.
——— 1950. The distribution of ranges from a discrete rectangular distribution, Proc. of the International Congress of Mathematics, I:583. (Published by the Am. Math. Soc., 1952.)
——— 1951a. The distribution of the quotient of ranges in samples from a rectangular population, J. Am. Stat. Assn., 46:502.
——— 1951b. The distribution of the range in samples from a discrete rectangular population, J. Am. Stat. Assn., 46:375.
——— 1953. The distribution of the product of ranges in samples from a rectangular population, J. Am. Stat. Assn., 48:546.
——— 1955. The distribution of the product of maximum values in samples from a rectangular distribution, J. Am. Stat. Assn., 50:1142.
——— 1956. The midrange of a sample as an estimator of the population midrange (Abstract), Ann. Math. Stat., 27:204.
Rietz, H. L. 1922. Frequency distributions obtained by certain transformations of normally distributed variables, Ann. Math. (Ser. 2), 23:292.
——— 1936. On the frequency distribution of certain ratios, Ann. Math. Stats., 7:145.
Robbins, H. E. 1944a. On the expected values of two statistics, Ann. Math. Stats., 15:321.
——— 1944b. On distribution-free tolerance limits in random samples, Ann. Math. Stats., 15:214.
——— 1948. On the asymptotic distribution of the sum of a random number of random variables, Proc. Nat. Acad. Sc., 34:162.
Robinson, W. H., and G. L. Bodhaine. 1952. Floods in Western Washington, Frequency and Magnitude in Relation to Drainage Basin Characteristics. U. S. Geol. Survey Circ. 192.
Romanovsky, V. 1933. On a property of the mean ranges in samples from a normal population and on some integrals of Professor T. Hojo, Biometrika, 25:195.
Ruben, H. 1956. On the moments of the range and product moments of extreme order statistics in normal samples, Biometrika, 43:458.
St.-Pierre, J., and A. Zanger. 1956. The null distribution of the difference between two sample values, Ann. Math. Stats., 27:849.
Sakamoto, H. 1943. On the distribution of the product and the quotient of independent and uniformly distributed random variables, Tohoku Math. J., 49:243.
Sarhan, A. E., and B. G. Greenberg. 1956. Estimation of location and scale parameters by order statistics from singly and doubly censored samples, Ann. Math. Stats., 27:427.
——— 1957. Tables for best linear estimates by order statistics of the parameters of single exponential distributions from singly and doubly censored samples, J. Am. Stat. Assn., 52:58.

Saunder, S. A. 1902-3. Note on the use of Peirce's criterion for the rejection of doubtful observations, Roy. Astron. Soc., Monthly Notices, 63:432.

Savage, J. R. 1953. Bibliography of nonparametric statistics and related topics, J. Am. Stat. Assn., 48:844 (Part L).

Saville, Thorndike. 1936. A study of methods of estimating flood flows applied to the Tennessee River. U. S. Geol. Survey, Water Supply Paper 771.

Scheffé, H., and J. W. Tukey. 1944. A formula for sample sizes for population tolerance limits, Ann. Math. Stats., 15:217.

Schmetterer, L. 1952. Über ein Beispiel aus der Statistik, Zeitschr. für angewandte Math. u. Mech., 32:281.

Schott, C. A. 1877. On Peirce's criterion, Proc. Am. Acad. Arts Sci. (New Series), 5:350.

Schultz, H. 1930. The standard error of a forecast from a curve, J. Am. Stat. Assn., 25:139.

Schützenberger, Marcel-Paul. 1948. An ABAC for the sample range, Psychometrika, 13:95.

Schwob, H. H. 1953. Iowa Floods Magnitude and Frequency. Iowa Highway Research Board, Bull. No. 1.

Severo, Norman C., and Edwin G. Olds. 1956. A comparison of tests on the mean of a logarithmico-normal distribution with known variance, Ann. Math. Stats., 27:670.

Sherman, B. 1950. A random variable related to the spacing of sample values, Ann. Math. Stats., 21:339.

—— 1957. Percentiles of the ω_n statistic, Ann. Math. Stats. 28:259.

Sherman, L. K. 1932. Stream-flow from rainfall by the unit-graph method, Eng. News-Record, 108:501.

Shimada, Shozo. 1954. Power of R-Charts, Rep. Stat. Appl. Res. JUSE (Japan), 3:14.

—— 1957. Moments of order statistics drawn from exponential distribution, Rep. Stat. Appl. Res. JUSE, 4:43.

Shone, K. J. 1949. Relation between the standard deviation and the distribution of range in non-normal populations, J. Roy. Stat. Soc., B, 11:85.

Shuh-Chai Lee. 1952. The Return Period and Variability of Floods in China. In: Essays and Papers in Memory of Late President Fu Ssu-nien. Taipan.

Simon, L. E. 1941. An Engineer's Manual of Statistical Methods. New York, Wiley.

Sibyua, M., and H. Toda. 1957. Tables of the probability density function of range in normal samples, Ann. Inst. Stat. Math. (Japan), 8:155.

Siotani, Minoru. 1956. Order statistics for discrete case with a numerical application to the binomial distribution, Ann. Inst. Stat. Math., Tokyo, 8:95.

Slade, J. J. 1936a. The reliability of statistical methods in the determination of flood frequencies. U. S. Geol. Survey, Water Supply Paper 771. P. 421.

—— 1936b. An asymmetric probability function, Trans. Am. Soc. Civ. Eng., 101:35

Smirnov, N. V. 1933. On the probability of large deviations, Math. Sbornik, 40:443 (Russian).

—— 1935. Über die Verteilung des allgemeinen Gliedes in der Variationsreihe, Metron, 12:59.

—— 1936. Sur la distribution de ω^2. Compt. rend., Ac. Sc. 202:449.

—— 1939. Sur les écarts de la courbe de distribution empirique, Matematicheskij Sbornik N.S. 6:3 (Russian).

—— 1941. On the estimation of the maximum term in a series of observations, Doklady Akad. Nauk SSSR (N.S.), 33:346.

—— 1952. Limit Distributions for the Terms of a Variational Series. Am. Math. Soc., Translation No. 67.

Standish, Hall. 1942. Discussion on statistical analysis in hydrology, Proc. Am. Soc. Civ. Eng., 68:458.

Stange, K. 1955. Zur Ermittlung der Abgangslinie für wirtschaftliche und technische Gesamtheiten, Mitteilungsblatt für Math. Stat., 7:113.

Steffensen, J. F. 1941. On the ω^2 test of dependence between statistical variables, Skand. Akt., Tidsk., 24:13.

Stewart, R. M. 1920a. Peirce's criterion, Pop. Astron., 27:2.
────── 1920b. The treatment of discordant observations, Pop. Astron., 28:4.
Stone, E. J. 1867–68. On the rejection of discordant observations, Roy. Astron. Soc., Monthly Notices, 28:165
────── 1873–74. On the rejection of discordant observations, Roy. Astron. Soc., Monthly Notices, 34:9.
────── 1874–75. Note on a discussion relating to the rejection of discordant observations, Roy. Astron. Soc., Monthly Notices, 35:107.
Streiff, A. 1927. Straight line plotting of skew frequency data, Trans. Am. Soc. Civ. Eng., 91:46.
"Student." 1927. Errors of routine analysis, Biometrika, 19:151.
Sukhatme, P. V. 1936. On the analysis of k samples from exponential populations with especial reference to the problem of random intervals. In: Statistical Research Memoirs, ed. by J. Neyman and E. S. Pearson. London, Cambridge Univ. Press. Vol. I, p. 94.
Tanenhaus, Seaman J. 1947. The median as a typical value for Walker yarn abrader data, Textile Res. J., 17:281.
Teichroew, D. 1956. Tables of expected values of order statistics and products of order statistics for samples of size twenty and less from the normal distribution, Ann. Math. Stats., 27:410.
Thom, H. C. S. 1954. Frequency of maximum wind speeds, Proc. Am. Soc. Civ. Eng., 80: Sep. No. 539.
────── 1956. Revised winter outside design temperatures, Heating, Piping and Air Conditioning, Nov., 1956, p. 137.
Thomas, George W. 1955. Bounds for the ratio of range to standard deviation, Biometrika, 43:268.
Thomas, H. A., jr. 1948. Frequency of minor floods, J. Boston Soc. Civ. Eng., 35:425; Grad. School of Eng. Pub. No. 466, Harvard Univ., Cambridge, Mass.
Thompson, Catherine M. 1941. Tables of percentage points of the Chi-square distribution, Biometrika, 32:187.
Thompson, W. R. 1935. On a criterion for the rejection of observations and the distribution of the ratio of deviation to sample standard deviation, Ann. Math. Stats., 6:214.
────── 1936. On confidence ranges for the median and other expectation distributions for populations of unknown distribution form, Ann. Math. Stats., 7:122.
────── 1938. Biological applications of normal range and associated significance tests in ignorance of original distribution forms, Ann. Math. Stats., 9:281.
Thomson, G. W. 1956. Bounds for the ratio of range to standard deviation, Biometrika, 42:268.
Tiago de Oliveira, J. 1952. Distribution-free Methods for the Statistical Analysis in Geophysics, Publicações do Sindicato Nacional dos Engenheiros Geógrafos, No. 3.
────── 1955. Distribution-free tests of goodness of fitting for distribution functions, Fac. de Ciencias de Lisboa, V:113.
────── 1957. Estimators and tests for continuous populations with locations and dispersion parameters, Revista da Faculdade de Ciéncias de Lisboa, 2.a Série-A, 6:121.
Tippett, L. H. C. 1925. On the extreme individuals and the range of samples taken from a normal population, Biometrika, 17:364.
Todd, David K. 1953. Stream-flow frequency distributions in California, Trans. Am. Geophys. Union, 34:897.
Todhunter, J. 1949. A History of the Mathematical Theory of Probability from the Time of Pascal to that of Laplace. (Reprint of 1865 ed.) New York, Chelsea.
Torabella, L. V. 1952. The distribution of the mth variate in certain chains of serially dependent populations, Ann. Math. Stats., 23:646.
Torroja, E. 1949. Notes sur le coefficient de sécurité. Assoc. Int. Ponts et Charpentes.
Trautwine, John C. 1872. The Civil Engineer's Pocket-book. Philadelphia, Claxton, Remsen and Haffelfinger.

Tricomi, F. 1933. Determinazione del valore asintotico di un certo integrale, Rendiconti A. dei Lincei, 17:116.
Tsao, Chia Kuei. 1954. An extension of Massey's distribution of the maximum deviation between two-sample cumulative step functions, Ann. Math. Stats., 25:587.
Tucker, J. 1941. Statistical theory of the effect of dimensions and method of loading upon the modulus of rupture of beams, Proc. Am. Soc. Testing Materials, 41:1072.
——— 1945a. Effect of dimensions of specimens upon the precision of strength data, Proc. Am. Soc. Testing Materials, 45:952.
——— 1945b. The maximum stresses present at failure of brittle materials, Proc. Am. Soc. Testing Materials, 45:961.
——— 1945c. Effect of length on the strength of compression test specimens, Proc. Am. Soc. Testing Materials, 45:976.
Tukey, J. W. 1946. An inequality for deviations from medians, Ann. Math. Stats., 17:75.
——— 1955. Interpolations and approximations related to the normal range, Biometrika, 42:480.
Tukey, J. W., and W. G. Ran. 1956. A rejection criterion based upon the range, Biometrika, 43:418.
U. S. Dept. of Commerce. Weather Bureau. 1947. Generalized Estimates, Maximum Possible Precipitation over the United States, East of the 105th Meridian, Hydrometeorological Report No. 23.
U. S. Geological Survey. 1952. Floods in Youghiogheny and Kiskiminetas River Basins, Pennsylvania and Maryland, Frequency and Magnitude. Circular 204.
Urquhart, Leonard Church, et al. 1940. Civil Engineering Handbook. New York, McGraw-Hill.
Uzgoeren, N. T. 1954. The Asymptotic Development of the Distribution of the Extreme Values of a Sample. In: Studies in Mathematics and Mechanics Presented to R. Von Mises. New York, Academic Press.
Velz, C. J. 1950a. Utilization of natural purification capacity in sewage and industrial waste disposal, Sewage and Industrial Wastes, 22:1601.
——— 1950b. Graphical approach to statistics, part III. Use of skewed probability paper, Water and Sewage Works, 97:393.
Velz, C. J., and J. J. Gannon. 1953. Low flow characteristics of streams, Ohio State Univ. Studies, Eng. Ser., 22:138.
Verhulst, P. F. 1845. Recherches mathématiques sur la loi d'accroisement de la population, Nouveaux mémoires de l'ac. de Bruxelles. Vol. 18.
Vincent, Paul. 1950. Quelques problèmes soulevés par l'étude de la mortalité aux âges élevés, Inst. Inter. Stat. Bull., 32:377.
——— 1951. La mortalité des vieillards, Population (Paris), 6:181.
Vora, S. A. 1951. Bounds on the distribution of Chi-square, Sankhyā, 11:365.
Waerden, B. L. van der, and E. Nievergelt. 1956. Tables for Comparing Two Samples by X-test and Sign Test. Berlin, Springer.
Wald, A. 1943. An extension of Wilks' method of setting tolerance limits, Ann. Math. Stats., 14:45.
——— 1947. Limit distribution of the maximum and minimum of successive cumulative sums of random variables, Bull. Am. Math. Soc., 53:142.
Wald, A., and J. Wolfowitz. 1946. Tolerance limits for a normal distribution, Ann. Math. Stats., 17:208.
Walsh, John E. 1946a. Some significance tests based on order statistics, Ann. Math. Stats., 17:44.
——— 1946b. Some order statistic distributions for samples of size four, Ann. Math. Stats., 17:246.
——— 1948. Some non-parametric tests of whether the largest observations of a set are too large (Prelim. Report), Bull. Am. Math. Soc., 54:1080.
——— 1949a. On the range-midrange test and some tests with bounded significance levels, Ann. Math. Stats., 20:257.

―――― 1949b. Some non-parametric tests of whether the largest observations of a set are too large or too small, Ann. Math. Stats., 21:583.

―――― 1950. Some estimates and tests based on the r smallest values in a sample, Ann. Math. Stats., 21:386.

Water Supply Commission of Pennsylvania. 1947. Floods. Water Resources Inventory Report, Pt. VIII.

Watson, G. N. 1948. A Treatise on the Theory of Bessel Functions. London, Cambridge Univ. Press.

Watson, G. S. 1952. Extreme value theory for m-dependent stationary sequences of continuous random variables, Ann. Math. Stats., 23:644.

―――― 1954. Extreme values in samples from m-dependent stationary stochastic processes, Ann. Math. Stats., 25:798.

Weber, K. H., and H. S. Endicott. 1956. Area effect and its extremal basis for the electric breakdown of transformer oil. Power Apparatus and Systems, 75:371.

―――― 1957a. Electrode area effect for the impulse breakdown of transformer oil, AIEE Transactions, Power Apparatus and Systems, 76:393.

―――― 1957b. Extremal area effect for large area electrodes for the electric breakdown of transformer oil, AIEE Transactions, 76:1051.

Weibull, W. 1939a. A statistical theory of strength of materials, Ing. Vet. Ak. Handl., no. 151. Stockholm.

―――― 1939b. The phenomenon of rupture in solids, Ing. Vet. Ak. Handl., no. 153. Stockholm.

―――― 1949. A statistical representation of fatigue failures in solids, Trans. Roy. Inst. Tech. (Stockholm), no. 27.

―――― 1951. A statistical distribution function of wide applicability, J. Appl. Mech., 18:293.

―――― 1952a. A survey of "statistical effects" in the field of material failure, Appl. Mech. Rev., 5:449.

―――― 1952b. Statistical design of fatigue experiments, Journal of Applied Mechanics, 19:109.

―――― 1955. New methods for computing parameters of complete or truncated distributions, Flygtekn. Försöksanstalt, Rep. 58.

Weiss, Leonard L. 1955. A nomogram based on the theory of extreme values for determining values for various return periods, Monthly Weather Rev., 83:69.

―――― 1957. Nomogram for log-normal frequency analysis, Trans. Am. Geophys. Union, 38:33

Welch, B. L. 1936. Specification of rules for rejecting too variable a product, with particular reference to an electric lamp, J. Roy. Stat. Soc., Supp. III:29.

Wemelsfelder, P. J. 1939. Wetmatigheden in het optreden van stormvloeden, De Ingenieur, 54:31.

Westenberg, J. 1948. Significance test for median and interquartile range in samples from continuous populations of any form, Wetenschappen Proc., 51:252.

Wilks, S. S. 1940. Confidence limits and critical differences between percentages, Public Opinion Quarterly, 4:332.

―――― 1941. Determination of sample sizes for setting tolerance limits, Ann. Math. Stats., 12:91.

―――― 1942. Statistical prediction with special reference to the problem of tolerance limits, Ann. Math. Stats., 13:400.

―――― 1943. Mathematical Statistics. Princeton, Princeton Univ. Press.

―――― 1948. Order statistics, Bull. Am. Math. Soc., 54:6.

Williams, Arthur, jr. 1950. On the choice of the number and width of classes for the Chi-square test of goodness of fit, J. Am. Stat. Assn., 45:77.

Wilson, E. B. 1923. First and second laws of error, Quart. Publ. Am. Stat. Assn., 18:841.

Wing, S. P., W. H. Price, and C. T. Douglass. 1944. Precision indices for compression tests of companion concrete cylinders, Proc. Am. Soc. Testing Materials, 44:839.

Winlock, J. 1856. On Professor Airy's objections to Peirce's criterion, Astron. J., 4:145.
Winston, C. B. 1946. Inequalities in terms of mean range, Biometrika, 33:283.
Wisler, C. D., and E. F. Brater. 1949. Hydrology. New York, Wiley.
Wolfowitz, J. 1946. Confidence limits for the fraction of a normal population which lies between two given limits, Ann. Math. Stats., 17:483.
Woodruff, Ralph S. 1952. Confidence intervals for medians and other position measures, J. Am. Stat. Assn., 47:635.
Working, Holbrook, and Harold Hotelling. 1929. The application of the theory of error to the interpretation of trends, J. Am. Stat. Assn., 24:73.
Yang, S. 1933. On partition values, J. Am. Stat. Assn., 28:184.
Yarnell, D. L. 1935. Rainfall intensity-frequency data. U. S. Dept. of Agric., Misc. pub. 204.
Zeigler, R. K. 1950. A note on the asymptotic simultaneous distribution of the sample median and the mean deviation from the sample median, Ann. Math. Stats., 21:542.

INDEX

Absolute extreme values, 94
Adyanthaya, 109
Aeronautics, 245
Anderson, 37
Asymptote, 156
Bachelier, 129
Barricelli, 184, 209, 241
Benard, 46
Benham, 237
Benson, 238
Berkson, 127
Bernoulli, D., 58, 63, 73
Bernoulli, N., 2
Bertrand, 94
Bessel functions, 3, 314
Birkeland, 241, 340
Bodhaine, 213, 237
Bortkiewicz, 2, 75, 97
Breakdown voltage, 249
Breaking strength, 148, 248, 302
Brownian movement, 148
Bulbs, 333
Cadwell, 338
Calculated risk, 24
Carlson, 36, 102
Carter, 213, 237
Cauchy Type, 149, 152, 155, 162, 326, 330, 346
Center, 109
Central value, 42
Chandra Sekar, 141
Characteristic largest (smallest) deviation, 95, 96
Characteristic largest (smallest) value, 82
Characteristic oldest age at death, 247
Characteristic product, 125
Characteristic range, 309
Charlier, 94
Chauvenet, 2, 83
Chi square test, 28
Chow, 179
Classes of exponential type, 122, 125
Control band, 52, 54, 216
Control interval, 24, 215
Court, 24, 244
Courtagne, 178
Cramér, 110, 139, 170, 194
Critical quotient Q, 119, 123, 127, 146, 149 150

Critical quotient q, 185
Cross, 179, 236
Dalrymple, 238
Daniels, 148
Darling, 37
Darwin, 198, 218
D.derick, 97
Design factor, 25
Design flood, 238
Design value, 24
Deviate, 140
Deviation, 94, 121
Dew point, 340
Dick, 218
Discharge, 40
Distance, 55, 197
Distribution, 7
Distribution of frequencies, 46
Distribution of repeated occurrences, 21
Dodd, 3, 75
Double exponential distribution, 166, 169
Drought, 299
Dutka, 2
Dutta, 243
Elfving, 3, 5, 312, 316
Endicott, 250
Epstein, 60, 249
Estimation of standard deviation, 338
Euler, 15
Exceedance, 21, 58
Exponential distribution, 113, 115, 219
Exponential type, 6, 113, 120, 149
Extrapolation, 67
Extremal intensity, 84
Extremal quotient, 111, 324, 325
Extremal statistic, 94, 306, 330
Extreme control band, 218
Extreme deviate, 140
Extreme range, 321
Fatigue failure, 302
Fermat, 23
Finetti, 75, 131
First double exponential distribution, 169
Fisher, 3, 5, 89, 136, 157, 173, 221, 270, 279, 298
Flood, 4, 177, 212, 236, 240, 272
Frank, 139
Fréchet, 3, 5, 29, 110, 158, 255, 272
Frequent exceedances, 72

Freudenthal, 302
Frogner, 340
Fuller, 178
Fung, 245
Galton, 16, 56
Galton's ratio, 56
Gamma distribution, 143, 182
Geological Survey, 177, 213, 236
Geology, 254
Geometric range, 324, 327
Gibrat, 16
Gnedenko, 3, 163, 169, 270, 276, 279, 289
Gompertz, 246
Grassberger, 16
Greenwood, 189
Grubbs, 141, 143, 222
Gumbel, 64, 75, 189, 238, 272, 300, 302, 325
Gurland, 166
Hagstroem, 23
Harley, 336
Harris, 67
Hartley, 99, 336, 338, 339
Hastings, 49
Hazen, 16, 32, 44, 206, 238
Herbach, 325
Homogeneity test, 212
Hotelling, 235
Independence of extremes, 110
Intensity function, 5, 20, 26, 75, 84
Interdependence, 164
Irwin, 55
Jasper, 252
Jenkinson, 163, 178
Johnson, A., 3
Johnson, L. G., 46
Kase, 249
Kawata, 49, 195
Kendall, 53, 136, 306
Kimball, 201, 210, 219, 229, 231, 233, 234, 276
Kolmogoroff, 16
Krumbein, 254
Lagrange, 90, 106
Landsberg, 343
Langbein, 79, 213
Laplace, 13, 15, 29, 47, 52, 119, 195
Largest deviation, 95
Leme, 3, 285
Lévy, 148
L'Hôpital's Rule, 118, 126, 151, 167, 171
Lieblein, 212, 224, 226, 229, 254
Logarithmic normal distribution, 16, 146, 181

Logarithmic trend, 123
Logistic distribution, 126, 220, 311, 330
Longuet-Higgins, 251, 253
Lower tolerance limit, 65
Markoff, 1
Maximum, 201, 288
Maximum likelihood, 231, 268, 296
Maximum mean largest value, 89
Maximum mean range, 106
McKay, 98, 140
McMillan, 307
Méré, 23
Midrange, 108, 311, 319, 331
Mill's ratio, 20
Minimum, 203, 288, 307
Minimum life, 303
Mises, 2, 3, 5, 37, 75, 149, 170, 262, 273, 279
Mitra, 106
Modal largest range, 323
Mode in the two sample problem, 199
Moment quotient, 136
Moriguti, 93, 106
Mosteller, 54
Most probable rank, 46
Mutual symmetry, 9
Nag, 243
Nair, 141, 222
National Bureau of Standards, 3, 169, 189, 195, 202, 204, 205, 212, 215, 318, 334
Naval Engineering, 251
Naval Proving Ground, Dahlgren, Virginia, 318
Newman, 212
Newton, 2
New Zealand Ministry of Works, 237
Neyman, 75
Nickel, 303
Normal distribution, 30, 38, 50, 54, 129, 147, 180, 219, 298, 336
Normal exceedances, 70
Normal extremes, 129, 298
Normal extreme deviate, 140
Normal range, 336
Observed distribution, 28
Occurrence interval, 44, 134, 202, 269
Ohio Water Resources Board, 236
Oldest ages, 246
Olmstead, 23
Order statistic, 42, 116, 187, 201, 223
Pareto distribution, 45, 151, 157
Pareto Type, 149, 152, 259, 327
Pearl, 126
Pearson, E. S., 3, 98, 109, 131, 141, 336, 338, 339

INDEX

Pearson, K., 20, 43, 56, 67, 97, 100, 129, 133, 182, 194, 245
Peirce, 2
Penultimate distribution, 279
Plackett, 89, 106
Plotting position, 29, 206
Poisson, 2, 72, 74, 143, 189, 345
Pólya, 1
Potter, 237
Powell, 176
Precipitation, 243
Press, 245
Pressure, 241, 272
Probability paper, 28, 31, 94, 176, 184, 250, 261, 277, 301, 333, 337, 341
Pseudosymmetry, 285, 293
Purcell, 331
Quantile, 79
Rainfall, 272
Range, 97, 306
Range of minima, 307
Range of range, 324
Rantz, 179, 236
Rare exceedance, 69
Ravilly, 302
Rayleigh, 251, 285
Reciprocal moments, 10, 267
Reciprocal transformation, 8
Reduced mth largest value, 168
Reduced range, 309
Reduced variate, 7
Return period, 5, 20, 23, 78, 80, 82, 176, 178, 180, 215, 239, 269, 286
Rider, 2
Riggs, 236
Robbins, 101
Robinson, 213, 237
Romanowsky, 87, 102
Rubber, 250
St. Pierre, 55
Schelling, 64, 174, 320
Schwarz, 93, 106
Second double exponential distribution, 169
Sherman, 37

Shuh, 236
Sibuya, 99
Siotani, 76, 101
Smirnov, 37
Spence, 212
Stability, 117, 157
Stirling's formula, 16, 65, 67, 70, 73
"Student," 97
Sufficient estimation function, 229
Symmetry, 8
Symmetry principle, 76
Temperature, 4, 242, 244, 272, 340
Thom, 244, 271
Thomas, 59
Thompson, 141
Tiago de Oliveira, 37
Tippett, 3, 5, 75, 97, 100, 129, 131, 136, 157, 173, 219, 270, 279, 298, 336, 339
Toda, 99
Todhunter, 2
Tolerance limit, 64, 103
Tricome, 75, 129
Two sample problem, 198
Type of initial distribution, 163
Uniform distribution, 45, 276
U.S. Geological Survey, 177, 213, 236
U.S. Soil Conservation Service, 237
Uzgoeren, 171, 222
Variability index, 13
Variance of forecast, 234
Verhulst, 126
Waerden, 40
Water year, 5
Watson, G. N., 314, 315
Watson, G. S., 164
Watson Laboratories, 226
Weaver, 225
Weber, 250
Weibull, 47, 279, 302
Weiss, 227
Wilks, 55, 58, 65, 103
Wilson, 29
Wind speed, 244, 272
Zanger, 55

A CATALOG OF SELECTED
DOVER BOOKS
IN SCIENCE AND MATHEMATICS

CATALOG OF DOVER BOOKS

Astronomy

BURNHAM'S CELESTIAL HANDBOOK, Robert Burnham, Jr. Thorough guide to the stars beyond our solar system. Exhaustive treatment. Alphabetical by constellation: Andromeda to Cetus in Vol. 1; Chamaeleon to Orion in Vol. 2; and Pavo to Vulpecula in Vol. 3. Hundreds of illustrations. Index in Vol. 3. 2,000pp. 6⅛ x 9¼.
Vol. I: 23567-X
Vol. II: 23568-8
Vol. III: 23673-0

EXPLORING THE MOON THROUGH BINOCULARS AND SMALL TELESCOPES, Ernest H. Cherrington, Jr. Informative, profusely illustrated guide to locating and identifying craters, rills, seas, mountains, other lunar features. Newly revised and updated with special section of new photos. Over 100 photos and diagrams. 240pp. 8¼ x 11. 24491-1

THE EXTRATERRESTRIAL LIFE DEBATE, 1750–1900, Michael J. Crowe. First detailed, scholarly study in English of the many ideas that developed from 1750 to 1900 regarding the existence of intelligent extraterrestrial life. Examines ideas of Kant, Herschel, Voltaire, Percival Lowell, many other scientists and thinkers. 16 illustrations. 704pp. 5⅜ x 8½. 40675-X

THEORIES OF THE WORLD FROM ANTIQUITY TO THE COPERNICAN REVOLUTION, Michael J. Crowe. Newly revised edition of an accessible, enlightening book recreates the change from an earth-centered to a sun-centered conception of the solar system. 242pp. 5⅜ x 8½. 41444-2

A HISTORY OF ASTRONOMY, A. Pannekoek. Well-balanced, carefully reasoned study covers such topics as Ptolemaic theory, work of Copernicus, Kepler, Newton, Eddington's work on stars, much more. Illustrated. References. 521pp. 5⅜ x 8½.
65994-1

A COMPLETE MANUAL OF AMATEUR ASTRONOMY: Tools and Techniques for Astronomical Observations, P. Clay Sherrod with Thomas L. Koed. Concise, highly readable book discusses: selecting, setting up and maintaining a telescope; amateur studies of the sun; lunar topography and occultations; observations of Mars, Jupiter, Saturn, the minor planets and the stars; an introduction to photoelectric photometry; more. 1981 ed. 124 figures. 26 halftones. 37 tables. 335pp. 6½ x 9¼.
42820-6

AMATEUR ASTRONOMER'S HANDBOOK, J. B. Sidgwick. Timeless, comprehensive coverage of telescopes, mirrors, lenses, mountings, telescope drives, micrometers, spectroscopes, more. 189 illustrations. 576pp. 5⅝ x 8¼. (Available in U.S. only.)
24034-7

STARS AND RELATIVITY, Ya. B. Zel'dovich and I. D. Novikov. Vol. 1 of *Relativistic Astrophysics* by famed Russian scientists. General relativity, properties of matter under astrophysical conditions, stars, and stellar systems. Deep physical insights, clear presentation. 1971 edition. References. 544pp. 5⅝ x 8¼. 69424-0

Chemistry

THE SCEPTICAL CHYMIST: The Classic 1661 Text, Robert Boyle. Boyle defines the term "element," asserting that all natural phenomena can be explained by the motion and organization of primary particles. 1911 ed. viii+232pp. 5⅜ x 8½. 42825-7

RADIOACTIVE SUBSTANCES, Marie Curie. Here is the celebrated scientist's doctoral thesis, the prelude to her receipt of the 1903 Nobel Prize. Curie discusses establishing atomic character of radioactivity found in compounds of uranium and thorium; extraction from pitchblende of polonium and radium; isolation of pure radium chloride; determination of atomic weight of radium; plus electric, photographic, luminous, heat, color effects of radioactivity. ii+94pp. 5⅜ x 8½. 42550-9

CHEMICAL MAGIC, Leonard A. Ford. Second Edition, Revised by E. Winston Grundmeier. Over 100 unusual stunts demonstrating cold fire, dust explosions, much more. Text explains scientific principles and stresses safety precautions. 128pp. 5⅜ x 8½. 67628-5

THE DEVELOPMENT OF MODERN CHEMISTRY, Aaron J. Ihde. Authoritative history of chemistry from ancient Greek theory to 20th-century innovation. Covers major chemists and their discoveries. 209 illustrations. 14 tables. Bibliographies. Indices. Appendices. 851pp. 5⅜ x 8½. 64235-6

CATALYSIS IN CHEMISTRY AND ENZYMOLOGY, William P. Jencks. Exceptionally clear coverage of mechanisms for catalysis, forces in aqueous solution, carbonyl- and acyl-group reactions, practical kinetics, more. 864pp. 5⅜ x 8½. 65460-5

ELEMENTS OF CHEMISTRY, Antoine Lavoisier. Monumental classic by founder of modern chemistry in remarkable reprint of rare 1790 Kerr translation. A must for every student of chemistry or the history of science. 539pp. 5⅜ x 8½. 64624-6

THE HISTORICAL BACKGROUND OF CHEMISTRY, Henry M. Leicester. Evolution of ideas, not individual biography. Concentrates on formulation of a coherent set of chemical laws. 260pp. 5⅜ x 8½. 61053-5

A SHORT HISTORY OF CHEMISTRY, J. R. Partington. Classic exposition explores origins of chemistry, alchemy, early medical chemistry, nature of atmosphere, theory of valency, laws and structure of atomic theory, much more. 428pp. 5⅜ x 8½. (Available in U.S. only.) 65977-1

GENERAL CHEMISTRY, Linus Pauling. Revised 3rd edition of classic first-year text by Nobel laureate. Atomic and molecular structure, quantum mechanics, statistical mechanics, thermodynamics correlated with descriptive chemistry. Problems. 992pp. 5⅜ x 8½. 65622-5

FROM ALCHEMY TO CHEMISTRY, John Read. Broad, humanistic treatment focuses on great figures of chemistry and ideas that revolutionized the science. 50 illustrations. 240pp. 5⅜ x 8½. 28690-8

CATALOG OF DOVER BOOKS

Engineering

DE RE METALLICA, Georgius Agricola. The famous Hoover translation of greatest treatise on technological chemistry, engineering, geology, mining of early modern times (1556). All 289 original woodcuts. 638pp. 6¾ x 11. 60006-8

FUNDAMENTALS OF ASTRODYNAMICS, Roger Bate et al. Modern approach developed by U.S. Air Force Academy. Designed as a first course. Problems, exercises. Numerous illustrations. 455pp. 5⅜ x 8½. 60061-0

DYNAMICS OF FLUIDS IN POROUS MEDIA, Jacob Bear. For advanced students of ground water hydrology, soil mechanics and physics, drainage and irrigation engineering, and more. 335 illustrations. Exercises, with answers. 784pp. 6⅛ x 9¼. 65675-6

THEORY OF VISCOELASTICITY (Second Edition), Richard M. Christensen. Complete, consistent description of the linear theory of the viscoelastic behavior of materials. Problem-solving techniques discussed. 1982 edition. 29 figures. xiv+364pp. 6⅛ x 9¼. 42880-X

MECHANICS, J. P. Den Hartog. A classic introductory text or refresher. Hundreds of applications and design problems illuminate fundamentals of trusses, loaded beams and cables, etc. 334 answered problems. 462pp. 5⅜ x 8½. 60754-2

MECHANICAL VIBRATIONS, J. P. Den Hartog. Classic textbook offers lucid explanations and illustrative models, applying theories of vibrations to a variety of practical industrial engineering problems. Numerous figures. 233 problems, solutions. Appendix. Index. Preface. 436pp. 5⅜ x 8½. 64785-4

STRENGTH OF MATERIALS, J. P. Den Hartog. Full, clear treatment of basic material (tension, torsion, bending, etc.) plus advanced material on engineering methods, applications. 350 answered problems. 323pp. 5⅜ x 8½. 60755-0

A HISTORY OF MECHANICS, René Dugas. Monumental study of mechanical principles from antiquity to quantum mechanics. Contributions of ancient Greeks, Galileo, Leonardo, Kepler, Lagrange, many others. 671pp. 5⅜ x 8½. 65632-2

STABILITY THEORY AND ITS APPLICATIONS TO STRUCTURAL MECHANICS, Clive L. Dym. Self-contained text focuses on Koiter postbuckling analyses, with mathematical notions of stability of motion. Basing minimum energy principles for static stability upon dynamic concepts of stability of motion, it develops asymptotic buckling and postbuckling analyses from potential energy considerations, with applications to columns, plates, and arches. 1974 ed. 208pp. 5⅜ x 8½. 42541-X

METAL FATIGUE, N. E. Frost, K. J. Marsh, and L. P. Pook. Definitive, clearly written, and well-illustrated volume addresses all aspects of the subject, from the historical development of understanding metal fatigue to vital concepts of the cyclic stress that causes a crack to grow. Includes 7 appendixes. 544pp. 5⅜ x 8½. 40927-9

CATALOG OF DOVER BOOKS

ROCKETS, Robert Goddard. Two of the most significant publications in the history of rocketry and jet propulsion: "A Method of Reaching Extreme Altitudes" (1919) and "Liquid Propellant Rocket Development" (1936). 128pp. 5⅜ x 8½. 42537-1

STATISTICAL MECHANICS: Principles and Applications, Terrell L. Hill. Standard text covers fundamentals of statistical mechanics, applications to fluctuation theory, imperfect gases, distribution functions, more. 448pp. 5⅜ x 8½. 65390-0

ENGINEERING AND TECHNOLOGY 1650–1750: Illustrations and Texts from Original Sources, Martin Jensen. Highly readable text with more than 200 contemporary drawings and detailed engravings of engineering projects dealing with surveying, leveling, materials, hand tools, lifting equipment, transport and erection, piling, bailing, water supply, hydraulic engineering, and more. Among the specific projects outlined–transporting a 50-ton stone to the Louvre, erecting an obelisk, building timber locks, and dredging canals. 207pp. 8⅜ x 11¼. 42232-1

THE VARIATIONAL PRINCIPLES OF MECHANICS, Cornelius Lanczos. Graduate level coverage of calculus of variations, equations of motion, relativistic mechanics, more. First inexpensive paperbound edition of classic treatise. Index. Bibliography. 418pp. 5⅜ x 8½. 65067-7

PROTECTION OF ELECTRONIC CIRCUITS FROM OVERVOLTAGES, Ronald B. Standler. Five-part treatment presents practical rules and strategies for circuits designed to protect electronic systems from damage by transient overvoltages. 1989 ed. xxiv+434pp. 6⅛ x 9¼. 42552-5

ROTARY WING AERODYNAMICS, W. Z. Stepniewski. Clear, concise text covers aerodynamic phenomena of the rotor and offers guidelines for helicopter performance evaluation. Originally prepared for NASA. 537 figures. 640pp. 6⅛ x 9¼. 64647-5

INTRODUCTION TO SPACE DYNAMICS, William Tyrrell Thomson. Comprehensive, classic introduction to space-flight engineering for advanced undergraduate and graduate students. Includes vector algebra, kinematics, transformation of coordinates. Bibliography. Index. 352pp. 5⅜ x 8½. 65113-4

HISTORY OF STRENGTH OF MATERIALS, Stephen P. Timoshenko. Excellent historical survey of the strength of materials with many references to the theories of elasticity and structure. 245 figures. 452pp. 5⅜ x 8½. 61187-6

ANALYTICAL FRACTURE MECHANICS, David J. Unger. Self-contained text supplements standard fracture mechanics texts by focusing on analytical methods for determining crack-tip stress and strain fields. 336pp. 6⅛ x 9¼. 41737-9

STATISTICAL MECHANICS OF ELASTICITY, J. H. Weiner. Advanced, self-contained treatment illustrates general principles and elastic behavior of solids. Part 1, based on classical mechanics, studies thermoelastic behavior of crystalline and polymeric solids. Part 2, based on quantum mechanics, focuses on interatomic force laws, behavior of solids, and thermally activated processes. For students of physics and chemistry and for polymer physicists. 1983 ed. 96 figures. 496pp. 5⅜ x 8½. 42260-7

Mathematics

FUNCTIONAL ANALYSIS (Second Corrected Edition), George Bachman and Lawrence Narici. Excellent treatment of subject geared toward students with background in linear algebra, advanced calculus, physics, and engineering. Text covers introduction to inner-product spaces, normed, metric spaces, and topological spaces; complete orthonormal sets, the Hahn-Banach Theorem and its consequences, and many other related subjects. 1966 ed. 544pp. 6⅛ x 9¼. 40251-7

ASYMPTOTIC EXPANSIONS OF INTEGRALS, Norman Bleistein & Richard A. Handelsman. Best introduction to important field with applications in a variety of scientific disciplines. New preface. Problems. Diagrams. Tables. Bibliography. Index. 448pp. 5⅜ x 8½. 65082-0

VECTOR AND TENSOR ANALYSIS WITH APPLICATIONS, A. I. Borisenko and I. E. Tarapov. Concise introduction. Worked-out problems, solutions, exercises. 257pp. 5⅝ x 8¼. 63833-2

THE ABSOLUTE DIFFERENTIAL CALCULUS (CALCULUS OF TENSORS), Tullio Levi-Civita. Great 20th-century mathematician's classic work on material necessary for mathematical grasp of theory of relativity. 452pp. 5⅝ x 8¼. 63401-9

AN INTRODUCTION TO ORDINARY DIFFERENTIAL EQUATIONS, Earl A. Coddington. A thorough and systematic first course in elementary differential equations for undergraduates in mathematics and science, with many exercises and problems (with answers). Index. 304pp. 5⅜ x 8½. 65942-9

FOURIER SERIES AND ORTHOGONAL FUNCTIONS, Harry F. Davis. An incisive text combining theory and practical example to introduce Fourier series, orthogonal functions and applications of the Fourier method to boundary-value problems. 570 exercises. Answers and notes. 416pp. 5⅜ x 8½. 65973-9

COMPUTABILITY AND UNSOLVABILITY, Martin Davis. Classic graduate-level introduction to theory of computability, usually referred to as theory of recurrent functions. New preface and appendix. 288pp. 5⅜ x 8½. 61471-9

ASYMPTOTIC METHODS IN ANALYSIS, N. G. de Bruijn. An inexpensive, comprehensive guide to asymptotic methods–the pioneering work that teaches by explaining worked examples in detail. Index. 224pp. 5⅜ x 8½ 64221-6

APPLIED COMPLEX VARIABLES, John W. Dettman. Step-by-step coverage of fundamentals of analytic function theory–plus lucid exposition of five important applications: Potential Theory; Ordinary Differential Equations; Fourier Transforms; Laplace Transforms; Asymptotic Expansions. 66 figures. Exercises at chapter ends. 512pp. 5⅜ x 8½. 64670-X

INTRODUCTION TO LINEAR ALGEBRA AND DIFFERENTIAL EQUATIONS, John W. Dettman. Excellent text covers complex numbers, determinants, orthonormal bases, Laplace transforms, much more. Exercises with solutions. Undergraduate level. 416pp. 5⅜ x 8½. 65191-6

CATALOG OF DOVER BOOKS

CALCULUS OF VARIATIONS WITH APPLICATIONS, George M. Ewing. Applications-oriented introduction to variational theory develops insight and promotes understanding of specialized books, research papers. Suitable for advanced undergraduate/graduate students as primary, supplementary text. 352pp. 5⅜ x 8½.
64856-7

COMPLEX VARIABLES, Francis J. Flanigan. Unusual approach, delaying complex algebra till harmonic functions have been analyzed from real variable viewpoint. Includes problems with answers. 364pp. 5⅜ x 8½.
61388-7

AN INTRODUCTION TO THE CALCULUS OF VARIATIONS, Charles Fox. Graduate-level text covers variations of an integral, isoperimetrical problems, least action, special relativity, approximations, more. References. 279pp. 5⅜ x 8½.
65499-0

COUNTEREXAMPLES IN ANALYSIS, Bernard R. Gelbaum and John M. H. Olmsted. These counterexamples deal mostly with the part of analysis known as "real variables." The first half covers the real number system, and the second half encompasses higher dimensions. 1962 edition. xxiv+198pp. 5⅜ x 8½.
42875-3

CATASTROPHE THEORY FOR SCIENTISTS AND ENGINEERS, Robert Gilmore. Advanced-level treatment describes mathematics of theory grounded in the work of Poincaré, R. Thom, other mathematicians. Also important applications to problems in mathematics, physics, chemistry, and engineering. 1981 edition. References. 28 tables. 397 black-and-white illustrations. xvii+666pp. 6⅛ x 9¼.
67539-4

INTRODUCTION TO DIFFERENCE EQUATIONS, Samuel Goldberg. Exceptionally clear exposition of important discipline with applications to sociology, psychology, economics. Many illustrative examples; over 250 problems. 260pp. 5⅜ x 8½.
65084-7

NUMERICAL METHODS FOR SCIENTISTS AND ENGINEERS, Richard Hamming. Classic text stresses frequency approach in coverage of algorithms, polynomial approximation, Fourier approximation, exponential approximation, other topics. Revised and enlarged 2nd edition. 721pp. 5⅜ x 8½.
65241-6

INTRODUCTION TO NUMERICAL ANALYSIS (2nd Edition), F. B. Hildebrand. Classic, fundamental treatment covers computation, approximation, interpolation, numerical differentiation and integration, other topics. 150 new problems. 669pp. 5⅜ x 8½.
65363-3

THREE PEARLS OF NUMBER THEORY, A. Y. Khinchin. Three compelling puzzles require proof of a basic law governing the world of numbers. Challenges concern van der Waerden's theorem, the Landau-Schnirelmann hypothesis and Mann's theorem, and a solution to Waring's problem. Solutions included. 64pp. 5⅜ x 8½.
40026-3

THE PHILOSOPHY OF MATHEMATICS: An Introductory Essay, Stephan Körner. Surveys the views of Plato, Aristotle, Leibniz & Kant concerning propositions and theories of applied and pure mathematics. Introduction. Two appendices. Index. 198pp. 5⅜ x 8½.
25048-2

CATALOG OF DOVER BOOKS

INTRODUCTORY REAL ANALYSIS, A.N. Kolmogorov, S. V. Fomin. Translated by Richard A. Silverman. Self-contained, evenly paced introduction to real and functional analysis. Some 350 problems. 403pp. 5⅜ x 8½. 61226-0

APPLIED ANALYSIS, Cornelius Lanczos. Classic work on analysis and design of finite processes for approximating solution of analytical problems. Algebraic equations, matrices, harmonic analysis, quadrature methods, more. 559pp. 5⅜ x 8½. 65656-X

AN INTRODUCTION TO ALGEBRAIC STRUCTURES, Joseph Landin. Superb self-contained text covers "abstract algebra": sets and numbers, theory of groups, theory of rings, much more. Numerous well-chosen examples, exercises. 247pp. 5⅜ x 8½. 65940-2

QUALITATIVE THEORY OF DIFFERENTIAL EQUATIONS, V. V. Nemytskii and V.V. Stepanov. Classic graduate-level text by two prominent Soviet mathematicians covers classical differential equations as well as topological dynamics and ergodic theory. Bibliographies. 523pp. 5⅜ x 8½. 65954-2

THEORY OF MATRICES, Sam Perlis. Outstanding text covering rank, nonsingularity and inverses in connection with the development of canonical matrices under the relation of equivalence, and without the intervention of determinants. Includes exercises. 237pp. 5⅜ x 8½. 66810-X

INTRODUCTION TO ANALYSIS, Maxwell Rosenlicht. Unusually clear, accessible coverage of set theory, real number system, metric spaces, continuous functions, Riemann integration, multiple integrals, more. Wide range of problems. Undergraduate level. Bibliography. 254pp. 5⅜ x 8½. 65038-3

MODERN NONLINEAR EQUATIONS, Thomas L. Saaty. Emphasizes practical solution of problems; covers seven types of equations. ". . . a welcome contribution to the existing literature. . . . "–*Math Reviews*. 490pp. 5⅜ x 8½. 64232-1

MATRICES AND LINEAR ALGEBRA, Hans Schneider and George Phillip Barker. Basic textbook covers theory of matrices and its applications to systems of linear equations and related topics such as determinants, eigenvalues, and differential equations. Numerous exercises. 432pp. 5⅜ x 8½. 66014-1

MATHEMATICS APPLIED TO CONTINUUM MECHANICS, Lee A. Segel. Analyzes models of fluid flow and solid deformation. For upper-level math, science, and engineering students. 608pp. 5⅜ x 8½. 65369-2

ELEMENTS OF REAL ANALYSIS, David A. Sprecher. Classic text covers fundamental concepts, real number system, point sets, functions of a real variable, Fourier series, much more. Over 500 exercises. 352pp. 5⅜ x 8½. 65385-4

SET THEORY AND LOGIC, Robert R. Stoll. Lucid introduction to unified theory of mathematical concepts. Set theory and logic seen as tools for conceptual understanding of real number system. 496pp. 5⅜ x 8¼. 63829-4

CATALOG OF DOVER BOOKS

TENSOR CALCULUS, J.L. Synge and A. Schild. Widely used introductory text covers spaces and tensors, basic operations in Riemannian space, non-Riemannian spaces, etc. 324pp. 5⅜ x 8¼. 63612-7

ORDINARY DIFFERENTIAL EQUATIONS, Morris Tenenbaum and Harry Pollard. Exhaustive survey of ordinary differential equations for undergraduates in mathematics, engineering, science. Thorough analysis of theorems. Diagrams. Bibliography. Index. 818pp. 5⅜ x 8½. 64940-7

INTEGRAL EQUATIONS, F. G. Tricomi. Authoritative, well-written treatment of extremely useful mathematical tool with wide applications. Volterra Equations, Fredholm Equations, much more. Advanced undergraduate to graduate level. Exercises. Bibliography. 238pp. 5⅜ x 8½. 64828-1

FOURIER SERIES, Georgi P. Tolstov. Translated by Richard A. Silverman. A valuable addition to the literature on the subject, moving clearly from subject to subject and theorem to theorem. 107 problems, answers. 336pp. 5⅜ x 8½. 63317-9

INTRODUCTION TO MATHEMATICAL THINKING, Friedrich Waismann. Examinations of arithmetic, geometry, and theory of integers; rational and natural numbers; complete induction; limit and point of accumulation; remarkable curves; complex and hypercomplex numbers, more. 1959 ed. 27 figures. xii+260pp. 5⅜ x 8½. 42804-4

POPULAR LECTURES ON MATHEMATICAL LOGIC, Hao Wang. Noted logician's lucid treatment of historical developments, set theory, model theory, recursion theory and constructivism, proof theory, more. 3 appendixes. Bibliography. 1981 ed. ix+283pp. 5⅜ x 8½. 67632-3

CALCULUS OF VARIATIONS, Robert Weinstock. Basic introduction covering isoperimetric problems, theory of elasticity, quantum mechanics, electrostatics, etc. Exercises throughout. 326pp. 5⅜ x 8½. 63069-2

THE CONTINUUM: A Critical Examination of the Foundation of Analysis, Hermann Weyl. Classic of 20th-century foundational research deals with the conceptual problem posed by the continuum. 156pp. 5⅜ x 8½. 67982-9

CHALLENGING MATHEMATICAL PROBLEMS WITH ELEMENTARY SOLUTIONS, A. M. Yaglom and I. M. Yaglom. Over 170 challenging problems on probability theory, combinatorial analysis, points and lines, topology, convex polygons, many other topics. Solutions. Total of 445pp. 5⅜ x 8½. Two-vol. set.
Vol. I: 65536-9 Vol. II: 65537-7

INTRODUCTION TO PARTIAL DIFFERENTIAL EQUATIONS WITH APPLICATIONS, E. C. Zachmanoglou and Dale W. Thoe. Essentials of partial differential equations applied to common problems in engineering and the physical sciences. Problems and answers. 416pp. 5⅜ x 8½. 65251-3

THE THEORY OF GROUPS, Hans J. Zassenhaus. Well-written graduate-level text acquaints reader with group-theoretic methods and demonstrates their usefulness in mathematics. Axioms, the calculus of complexes, homomorphic mapping, p-group theory, more. 276pp. 5⅜ x 8½. 40922-8

Math–Decision Theory, Statistics, Probability

ELEMENTARY DECISION THEORY, Herman Chernoff and Lincoln E. Moses. Clear introduction to statistics and statistical theory covers data processing, probability and random variables, testing hypotheses, much more. Exercises. 364pp. 5⅜ x 8½. 65218-1

STATISTICS MANUAL, Edwin L. Crow et al. Comprehensive, practical collection of classical and modern methods prepared by U.S. Naval Ordnance Test Station. Stress on use. Basics of statistics assumed. 288pp. 5⅜ x 8½. 60599-X

SOME THEORY OF SAMPLING, William Edwards Deming. Analysis of the problems, theory, and design of sampling techniques for social scientists, industrial managers, and others who find statistics important at work. 61 tables. 90 figures. xvii +602pp. 5⅜ x 8½. 64684-X

LINEAR PROGRAMMING AND ECONOMIC ANALYSIS, Robert Dorfman, Paul A. Samuelson and Robert M. Solow. First comprehensive treatment of linear programming in standard economic analysis. Game theory, modern welfare economics, Leontief input-output, more. 525pp. 5⅜ x 8½. 65491-5

PROBABILITY: An Introduction, Samuel Goldberg. Excellent basic text covers set theory, probability theory for finite sample spaces, binomial theorem, much more. 360 problems. Bibliographies. 322pp. 5⅜ x 8½. 65252-1

GAMES AND DECISIONS: Introduction and Critical Survey, R. Duncan Luce and Howard Raiffa. Superb nontechnical introduction to game theory, primarily applied to social sciences. Utility theory, zero-sum games, n-person games, decision-making, much more. Bibliography. 509pp. 5⅜ x 8½. 65943-7

INTRODUCTION TO THE THEORY OF GAMES, J. C. C. McKinsey. This comprehensive overview of the mathematical theory of games illustrates applications to situations involving conflicts of interest, including economic, social, political, and military contexts. Appropriate for advanced undergraduate and graduate courses; advanced calculus a prerequisite. 1952 ed. x+372pp. 5⅜ x 8½. 42811-7

FIFTY CHALLENGING PROBLEMS IN PROBABILITY WITH SOLUTIONS, Frederick Mosteller. Remarkable puzzlers, graded in difficulty, illustrate elementary and advanced aspects of probability. Detailed solutions. 88pp. 5⅜ x 8½. 65355-2

PROBABILITY THEORY: A Concise Course, Y. A. Rozanov. Highly readable, self-contained introduction covers combination of events, dependent events, Bernoulli trials, etc. 148pp. 5⅜ x 8¼. 63544-9

STATISTICAL METHOD FROM THE VIEWPOINT OF QUALITY CONTROL, Walter A. Shewhart. Important text explains regulation of variables, uses of statistical control to achieve quality control in industry, agriculture, other areas. 192pp. 5⅜ x 8½. 65232-7

Math–Geometry and Topology

ELEMENTARY CONCEPTS OF TOPOLOGY, Paul Alexandroff. Elegant, intuitive approach to topology from set-theoretic topology to Betti groups; how concepts of topology are useful in math and physics. 25 figures. 57pp. 5⅜ x 8½. 60747-X

COMBINATORIAL TOPOLOGY, P. S. Alexandrov. Clearly written, well-organized, three-part text begins by dealing with certain classic problems without using the formal techniques of homology theory and advances to the central concept, the Betti groups. Numerous detailed examples. 654pp. 5⅜ x 8½. 40179-0

EXPERIMENTS IN TOPOLOGY, Stephen Barr. Classic, lively explanation of one of the byways of mathematics. Klein bottles, Moebius strips, projective planes, map coloring, problem of the Koenigsberg bridges, much more, described with clarity and wit. 43 figures. 210pp. 5⅜ x 8½. 25933-1

CONFORMAL MAPPING ON RIEMANN SURFACES, Harvey Cohn. Lucid, insightful book presents ideal coverage of subject. 334 exercises make book perfect for self-study. 55 figures. 352pp. 5⅝ x 8¼. 64025-6

THE GEOMETRY OF RENÉ DESCARTES, René Descartes. The great work founded analytical geometry. Original French text, Descartes's own diagrams, together with definitive Smith-Latham translation. 244pp. 5⅜ x 8½. 60068-8

PRACTICAL CONIC SECTIONS: The Geometric Properties of Ellipses, Parabolas and Hyperbolas, J. W. Downs. This text shows how to create ellipses, parabolas, and hyperbolas. It also presents historical background on their ancient origins and describes the reflective properties and roles of curves in design applications. 1993 ed. 98 figures. xii+100pp. 6½ x 9¼. 42876-1

THE THIRTEEN BOOKS OF EUCLID'S ELEMENTS, translated with introduction and commentary by Thomas L. Heath. Definitive edition. Textual and linguistic notes, mathematical analysis. 2,500 years of critical commentary. Unabridged. 1,414pp. 5⅜ x 8½. Three-vol. set. Vol. I: 60088-2 Vol. II: 60089-0 Vol. III: 60090-4

GEOMETRY OF COMPLEX NUMBERS, Hans Schwerdtfeger. Illuminating, widely praised book on analytic geometry of circles, the Moebius transformation, and two-dimensional non-Euclidean geometries. 200pp. 5⅝ x 8¼. 63830-8

DIFFERENTIAL GEOMETRY, Heinrich W. Guggenheimer. Local differential geometry as an application of advanced calculus and linear algebra. Curvature, transformation groups, surfaces, more. Exercises. 62 figures. 378pp. 5⅜ x 8½. 63433-7

CURVATURE AND HOMOLOGY: Enlarged Edition, Samuel I. Goldberg. Revised edition examines topology of differentiable manifolds; curvature, homology of Riemannian manifolds; compact Lie groups; complex manifolds; curvature, homology of Kaehler manifolds. New Preface. Four new appendixes. 416pp. 5⅜ x 8½. 40207-X

CATALOG OF DOVER BOOKS

History of Math

THE WORKS OF ARCHIMEDES, Archimedes (T. L. Heath, ed.). Topics include the famous problems of the ratio of the areas of a cylinder and an inscribed sphere; the measurement of a circle; the properties of conoids, spheroids, and spirals; and the quadrature of the parabola. Informative introduction. clxxxvi+326pp; supplement, 52pp. 5⅜ x 8½. 42084-1

A SHORT ACCOUNT OF THE HISTORY OF MATHEMATICS, W. W. Rouse Ball. One of clearest, most authoritative surveys from the Egyptians and Phoenicians through 19th-century figures such as Grassman, Galois, Riemann. Fourth edition. 522pp. 5⅜ x 8½. 20630-0

THE HISTORY OF THE CALCULUS AND ITS CONCEPTUAL DEVELOPMENT, Carl B. Boyer. Origins in antiquity, medieval contributions, work of Newton, Leibniz, rigorous formulation. Treatment is verbal. 346pp. 5⅜ x 8½. 60509-4

THE HISTORICAL ROOTS OF ELEMENTARY MATHEMATICS, Lucas N. H. Bunt, Phillip S. Jones, and Jack D. Bedient. Fundamental underpinnings of modern arithmetic, algebra, geometry, and number systems derived from ancient civilizations. 320pp. 5⅜ x 8½. 25563-8

A HISTORY OF MATHEMATICAL NOTATIONS, Florian Cajori. This classic study notes the first appearance of a mathematical symbol and its origin, the competition it encountered, its spread among writers in different countries, its rise to popularity, its eventual decline or ultimate survival. Original 1929 two-volume edition presented here in one volume. xxviii+820pp. 5⅜ x 8½. 67766-4

GAMES, GODS & GAMBLING: A History of Probability and Statistical Ideas, F. N. David. Episodes from the lives of Galileo, Fermat, Pascal, and others illustrate this fascinating account of the roots of mathematics. Features thought-provoking references to classics, archaeology, biography, poetry. 1962 edition. 304pp. 5⅜ x 8½. (Available in U.S. only.) 40023-9

OF MEN AND NUMBERS: The Story of the Great Mathematicians, Jane Muir. Fascinating accounts of the lives and accomplishments of history's greatest mathematical minds–Pythagoras, Descartes, Euler, Pascal, Cantor, many more. Anecdotal, illuminating. 30 diagrams. Bibliography. 256pp. 5⅜ x 8½. 28973-7

HISTORY OF MATHEMATICS, David E. Smith. Nontechnical survey from ancient Greece and Orient to late 19th century; evolution of arithmetic, geometry, trigonometry, calculating devices, algebra, the calculus. 362 illustrations. 1,355pp. 5⅜ x 8½. Two-vol. set. Vol. I: 20429-4 Vol. II: 20430-8

A CONCISE HISTORY OF MATHEMATICS, Dirk J. Struik. The best brief history of mathematics. Stresses origins and covers every major figure from ancient Near East to 19th century. 41 illustrations. 195pp. 5⅜ x 8½. 60255-9

CATALOG OF DOVER BOOKS

Physics

OPTICAL RESONANCE AND TWO-LEVEL ATOMS, L. Allen and J. H. Eberly. Clear, comprehensive introduction to basic principles behind all quantum optical resonance phenomena. 53 illustrations. Preface. Index. 256pp. 5⅜ x 8½. 65533-4

QUANTUM THEORY, David Bohm. This advanced undergraduate-level text presents the quantum theory in terms of qualitative and imaginative concepts, followed by specific applications worked out in mathematical detail. Preface. Index. 655pp. 5⅜ x 8½. 65969-0

ATOMIC PHYSICS: 8th edition, Max Born. Nobel laureate's lucid treatment of kinetic theory of gases, elementary particles, nuclear atom, wave-corpuscles, atomic structure and spectral lines, much more. Over 40 appendices, bibliography. 495pp. 5⅜ x 8½. 65984-4

A SOPHISTICATE'S PRIMER OF RELATIVITY, P. W. Bridgman. Geared toward readers already acquainted with special relativity, this book transcends the view of theory as a working tool to answer natural questions: What is a frame of reference? What is a "law of nature"? What is the role of the "observer"? Extensive treatment, written in terms accessible to those without a scientific background. 1983 ed. xlviii+172pp. 5⅜ x 8½. 42549-5

AN INTRODUCTION TO HAMILTONIAN OPTICS, H. A. Buchdahl. Detailed account of the Hamiltonian treatment of aberration theory in geometrical optics. Many classes of optical systems defined in terms of the symmetries they possess. Problems with detailed solutions. 1970 edition. xv+360pp. 5⅜ x 8½. 67597-1

PRIMER OF QUANTUM MECHANICS, Marvin Chester. Introductory text examines the classical quantum bead on a track: its state and representations; operator eigenvalues; harmonic oscillator and bound bead in a symmetric force field; and bead in a spherical shell. Other topics include spin, matrices, and the structure of quantum mechanics; the simplest atom; indistinguishable particles; and stationary-state perturbation theory. 1992 ed. xiv+314pp. 6⅛ x 9¼. 42878-8

LECTURES ON QUANTUM MECHANICS, Paul A. M. Dirac. Four concise, brilliant lectures on mathematical methods in quantum mechanics from Nobel Prize–winning quantum pioneer build on idea of visualizing quantum theory through the use of classical mechanics. 96pp. 5⅜ x 8½. 41713-1

THIRTY YEARS THAT SHOOK PHYSICS: The Story of Quantum Theory, George Gamow. Lucid, accessible introduction to influential theory of energy and matter. Careful explanations of Dirac's anti-particles, Bohr's model of the atom, much more. 12 plates. Numerous drawings. 240pp. 5⅜ x 8½. 24895-X

ELECTRONIC STRUCTURE AND THE PROPERTIES OF SOLIDS: The Physics of the Chemical Bond, Walter A. Harrison. Innovative text offers basic understanding of the electronic structure of covalent and ionic solids, simple metals, transition metals and their compounds. Problems. 1980 edition. 582pp. 6⅛ x 9¼. 66021-4

CATALOG OF DOVER BOOKS

HYDRODYNAMIC AND HYDROMAGNETIC STABILITY, S. Chandrasekhar. Lucid examination of the Rayleigh-Benard problem; clear coverage of the theory of instabilities causing convection. 704pp. 5⅜ x 8¼. 64071-X

INVESTIGATIONS ON THE THEORY OF THE BROWNIAN MOVEMENT, Albert Einstein. Five papers (1905–8) investigating dynamics of Brownian motion and evolving elementary theory. Notes by R. Fürth. 122pp. 5⅜ x 8½. 60304-0

THE PHYSICS OF WAVES, William C. Elmore and Mark A. Heald. Unique overview of classical wave theory. Acoustics, optics, electromagnetic radiation, more. Ideal as classroom text or for self-study. Problems. 477pp. 5⅜ x 8½. 64926-1

PHYSICAL PRINCIPLES OF THE QUANTUM THEORY, Werner Heisenberg. Nobel Laureate discusses quantum theory, uncertainty, wave mechanics, work of Dirac, Schroedinger, Compton, Wilson, Einstein, etc. 184pp. 5⅜ x 8½. 60113-7

ATOMIC SPECTRA AND ATOMIC STRUCTURE, Gerhard Herzberg. One of best introductions; especially for specialist in other fields. Treatment is physical rather than mathematical. 80 illustrations. 257pp. 5⅜ x 8½. 60115-3

AN INTRODUCTION TO STATISTICAL THERMODYNAMICS, Terrell L. Hill. Excellent basic text offers wide-ranging coverage of quantum statistical mechanics, systems of interacting molecules, quantum statistics, more. 523pp. 5⅜ x 8½. 65242-4

THEORETICAL PHYSICS, Georg Joos, with Ira M. Freeman. Classic overview covers essential math, mechanics, electromagnetic theory, thermodynamics, quantum mechanics, nuclear physics, other topics. xxiii+885pp. 5⅜ x 8½. 65227-0

PROBLEMS AND SOLUTIONS IN QUANTUM CHEMISTRY AND PHYSICS, Charles S. Johnson, Jr. and Lee G. Pedersen. Unusually varied problems, detailed solutions in coverage of quantum mechanics, wave mechanics, angular momentum, molecular spectroscopy, more. 280 problems, 139 supplementary exercises. 430pp. 6½ x 9¼. 65236-X

THEORETICAL SOLID STATE PHYSICS, Vol. I: Perfect Lattices in Equilibrium; Vol. II: Non-Equilibrium and Disorder, William Jones and Norman H. March. Monumental reference work covers fundamental theory of equilibrium properties of perfect crystalline solids, non-equilibrium properties, defects and disordered systems. Total of 1,301pp. 5⅜ x 8½. Vol. I: 65015-4 Vol. II: 65016-2

WHAT IS RELATIVITY? L. D. Landau and G. B. Rumer. Written by a Nobel Prize physicist and his distinguished colleague, this compelling book explains the special theory of relativity to readers with no scientific background, using such familiar objects as trains, rulers, and clocks. 1960 ed. vi+72pp. 23 b/w illustrations. 5⅜ x 8½.
42806-0 $6.95

A TREATISE ON ELECTRICITY AND MAGNETISM, James Clerk Maxwell. Important foundation work of modern physics. Brings to final form Maxwell's theory of electromagnetism and rigorously derives his general equations of field theory. 1,084pp. 5⅜ x 8½. Two-vol. set. Vol. I: 60636-8 Vol. II: 60637-6

CATALOG OF DOVER BOOKS

QUANTUM MECHANICS: Principles and Formalism, Roy McWeeny. Graduate student–oriented volume develops subject as fundamental discipline, opening with review of origins of Schrödinger's equations and vector spaces. Focusing on main principles of quantum mechanics and their immediate consequences, it concludes with final generalizations covering alternative "languages" or representations. 1972 ed. 15 figures. xi+155pp. 5⅜ x 8½. 42829-X

INTRODUCTION TO QUANTUM MECHANICS WITH APPLICATIONS TO CHEMISTRY, Linus Pauling & E. Bright Wilson, Jr. Classic undergraduate text by Nobel Prize winner applies quantum mechanics to chemical and physical problems. Numerous tables and figures enhance the text. Chapter bibliographies. Appendices. Index. 468pp. 5⅜ x 8½. 64871-0

METHODS OF THERMODYNAMICS, Howard Reiss. Outstanding text focuses on physical technique of thermodynamics, typical problem areas of understanding, and significance and use of thermodynamic potential. 1965 edition. 238pp. 5⅜ x 8½. 69445-3

TENSOR ANALYSIS FOR PHYSICISTS, J. A. Schouten. Concise exposition of the mathematical basis of tensor analysis, integrated with well-chosen physical examples of the theory. Exercises. Index. Bibliography. 289pp. 5⅜ x 8½. 65582-2

THE ELECTROMAGNETIC FIELD, Albert Shadowitz. Comprehensive undergraduate text covers basics of electric and magnetic fields, builds up to electromagnetic theory. Also related topics, including relativity. Over 900 problems. 768pp. 5⅜ x 8¼. 65660-8

GREAT EXPERIMENTS IN PHYSICS: Firsthand Accounts from Galileo to Einstein, Morris H. Shamos (ed.). 25 crucial discoveries: Newton's laws of motion, Chadwick's study of the neutron, Hertz on electromagnetic waves, more. Original accounts clearly annotated. 370pp. 5⅜ x 8½. 25346-5

RELATIVITY, THERMODYNAMICS AND COSMOLOGY, Richard C. Tolman. Landmark study extends thermodynamics to special, general relativity; also applications of relativistic mechanics, thermodynamics to cosmological models. 501pp. 5⅜ x 8½. 65383-8

STATISTICAL PHYSICS, Gregory H. Wannier. Classic text combines thermodynamics, statistical mechanics, and kinetic theory in one unified presentation of thermal physics. Problems with solutions. Bibliography. 532pp. 5⅜ x 8½. 65401-X

Paperbound unless otherwise indicated. Available at your book dealer, online at **www.doverpublications.com**, or by writing to Dept. GI, Dover Publications, Inc., 31 East 2nd Street, Mineola, NY 11501. For current price information or for free catalogs (please indicate field of interest), write to Dover Publications or log on to **www.doverpublications.com** and see every Dover book in print. Dover publishes more than 500 books each year on science, elementary and advanced mathematics, biology, music, art, literary history, social sciences, and other areas.